THE NEW NATURALIST LIBRARY

A SURVEY OF BRITISH NATURAL HISTORY

PEMBROKESHIRE

THE NEW NATURALIST LIBRARY

PEMBROKESHIRE

JONATHAN MULLARD

N

WILLIAM COLLINS

This edition published in 2020 by William Collins,
An imprint of HarperCollins Publishers

HarperCollins Publishers
1 London Bridge Street
London SE1 9GF

WilliamCollinsBooks.com

First published 2020

© Jonathan Mullard, 2020

Photographs © Individual copyright holders

A CIP catalogue record for this book is available
from the British Library.

Set in FF Nexus, designed and produced by
Tom Cabot/ketchup

Printed in China by RR Donnelley APS

Hardback
ISBN 978-0-00-811280-6

Paperback
ISBN 978-0-00-811282-0

Reading and imagining are two of the principal doorways – curiosity is the third – through which one gains a knowledge of things. If you don't fling open the doors of imagination, curiosity and reading – and by reading, I mean, of course, study – you won't get very far in understanding the world or yourself.

José Saramago

Contents

Editors' Preface

THE NATIONAL PARK AUTHORITIES FACE growing and sometimes conflicting demands to deliver conservation, public access, local employment and affordable housing, at a time when resources are being constrained. Whether it's the threat of climate change, the pressures of tourism or the challenge of balancing the needs of different groups, all of our National Parks need to address a number of difficult issues.

Pembrokeshire Coast National Park (Parc Cenedlaethol Arfordir Penfro) was established as an early member of the National Park family, under the National Parks and Access to the Countryside Act 1949, and of the fifteen National Parks in the UK, it was the fifth to be created (with the North York Moors) in 1952, and one of three in Wales. Uniquely, it is the only one in the United Kingdom to have been designated primarily because of its spectacular coastline, with nowhere more than 16 km from the sea, and with 300 km of coastal walks, full of wildlife and history, scenic beaches, coves and cliffs.

The National Park covers almost all of the Pembrokeshire coast, including every offshore island, as well as the Daugleddau estuary, Mynydd Preseli and Cwm Gwaun. Milford Haven Waterway is the second-deepest natural harbour in the world, after Sydney Harbour in Australia. While it is one of the smallest UK National Parks, it has one of the most diverse landscapes, with a greater variety of geological and landform scenery than almost any area of the same size in the British Isles. In addition, it is ecologically one of the richest and most diverse parts of Wales, and recognised as of international importance for a wide range of high-quality habitats and rare species. The Park includes thirteen Special Areas of Conservation (including three marine SACs), five Special Protection Areas, one Marine Conservation Zone, seven National Nature Reserves and sixty Sites of Special Scientific Interest. One-third of Britain's nesting pairs of choughs are in Pembrokeshire, Skomer is well known for its puffins, and Grassholm has one of the world's largest gannet colonies. But not all visitors will be prepared for the water buffalo in the Teifi Valley or the live-firing tank training exercises on the Castlemartin Range, currently celebrating its eightieth birthday.

We are guided through all this, and the wildlife in the rest of the county, by Jonathan Mullard, a great authority on the natural history of South Wales. How fortunate for our series that he was marooned on Skomer in May 2015 by high winds which prevented his scheduled departure. While waiting for the next boat he began to sketch out this present volume. We therefore have the benefit of his impressive knowledge and accessible writing skills in this latest New Naturalist.

All three National Parks in Wales are now covered by the New Naturalist series, for the older volume on Snowdonia (No. 47) was joined a few years ago by *Brecon Beacons* (No. 126), also written by Jonathan Mullard – who in addition wrote the volume on the Gower Area of Outstanding Natural Beauty (No. 99). Our coverage of the natural history of South Wales is therefore now impressive, and although – as Jonathan himself admits – this present volume has to be seen as an overview, his three contributions to the series have immeasurably enhanced our knowledge of the whole region. And not only that, for in his final chapter he issues a grave warning to us all about the lack of climatic and environmental awareness that persists even several decades after we entered the Anthropocene era. Environmental activists can draw attention to the issues, but education must surely be the key to greater understanding – and a volume such as this, with its striking illustrations and comprehensive text, will undoubtedly help disseminate information about the splendid biodiversity to be found within Pembrokeshire, and the pressing need to recognise its fragility.

Author's Foreword and Acknowledgements

No man may step twice into the same river; no naturalist revisiting Wales after twenty years or even less, could fail to realise that the fauna he saw around him was not quite the same as that he knew before.

Colin Matheson (1932)

T HIS BOOK HAS ITS ORIGIN in our annual family holidays in Pembrokeshire, a tradition that continues, with few exceptions, today. In contrast to my previous volumes in the series, which covered Gower and the Brecon Beacons, this book therefore derives not from a professional familiarity, but from a deep affection for Pembrokeshire and all it represents for the visiting naturalist. For over forty years I have been exploring the county, and I have seen many changes during this period – some positive and others not so welcome. The quiet and somewhat untidy landscape, seemingly full of wildlife, that I first encountered in the late 1970s has largely disappeared. New roads, more vehicles and the intensification of all kinds of activities, including farming, are not issues confined to Pembrokeshire but they are increasingly diluting its distinctive qualities. Retaining a sense of place will be an important issue in the coming years.

I started writing the book while marooned on Skomer in May 2015. The combination of small boats and high winds, in this case a force eight gale, had prevented my departure on the scheduled day. There was nothing to be done except wait for calmer seas before the *Dale Princess* could make the crossing from Martin's Haven on the mainland. Unsurprisingly, I began by sketching out the chapters on the islands – but, given the variety of geology and wildlife to be found in Pembrokeshire, covering all aspects of the county's natural history in one book presents a challenge to the author. Whole books have indeed been written on Skomer alone, but in the space available it is not possible to cover everything in depth. Although this book is therefore necessarily an overview, I hope it will provide a flavour of the wildlife to be found in this particular corner of Wales, and that the various specialists who have willingly provided their expertise and support will forgive me if I have not covered their interests in detail.

Another issue for a natural history author in these rapidly changing times is how to deal with the issue of ecosystem degradation and the climate crisis. To continually refer to these large-scale topics, which nevertheless affect individual plants and animals on the ground, could make the text unduly negative, especially when there is otherwise much of interest to be described. Because of this I decided to use the last chapter, as in my previous book, to describe the wider picture in more detail, while looking at the historical context of both habitats and species, as appropriate, throughout the text. In this I have been helped by the availability of a number of early sources, notably *The Description of Pembrokeshire* by George Owen, which dates to 1603. Publications such as this provide a startling reminder of how much we have lost over the centuries and, conversely, how much we have to gain by recovering some of the past glories.

Previously, I included a list of designated sites and nature reserves in the appendices of my books, since it was difficult to find the information, but times have changed. The website for National Resources Wales now contains many documents on designated areas in Pembrokeshire, and material on the nature reserves managed by other conservation organisations is also readily available, as is a list of Geological Review Sites. Details of Regionally Important Geological Sites (RIGS) in the National Park can be found on the Park Authority's website. So I have not repeated the material here. Although I used these sources in compiling this book, together with material from the Skokholm, Skomer and Ramsey and Pembrokeshire Birds blogs, I have not specifically referenced them, in contrast to knowledge gained from books and scientific papers. My aim has been to avoid unduly disrupting the flow of the text and to keep the bibliography, which is already long, relatively manageable. Copies of many of the articles and papers, especially on botany, can be found in the ever-expanding, and extremely useful, *Biodiversity Heritage Library*. I have also drawn on the equally important *Welsh Journals* website, developed by the National Library of Wales, which provides access to journals relating to Wales that were published between 1735 and 2007.

ACCESS AND GEOLOGY

The large-scale Ordnance Survey maps covering Pembrokeshire are a useful companion to this book and include many of the sites that I have described. It is not, however, intended as a field guide, and while a considerable area of the county is accessible to the public (with the exception of the military ranges), the description, or mention, of any site does not necessarily imply that there is access to it, or that a right of way exists. An account of an area or site, or its appearance in a photograph or on a map, should not be taken as an invitation to visit. Where necessary, you should obtain permission from landowners for access and field study.

Most of the coastline, and many inland locations, are covered by Sites of Special Scientific Interest (SSSIs), and the majority of these include primary, or secondary, geological features identified through the work of the Geological Conservation Review. All these sites are legally protected, so please avoid causing damage or disturbance to wildlife; both of which could potentially lead to prosecution. If you are interested in the geology of the county do not remove in-situ fossils, or sample, or core, rocks and minerals from significant exposures. Take photographs instead and only collect from loose material that would otherwise be lost. If in doubt, seek advice from Natural Resources Wales and pass on details of any important new findings that you make.

ACKNOWLEDGEMENTS

For a number of reasons, this book has taken longer to complete than originally agreed with my publishers. I must therefore thank Myles Archibald, Julia Koppitz and Hazel Eriksson, and Brian Short, my editorial board member, for being patient while I finished the manuscript. David Streeter, my previous editorial contact, has remained a welcome source of support and encouragement. My warmest thanks also to Hugh Brazier, who has expertly copy-edited all my books to date. Having spent many of his formative years in South Wales, he is familiar with the landscapes and wildlife I have described, a real advantage for an author. Melanie Francis, my wife, patiently read through a number of early chapter drafts and accompanied me on various field visits. Her comments and assistance were invaluable.

There are many other people I must acknowledge for their contributions. In relation to material on the early naturalists and their exploits, I am indebted to David Saunders, who has had a lifelong association with Pembrokeshire, for photographs and information on R. M. Lockley and the Reverend Murray Mathew; David Utting, Secretary of the Hertfordshire Natural History Society, for the details of Bertram Lloyd's visits; Andy Kendall, Publicity Officer for the Cardiff Naturalists' Society, for copies of the Society's publications and photographs relating to its involvement in Pembrokeshire; and Mark Lewis, Curator of Tenby Museum and Art Gallery, for details of the museum's founders. Lizelle de Jager of the Royal Museums Greenwich was of great assistance concerning the background to the figurehead of HMS *Gannet*.

Sid Howells, former Regional Earth Scientist for Natural Resources Wales, read the geological chapter and other relevant sections and made many useful suggestions which have greatly improved the text. Gareth George supplied information on another (unrelated) geologist, Thomas Neville George.

Regarding the islands, Greg and Lisa Morgan, RSPB wardens, kindly checked the text on Ramsey, Grassholm and the smaller isles. Likewise, Richard Brown and Giselle Eagle of the Wildlife Trust of South and West Wales provided a range of material on Skokholm. A former Skomer warden, Mike Alexander, checked the section on that island and its conservation history. Bob Haycock, the former Senior Reserves Manager for Pembrokeshire for the Countryside Council for Wales, and Annie Haycock, freelance naturalist and the county's Mammal Recorder, were also very helpful, commenting on several sections of the text and providing access to their extensive photographic library. Blaise Bullimore, former Skomer Marine Nature Reserve Officer and Project Manager for the Milford Haven Waterway Environmental Surveillance Group, similarly provided detailed guidance on the marine chapters, together with many excellent underwater photographs. Sue Burton, Pembrokeshire Marine Special Area of Conservation Officer, sent information on the SWEPT project. Matthew Green, Marine Monitoring Ecologist for Natural Resources Wales, explained the background to the finding of the fan mussel in Milford Haven.

Both James Parkin, Director of Countryside, Community and Visitor Services, and Sarah Mellor, Biodiversity Officer at the Pembrokeshire Coast National Park Authority, have been supportive of the project from the beginning, providing advice and reports. Jane Hodges, the former National Park Ecologist, collated the facts on Carew Castle and the species found there.

Stephen Jones, Senior Conservation and Design Services Manager for Cadw, arranged access to the ecology reports for their guardianship sites. Lynne Houlston, Ranger, and Steven Philipps-Harries, Deputy Training Safety Officer for the Defence Infrastructure Organisation on the Castlemartin Range, have both been extremely accommodating in providing information on, and tours of, the Training Area.

Fred Rumsey, Senior Curator for the Historical Collections in the Natural History Museum, kindly commented on the text relating to rock sea-lavenders and supplied his drawings of them, originally intended for a Botanical Society of Britain and Ireland (BSBI) guide, together with several photographs. Robyn Cowan at Kew Gardens provided useful advice on the genetics of this interesting group of plants. Stephen Evans, Vice County Recorder for the BSBI, also contributed his thoughts and expertise on the botany, as did Tim Rich, another well-known member of the BSBI. Trevor Theobald, Ecologist for Pembrokeshire County Council, was again very helpful, as was Ant Rogers, Biodiversity Implementation Officer with the Pembrokeshire Nature Partnership. In addition, Ant put me in contact with a number of other specialists, including David and Holly Harries of the Pembrokeshire Fungus Recording Network. Pat O'Reilly from First Nature supplied lists of fungi in north Pembrokeshire. Matt Sutton and Vicky Swann kindly showed me around their farm and explained the meadow restoration projects they have under way. Sam Bosanquet, Vegetation Ecologist for Natural Resources Wales, once again provided guidance on the bryophytes. Ivan Pedley of the British Lichen Society explained the factors affecting churchyard lichens, while Alan Orange also helped with information on the lichen flora.

Ben Rowson, Senior Curator for Terrestrial Mollusca at National Museum Wales, produced a summary of the molluscs found in the woodland at Ty Canol. Peter Hill, Dragonscapes Habitat Officer with Amphibian and Reptile Conservation, supplied material on adders, as did Sam Langdon, the Pembrokeshire Adder Coordinator.

In relation to insects, Karim Vahed, Professor of Entomology at the University of Derby, generously shared his papers on the scaly cricket, prior to their publication. Max Barclay, Senior Curator for Coleoptera in the Natural History Museum, and Mark Telfer, consultant entomologist, supplied information on the Stackpole seed-eater, and Bob Heckford, one of our leading lepidopterists, was of assistance regarding the Pembroke dwarf moth. Steven Falk made available material and photographs derived from his detailed recording of bees and other insects in Pembrokeshire. Mike Howe, Invertebrate Ecologist for Natural Resources Wales, advised on surveys of soft-cliff invertebrates. David Redhead, County Recorder for Butterflies, also provided information and advice.

Mary Chadwick, Secretary of the Pembrokeshire Bat Group, Tom McOwat and Den Vaughan submitted details of bat roosts and hibernation sites in Pembrokeshire. Dave Astins of West Coast Birdwatching allowed me to join one of his early autumn trips to Skokholm.

People who allowed me to use their photographs are also acknowledged in the appropriate locations in the book, but it would be remiss of me not to mention here the excellent support I had from Brian Southern, Graphic Services Manager for the Pembrokeshire Coast National Park Authority, who kindly allowed me to use a number of their photographs. Finally, Anna Malloy, Stakeholder Engagement and Communications Manager for the Port of Milford Haven, was very helpful in sourcing aerial views of the waterway. My apologies to anyone I have accidentally omitted from this list.

Despite the aim of providing a balanced account of the county's landscape and wildlife, writing a book such as this often exposes the personal thoughts of the author, intentionally or unintentionally. The views expressed in these pages are mine alone and do not necessarily represent the position of past, or current, employers, or people who have assisted me. I also take responsibility for any unintended errors in the text.

In conclusion, as Dawson Turner and Lewis Weston Dillwyn stated in the introduction to their *Botanist's Guide* (1805), I hope that 'at a period when pursuits in Natural History become daily more prevalent, the present work will not be deemed an unacceptable offering to the public'.

Map of Pembrokeshire with the area of the National Park marked in green.

(Pembrokeshire Coast National Park Authority)

CHAPTER 1

At the Land's End

As touching the form and fashion thereof by the topographical description it is
neither perfect square, long or round, but shaped with diverse corners, some sharp,
some obtuse, in some places concave in some convex, but in most places concave
and bending inward, as the moon does in her decreasing ...

George Owen (1603)

S URROUNDED BY THE SEA ON three sides, and by the counties
of Ceredigion to the north and Carmarthenshire to the east,
Pembrokeshire is the most westerly peninsula in Wales. It is also the
most attractive, at least according to Gerald de Barri, the cleric and historian
popularly known as Gerald of Wales or Giraldus Cambrensis – although he
was undoubtedly biased, since he was born in the south of the county, in
Manorbier Castle, in 1146. 'Penbroch', he wrote in 1191, 'is the finest part of the
province of Demetia, and Demetia ... is the most beautiful ... district of Wales.'
He does go on to say, however, that he hopes 'the author may be pardoned for
having thus extolled his native soil, his genial territory, with a profusion of
praise and admiration'. Some 800 years later, the great authority on the British
coast, Professor J. A. Steers, made no excuses for stating that 'No part of the
coastline of England and Wales is more beautiful or more interesting than that
of Pembrokeshire' (Steers 1953). Everyone who visits has their own favourite
location, but Whitesands Bay, to the south of St Davids Head, is one of the most
popular beaches in the county (Fig. 1).

According to George Owen of Henllys, the renowned Welsh Elizabethan antiquary and polymath, Pembrokeshire was originally Penbrokshire or Penfro (Owen 1603). This is derived from *pen*, meaning 'head', 'promontory' or 'end', and *bro*, 'region' or 'land'. *Pen bro* is softly mutated in Welsh to *Penfro*, and anglicised as *Pembroke* (Miles 1987). The county's name is therefore a Welsh version of Land's End.

Pembrokeshire was, for centuries, renowned as the ancient land of mystery and enchantment, *gwlad hud a lledrith*, described in *The Mabinogion*, the earliest prose stories of the literature of Britain. The legends, which are related to specific localities in the county, include those about St David, the patron saint of Wales, whose mother was said to be St Non, niece of King Arthur. St Non's Well, near the city of St Davids, marks the very spot where St David was apparently born. One of the many sites of Arthur's grave is said to be located at Bedd Arthur, a megalithic site high on Mynydd Preseli, although other legends state that he is asleep in a cave under Dinas Rock in the Brecon Beacons, waiting for a call to defend Britain (Mullard 2014). The nearby outcrop of Cerrigmarchogion, source of a number of the Stonehenge bluestones, marks the site where the mythical boar Twrch Twyth slew several of Arthur's men and turned them to stone. Other prehistoric sites include Pentre Ifan, a Neolithic dolmen near Newport, which is

FIG 1. Whitesands Bay is one of the most popular beaches in Pembrokeshire, especially with surfers. The prominent crags of Carn Llidi in the background are formed from intrusive igneous rocks. (Pembrokeshire Coast National Park Authority)

FIG 2. The huge capstone of Pentre Ifan is estimated to weigh 16 tonnes. George Owen noted that 'The stones whereon this is laid are so high that a man on horseback may well ride under it without stooping.' (Jonathan Mullard)

probably the most well-known prehistoric monument in Wales (Fig. 2). Studied by early travellers and antiquarians, it rapidly became famous because of the many engravings that were made of the 'romantic' stones.

The north of the county, where these sites are located, certainly has a different feel to the south – which is a part of Wales that has been English in language and culture for many centuries, despite the distance from the border. In his book *The History of Little England beyond Wales, and the Non-Kymric Colony settled in Pembrokeshire,* Edward Laws (1888) provides 'a sample of the vocabulary in daily use' at that time in the English-speaking part of the county, including common names of plants and animals. These include 'sea parrot' for the puffin *Fratercula arctica,* 'spur' for the common tern *Sterna hirundo,* 'anny' for the kittiwake *Rissa tridactyla* and 'cutty moorcock'[*] for the water rail *Rallus aquaticus.* See Appendix 1 for the full list.

---

[*] 'Cutty' means small. The female figurehead of the tea clipper *Cutty Sark* wears a 'cutty sark', an archaic Scottish name for a small nightdress. Cutty-sark was a nickname given to Nannie Dee, a witch created by Robert Burns in his poem *Tam o' Shanter,* referring to the garment she wore.

FIG 3. This little wooden cottage dressed with ribbons is a wren house made in Marloes in 1869. A dead wren was placed in it and carried around the community to symbolise the death of winter. (St Fagans National Museum of History)

Another small bird, the cutty wren *Troglodytes troglodytes*, once played a key role in the ritual year in south Pembrokeshire, forming part of the celebrations around the winter solstice. The wren is one of the most numerous Welsh birds, breeding in a wider variety of habitats than any other species. It has been estimated that there are 62,000 pairs in Pembrokeshire (Rees *et al.* 2009). Up until the late nineteenth century, in certain parts of the British Isles and Ireland, wrens were hunted on St Stephen's Day (26 December) or Twelfth Day (6 January) and killed to symbolise the death of winter. Placed in a garlanded box, they were then taken from door to door. At each house a song was sung and the occupants asked *'Please to see the King'* – the wren being considered the King of Birds. A wren house from Marloes, made in 1869, is preserved in the National Museum of History at St Fagans (Fig. 3). Despite Marloes being one of the last places in Wales where the wren hunt took place, the residents have traditionally been known as 'gulls', owing to the other historic village practice of harvesting seabirds' eggs from the nearby island of Skomer.

An early description of the custom is provided by Edward Lhuyd (1660–1709), the Welsh naturalist and antiquarian: 'They are accustomed in Pembrokeshire to carry a wren in a bier on Twelfth Night.' In Tenby:

> *Having procured a wren, and placed it in a small ornamented box, or paper house, with a square of glass at each end, two or four men would carry it about, elevated on four poles fixed to the corners, singing a long ditty … The men would enter the doorway, groaning under the weight of their burden, and looking as if they had just relieved Atlas of his shoulder-piece. (Owen 1978)*

In north Pembrokeshire, in Cwm Gwaun (the Gwaun valley), local people continue to follow the tradition of the pre-1752 Julian calendar and celebrate New Year (*Hen Galan*) on 13 January. Children walk from house to house to sing traditional songs, and in return householders provide gifts, or *calennig*, of sweets and money. The Gregorian calendar removed 11 days from September 1752 in order to correct a growing discrepancy between the dates of festivals and the actual seasons. The local community, however, ignored this decree and to this day have carried on using the older calendar. The Jabes Baptist Chapel in the valley is one of the few remaining chapels in Wales to have an outdoor baptistry, which is filled with water from the nearby Afon Gwaun.

LANDSCAPE AND VISITORS

For over two centuries Pembrokeshire has attracted visitors, lured mainly by its spectacular coastal scenery, especially the wide sandy beaches and the offshore islands. People began to visit Wales, rather than come on business, from 1700, but around 1770 the numbers of tourists increased significantly, partly due to wars on the continent making travel there difficult. The first English visitors knew very little about Wales and the early eighteenth-century publications, which described their visits, were often very offensive about the Welsh. As interest in 'the picturesque' grew, however, and the roads improved and published accounts of tours of Wales began to appear, the number of tourists increased, but they tended to follow in the footsteps of the first (Freeman 2019).

Between 1770 and 1815, no fewer than 80 descriptions of tours through Wales were published. Pembrokeshire though was the last county in Wales to have an Act, in 1771, for establishing turnpikes, so communications here remained poor for some time. St Davids was not a popular place, and even George Borrow, who visited in August 1857, did not include it in his book on *Wild Wales* (1862). Many of

FIG 4. Tenby has been a busy holiday resort since 1781, when John Jones, a medical man from Haverfordwest, promoted seawater bathing here. The town's many famous visitors have included Beatrix Potter. (Pembrokeshire Coast National Park Authority)

the tourists who came to Pembrokeshire went to Tenby (Fig. 4) and occasionally Pembroke and Milford Haven, and then returned eastwards, or travelled north to Cardigan, sometimes via Haverfordwest. Few took the road from Haverfordwest to St Davids and then to Fishguard and on to Cardigan, and even fewer followed the reverse of that route.

In 1781 John Jones, a medical man from Haverfordwest, promoted seawater bathing at Tenby – and ever since the town has been a busy holiday resort, with the rest of the county following its lead. One of the oldest Welsh poems surviving in the ninth-century *Book of Taliesin* is *Etmic Dinbych* ('In praise of Tenby'), which describes the town as 'a fine fortress of revel and tumult' (*aduwyn gaer yssyd ae gwna kyman*) – a statement which could be said to anticipate the town's role in tourism (Miles 1987). In 2011 *National Geographic Traveler* magazine voted Pembrokeshire the second-best coastal destination in the world for sustainable tourism, surpassed only by the Avalon peninsula in Newfoundland and Labrador.

One of the many visitors to Tenby was Beatrix Potter, who stayed there with her parents on at least two occasions in 1900 and 1901. She was apparently fascinated with the area, and her letters to the children of her former governess refer to her boat trip around the Tenby cliffs, the rabbit holes, the wallflowers and the many wild cabbages she saw in their little rabbit gardens. According to Potter, Mrs Rabbit cooked the cabbages for her family using the local coal. The Castle Rocks are the classic Pembrokeshire location for wild cabbage *Brassica oleracea* var. *oleracea* (Fig. 5). The first written record of the species here, by the Reverend John Holcombe, the greatest Pembrokeshire botanist of the eighteenth century, dates from 1775. Similarly, in 1853, Edwin Lees noted, 'about the Castle-rocks', 'the most exuberant growth of sea-cabbage (*Brassica oleracea*) and *Smyrnium olusatrum* [Alexanders] that ever met my view'. Lees went on to speculate that 'Whether this was so previous to the occupation of the Castle is, perhaps, worth the discussion of those who would pry curiously into the first immigration of plants considered to be "doubtfully wild", or "perhaps only escaped from cultivation".' While Alexanders is native to the Mediterranean, and was introduced to Britain by the Romans for use as a vegetable, wild cabbage is a native species, though at the northern edge of its range in Wales.

FIG 5. Wild cabbage, with Alexanders, on the Castle Rocks near Tenby bandstand. Wild cabbage was first recorded here by the Reverend John Holcombe some time before 1775. This is the classic Pembrokeshire location for the species. (Jonathan Mullard)

Visitors to Tenby today can still see rabbit holes, wild cabbages, wallflowers *Erysimum cheiri* and Alexanders along the cliffs. Later, when Potter became a children's author, it was these letters from Tenby that helped shape the story of one of her best-known characters, Peter Rabbit. Facsimile copies of the letters can be seen in Tenby Museum and Art Gallery. The *Tale of Peter Rabbit* includes a picture of a cat peering into a pond that had been in her painting of the garden of 2 The Croft, the house where she stayed on her visits (Potter 1902).

Beatrix Potter also recalled a trip to an island, almost certainly Caldey, to see the puffins – which would have been plentiful there at the time of her visit. The main offshore islands are certainly a draw for visitors, and Caldey, along with Skomer and Ramsey, is one of the easiest to visit, with daily boat trips from the mainland between Easter and October. Now that there are no longer puffins on Caldey, Skomer is the place to see them – and for many people these birds are the highlight of their visit (Fig. 6). At the magnificent inlet on the south side of the island, known as the Wick, the puffins completely ignore the visitors and will just

FIG 6. Skomer is the best place in Pembrokeshire to see puffins, an estimated total of 25,227 birds being resident on the island during the breeding season in 2018. (Mike Alexander)

FIG 7. The Pre-Raphaelite artist John Brett produced a detailed oil painting of Forest Cove, now known as Aberfforest, which lies to the east of Dinas Head. Paintings such as this can be used to estimate the rate of coastal change in a particular area. (National Museum Wales)

walk around them, or even over their shoes, to get to their chicks. It is important not to stray from the waymarked paths, as even slight trampling will damage the burrows in which the young are located.

The coastal light and the spectacular scenery attract many artists to Pembrokeshire. The Pre-Raphaelite artist John Brett, for instance, produced a detailed painting in oils of the coastal geology of Fishguard Bay in 1883. The Pre-Raphaelites wished to capture nature in an exact form through their works, often painting outside. The accurate images they created now provide an important record of conditions in the coastal zone in the mid to late nineteenth century (McInnes & Benstead 2013). Brett devoted his later years to touring and painting the British coastline but wrote that Pembrokeshire was the 'only one really satisfactory seaside place on the whole British coast' (Fig. 7).

Tenby itself was painted by many artists, including Alfred Robert Quinton, who produced several views of the town in the early twentieth century. His views show extensive exposures of rock which are not visible today, demonstrating how the level of the beach has risen over time. There is a rich resource of historical images such as these that can be used to illustrate the rate of coastal change over the last 200 years. A modern version has been developed by the Pembrokeshire Coast National Park Authority, which has installed fourteen fixed-point

FIG 8. A traditional field boundary, or *clawdd* (plural *cloddiau*), on the lane leading to Strumble Head. A *clawdd* consists of a drystone wall with a compacted earth core, and it provides a habitat for numerous species. Here, in this exposed location, lichens and thrift dominate. (Jonathan Mullard)

photography posts along the coast and encourages visitors to take pictures from these with their mobile phones and share the results.

Away from the coast, the mainland landscape consists, for the most part, of farmland and bare rounded hills, while craggy outcrops stretch from Frenni Fawr at the eastern end of Mynydd Preseli to St Davids and Ramsey Island in the west (Lloyd *et al.* 2010). The highest point, Foel Cwmcerwyn near Rosebush, only reaches 536 m (1,760 feet), but the mountains, whether the long ridge of Mynydd Preseli or the volcanic outcrops such as Carn Llidi to the west, dominate the north of the county. In contrast to other areas of Wales, the *ffridd*, that distinctive transition between the enclosed fields and the open hill, is largely absent, the landscape of small fields giving way abruptly to the uplands. Coastal winds have created the largely treeless landscape west of Fishguard, which is divided by the traditional field boundaries known as *cloddiau* (Fig. 8). Found in Pembrokeshire and other western coastal districts of Wales, these are a characteristic feature of the area. The lowland fields, particularly in the south of the county, are larger and generally heavily 'improved', with intensive arable and livestock production. Here and there, however, there are still remnants of an older, and richer, landscape.

Rivers in the north of the county are generally short but dramatic, the Afon Gwaun running down the valley separating Carningli from the main part of Mynydd Prescli being, perhaps, the most scenic. To the south, the landscape is dominated by the two Cleddaus which unite to form the Milford Haven Waterway. Both rivers drain the southern slopes of the hills, the course of the Western Cleddau running through the deep cut of Treffgarne Gorge to Haverfordwest, which is situated at its highest navigable point. In contrast, the Eastern Cleddau follows a wooded route, joining the Western Cleddau at Picton Point.

The industrialised shores of Milford Haven provide a stark contrast to the rural character of the rest of the county. The waterway supported a major naval dockyard during the nineteenth and twentieth centuries and an important naval

FIG 9. The Pembroke Refinery at Rhoscrowther first came on stream in 1964. It occupies a prominent position on the south bank of the Milford Haven Waterway, around a quarter of the site being within the National Park. The tower of Rhoscrowther church can just be seen in the wooded area in the centre of the photograph. (Jonathan Mullard)

base in both World Wars, but it was not until after the Second World War that the major developments occurred. Between 1955 and 1975 four oil refineries and an oil terminal were constructed and a new power station built at Pennar near Pembroke Dock. An inland reservoir, at Llys-y-frân (now a Country Park), was developed to provide water for industry, and the Cleddau Bridge was built to link the north and south shores. The industries and their associated supporting maritime businesses surrounding the Haven constantly change; three refineries have closed since 1980 and two liquid natural gas terminals opened in the 2000s, while a decommissioned oil-fired power station was replaced by a gas-fired station (Lock & Bullimore 2018). The largest natural gas pipeline ever built in Britain connects the two terminals to the national gas network in Gloucestershire, crossing the Brecon Beacons on the way (Mullard 2014). All have had an impact on the landscape, wildlife and local communities. A large part of Rhoscrowther, on the south shore of Milford Haven, described now as a 'strange and unhappy village', was evacuated in 1996–97 after an explosion in the nearby oil refinery (Fig. 9; Lloyd et al. 2010). The spectacular medieval church, which

FIG 10. The Castlemartin Range is one of thelargest training areas for armoured fighting vehicles in the United Kingdom but still supports an impressive variety of wildlife. These old tracked vehicles are used as targets. (Jonathan Mullard)

thankfully survived, despite damage to the roof, is now in the care of the Friends of Friendless Churches.

Around a quarter of the refinery is situated within the Pembrokeshire Coast National Park, as is the Castlemartin Range to the south, which covers an area of 2,429 ha. Stretching from Stackpole in the east to Freshwater West, this coastal plateau, with 14 km of cliffs, was requisitioned by the War Office in 1938 from the Cawdor Estate (Fig. 10). Today Castlemartin is one of the major live-fire training and manoeuvre facilities for armoured fighting vehicles in the United Kingdom and is intensively used. In 2015, for example, 153,566 'man training days' took place here, and that training involved the firing of more than three million rounds of ammunition. There is a coastal exclusion zone of between 3 and 12 nautical miles (6–22 km) offshore, depending on what type of weapons and ammunition are being used.

During non-firing periods the public have access to a section of the Pembrokeshire Coast Path, but much of the area is permanently off limits owing to the dangers of unexploded bullets and shells. During the summer, however, the National Park Authority organises guided walks through areas that are normally out of bounds, and this provides a fascinating insight into the range. Because of the limited access, the military training has preserved a spectacular coastal landscape that supports many rare, or uncommon, plants, insects and birds.

There are smaller areas of land used for military training at Manorbier and Penally. The air defence range at Manorbier is the only range on the mainland where high-velocity missiles, employed in an anti-aircraft role, can be fired. The range also hosts an air warfare centre that conducts a variety of aircraft-related trials. Penally rifle range was established in 1860 to cater for a need for musketry training following the Crimean War. Although there are nearby footpaths from which the whole of the site can be seen, there is no public access to Penally – but, as at Castlemartin, the Manorbier range is accessible through National Park Authority guided walks.

THE NATIONAL PARK

The Pembrokeshire Coast National Park was designated on 29 February 1952, not long after the National Parks and Access to the Countryside Act of 1949 was brought into force. Harold Abrahams, Secretary of the National Parks Commission, and Lord Merthyr, a member of the Commission and a former member of Pembrokeshire County Council, had met the County Planning

FIG 11. The razorbill is the symbol of the Pembrokeshire Coast National Park, more than 8,000 birds being present in the area during the breeding season. (Annie Haycock)

FIG 12. The National Park visitor centre and art gallery in St Davids is a striking modern building, the first part being built in 2001. (Pembrokeshire Coast National Park Authority)

Committee the previous month and obtained agreement to the setting up of a separate National Park Committee. The first National Park Officer, John Price, was the former County Planning Officer. Subsequently, the razorbill *Alca torda* was adopted as the emblem of the new designation (Fig. 11). At the same meeting Lord Merthyr put forward 'proposals for the establishment of a footpath round the Pembrokeshire coast as a long-distance route' and a preliminary survey was carried out by Ronald (R. M.) Lockley, the well-known Welsh naturalist. A formal decision to establish the path was taken by the County Council in July 1953 but 'its completion was long delayed by the obstinacy of a small number of landowners' (Miles 1987). The path was eventually declared open by Wynford Vaughan-Thomas, then President of the Council for the Protection of Rural Wales, at a ceremony held on Monkstone Point, near Tenby, on 16 May 1970.

The only National Park in the United Kingdom established primarily for its coastal landscape, the Pembrokeshire Coast is divided into four distinct, and separate, sections. The designation covers almost all the coast, every offshore island, the Daugleddau estuary and large areas of Mynydd Preseli and Cwm Gwaun but excludes the lower, industrialised, section of Milford Haven and the area around Fishguard. Despite occupying more than a third of the area of the county, some 612 km², it is one of our smallest National Parks, being only 16 km across at its widest point and 100 m at its narrowest. It is busy, however, with around 7.2 million day visits every year, 1.1 million of these from people on holiday. Many of these visitors walk at least part of the Pembrokeshire Coast Path. Oriel y Parc, the National Park visitor centre and art gallery in St Davids (Fig. 12) is an 'emphatically post-modern introduction to the ancient city' (Lloyd *et al.* 2010). Pembrokeshire Coast is one of the most densely populated National Parks, with approximately 22,542 people living within its boundaries. There are, however, only two large settlements inside the designated area, the city of St Davids and the town of Tenby; the majority of the population live in the small coastal villages and isolated farms and hamlets.

The National Park includes numerous sites of historic and archaeological importance and areas of national, or international, importance for their wildlife, including 7 Special Areas of Conservation, a Marine Conservation Zone, 6 National Nature Reserves and 75 Sites of Special Scientific Interest. One of the reasons we know these areas, and similar sites outside the National Park boundary, are special is that they have been extensively studied over many years by both amateur and professional naturalists. The activities of a number of these people are described in the following chapter.

CHAPTER 2

A Naturalist's Paradise?

*The county has been fortunate in having many all-round naturalists of the old
school living within its boundaries over the years. As a result, very few branches
of the natural sciences have remained unexplored in Pembrokeshire, and the
available literature is testament to this bounty.*

Jonathan Green and Owen Roberts (2004)

THIS BOOK IS MAINLY CONCERNED with the wildlife that can be found in
Pembrokeshire today, but it is difficult to describe the subject without
reference to those people who have done so much to raise the profile
of the area and its wildlife. This local knowledge has been supplemented by that of
visiting experts. Surprisingly, despite its popularity as a holiday destination, there
is not a particularly long history of wildlife recording in the county. Only a few
naturalists ventured this far west before the nineteenth century. The geologists
have their own separate story, which is described in the following chapter.

EARLIEST ACCOUNTS

Probably the earliest references to Pembrokeshire's wildlife come from Gerald
of Wales (Fig. 13). As a clerk to Henry II and two archbishops, he travelled widely
and wrote extensively. Gerald was selected to accompany the Archbishop of
Canterbury on a tour of Wales in 1188, a recruitment campaign for the Third
Crusade. His account of that journey, the *Itinerarium Cambriae* or *Journey through*

FIG 13. The statue of Gerald of Wales in St Davids Cathedral. His account of a journey through Wales, produced in 1191, was based initially on observation and eye-witness accounts and is a valuable historical document. (Jonathan Mullard)

Wales, produced in 1191, is a valuable historical document. His work was initially surprisingly accurate, based on observation and eye-witness accounts. Over time, however, and especially after Gerald retired from court in 1195, he added more classical and religious allegorical references (Raye 2016). For example, he gives a vivid and accurate description of the last beavers *Castor fiber* in Wales on the Afon Teifi, including many details which indicate that Gerald, or a correspondent, had observed the animals in the wild, but, at a later date, adds the myth that beavers castrate themselves to avoid danger. Among many other tales about the landscape and wildlife of Pembrokeshire, Gerald notes how a 'brood' of young weasels *Mustela nivalis* was found concealed in a fleece in someone's house, and how a young weasel was taken by an 'invidious kite' – which would have been a red kite *Milvus milvus*. Gerald died around 1223 and is buried in St Davids Cathedral.

After Gerald of Wales there seems to be a gap of a few centuries until the 1530s, when a short note by John Leland breaks the silence. Leland recorded that '*Bytwixt Whitland and Llanfeth is almost hethy yet on everi side good corne. But the ground is sumwhat baren of wood as al Penbrookshire almost is, except where a few parkes be'* (Toulmin Smith 1906). For more detailed information we have to wait until 1603, when George Owen of Henllys produced *The Description of Pembrokeshire*. His manuscript was a product of the new spirit of inquiry that developed in Elizabethan Britain, and it provides a fascinating insight into the county in the early seventeenth century. Anyone writing about the landscape, history or wildlife of Pembrokeshire does so in the shadow of George Owen. While his stated purpose was to place on record the results of his research, as a Tudor gentleman who had inherited a lordship, with feudal rights, he had other motives. His records of geology, landscape and wildlife provide a backdrop to this present book – and a sad reminder of how much we have lost over the succeeding centuries.

While we are lucky to have a copy of Owen's *First Booke*, a general history of the county, only a fragment of his *Second Booke*, which was going to be a detailed history of the county parish by parish, has survived, and it is doubtful whether Owen ever completed his ambitious scheme (Raye 2016). Tantalisingly, the antiquary Browne Willis, writing in 1717, mentions another manuscript, 'written in the latter End of Queen Elizabeth's Reign but, by whom I cannot certainly learn', given to him by 'an excellent Friend Mr Havard, Vicar of Aber-Gwily in Caermarthenshire'. Queen Elizabeth died in 1603, so this may have been a copy of Owen's manuscript – but Willis mentions him separately and the text he quotes does not match the published material.

Unfortunately, no portrait survives of this remarkable man, and perhaps none was painted. Neither Owen, nor any of his contemporaries, referred to his appearance and he left no diaries or letters that shed any light on the matter. All that is known is that he was lame in the right leg and suffered from gout in middle age. The pain was often so severe that he was unable to walk, but he still insisted on travelling and exploring, being lifted onto his horse by servants and carried to his bed. Despite this he never lost his sense of wonder, devoting a whole chapter of his book to the 'divers wonders' of the county, which he limited to nine, 'lest I become too wonder-full'!

Owen touched on many aspects of natural history. For instance, he ends his chapter 'Of the several sorts of fish' by marvelling at those that 'engender after the nature of beasts' – that is, the grey seals *Halichoerus grypus*, harbour porpoise *Phocoena phocoena* and thornpole (a dolphin?) – and he brings his discussion on birds to a close in amazement at the long-held belief that the barnacle goose *Branta leucopsis* was born from a barnacle attached to a ship's timbers (Owen 1603). In the same chapter he notes that 'in the bogs breed the crane'. So we have evidence that the common cranes *Grus grus* still bred in Pembrokeshire at this

FIG 14. Three cranes over Skokholm in March 2017. In time, they may return to breed once again in Pembrokeshire. (Richard Brown and Giselle Eagle)

date. Owen states that cranes can be found year-round, and while they breed in marshes, herons breed in tall trees, which is true. He notes, however, that 'the heronshaws are also found in many places of the sea cliffs' – and indeed, this was previously a favoured nesting place. Crane bones, dating to just before, or just after, the maximum of the last glaciation have been found in Little Hoyle Cave near Tenby, so this bird had a long history in Pembrokeshire.

Between 1994 and 2004 cranes were rare visitors to the county, only four individuals being recorded during the ten-year period (Green & Roberts 2004). In 2017 though a crane was seen at Goodwick in February, three individuals were recorded flying over Skokholm in March (Fig. 14), and one was sighted at Carnhedryn in April (Welsh Ornithological Society 2017). The previous year a pair of cranes bred successfully in Wales for the first time in around 400 years, successfully rearing a single chick on the Gwent Levels. The adults, like those seen in Pembrokeshire, came from the Great Crane Project reintroduction scheme which released 93 hand-reared cranes between 2010 and 2014 on the RSPB West Sedgemoor Reserve in Somerset. It seems likely therefore that, in time, cranes may return to breed in their old haunts.

LATER CENTURIES

In 1662 John Ray, the leading naturalist of the seventeenth century, visited St Davids, Haverfordwest, Tenby, Caldey and Pembroke with Francis Willughby, but most of their journey seems to have been unprofitable: 'Pembroke yielded nothing and Tenby only a long list of fishes.' But on Caldey Island and 'a little island between that and the mainland' (St Margaret's) there was 'great plenty of fowl' and 'some plants' (Raven 1942). They had decided to catalogue all living things in a methodical way for the first time, with Willughby concentrating on birds, mammals, fish and insects and Ray on plants. To achieve their aims they needed, wherever possible, to study at first hand the specimens in their natural habitat, which would have been an enormously difficult, and novel, undertaking at the time.

In later years, Ray was assisted in his botanical work by Edward Lhuyd (Fig. 15). In 1684 Lhuyd was appointed as assistant to Robert Plot, the Keeper of the Ashmolean Museum in Oxford, and replaced him as Keeper in 1690. In 1693–94

FIG 15. The statue of Edward Lhuyd (1660–1709), the famous Welsh naturalist, outside the University of Wales Centre for Advanced Welsh and Celtic Studies in Aberystwyth. (René and Peter van der Krogt)

he worked on revised descriptions of all the Welsh counties for a new edition of Camden's *Britannia* (a survey of Great Britain and Ireland first published in English in 1610) and this inspired him to plan an ambitious natural and human history of Wales, *Archaeologia Britannica*. The project led to an extended research tour in 1697–98 with his trained helpers William Jones, Robert Wynne and David Parry. They were regarded with suspicion almost everywhere, being looked upon as spies and conjurers in Pembrokeshire.

During his great tour Lhuyd was in Pembrokeshire for six months, staying near Tenby. A letter to him from the gardener at the Badminton Estate in Gloucestershire on 22 April 1698 states that the Duchess would like 'Sea Plants from Pembrokeshire, if she can have them within six weeks' (Gunther 1945). Unfortunately, we have no knowledge as to whether Lhuyd met this deadline, and if so, what plants he collected. His extensive collection of manuscript notes was sold at Sotheby's, London, in 1807 but it is believed that they were destroyed shortly afterwards in a fire at a bookbinder's workshop. They apparently included many records of both plants and animals, and we are much poorer today for their loss. Lhuyd was mainly looking for fossils, which he called 'formed stones', but writing from Swansea in September 1696 he states that:

> *This time two years [i.e. September 1694], there came a flock of birds (about a hundred) to a hemp-yard, at a place called Lhan Dewi Velfrey in Pembrokeshire; and in one afternoon destroyed all the Hempseed. They described the Cocks to be all over red as scarlet; the Hens greenish above, and red underneath; about as big, or little less than Blackbirds; with bills more stubbed and bigger than that of a Bullfinch.*

Lhuyd suspected them to be 'Virginia Nightgals' but we now know that they were crossbills *Loxia curvirostra*. Small numbers of crossbills are recorded in the county each summer and autumn on passage, although occasionally large flocks occur, as in this case (Green & Roberts 2004). The observer, and Lhuyd's correspondent, was the Reverend Nicholas Roberts of Llanddewi Velfrey, referred to as 'my ingenious Friend', who similarly provided him with information about the seabirds on Ramsey.

Between 25 June and 16 August 1773, a party including Sir Joseph Banks, Dr Daniel Solander, the Reverend John Lightfoot, Charles Greville, the artist Paul Sandby, and possibly the Swiss geologist and meteorologist Jean Andre de Luc, explored the coast of South Wales from Chepstow to St Davids. The purpose of the expedition was primarily botanical, but the travellers also had artistic aims. Their route was intended to retrace Ray and Willughby's itinerary, but they spent longer in Pembrokeshire than they had planned, being in the county from 11 to 26 July. The main reason for this was the presence of a keen local botanist and clergyman, the Reverend John Holcombe (1710–77), mentioned in Chapter 1. He was Rector of Saint Mary's Church in Tenby, and his grave inside the building is near a monument to Robert Recorde (1510–58), the Tenby mathematician who invented the equals sign – but there is unfortunately no plaque recording Holcombe's botanical expertise.[*]

Holcombe was a friend of Lightfoot, and corresponded with Sir John Cullum (who was working at the time on a new *Flora Anglicana*) and Sir Joseph Banks (Babington 1886). Judging by the available correspondence, both Banks and Cullum seem to have unfairly taken the credit for discoveries of plants made by Holcombe. In a letter sent to Cullum in 1775, for instance, Holcombe confirms that he was the first person to record several plants some years earlier, including '*Brassica oleracea* and *Lavatera arborea* near Tenby; *Convallaria polygonatum* in the very spot where Lightfoot saw it, the well-known locality for *Cyperus longus* at St David's – in a little gully about ¼ mile above Whitesand Bay' (Riddelsdell 1905). The four species mentioned by Holcombe are wild cabbage, tree mallow *Malva eriocalyx*, angular Solomon's seal (*Polygonatum odoratum*) and galingale *Cyperus longus* – the last of which had also been found in 1662 by Ray at Haverfordwest, 'by a little river, which is kept up by banks to drive a mill'. This scarce sedge no longer occurs near Whitesands, or Haverfordwest, the only location now being Cors Penally on the Penally rifle range (Fig. 16). Unfortunately, Lightfoot's record of the journey does not include a complete list of the plants the party found, as he aimed only to mention the rarer species, or criticise previous records.

Galingale was listed in the Pembrokeshire section of Dawson Turner and Lewis Weston Dillwyn's *Botanist's Guide*, published in 1805, again referencing Sir John Cullum. The localities of 48 species in the county were given, but as Turner and Dillwyn make clear, 'our responsibility is confined to the mere circumstance of having faithfully copied and properly arranged the materials before us' – so most plants were apparently not seen by the authors. Eight listings, however, were attributed to 'L.W.D.', including Danish scurvygrass *Cochlearia danica* 'about St Anne's Lighthouse and other places on the Coast'. So Dillwyn, at least, was familiar with south Pembrokeshire.

---

[*] Confusingly, the grave slab in St Mary's Church records that Holcombe died in 1770 instead of 1777, which appears to be the correct date, dying as he did in the '67th year of his age'. He was clearly alive in 1773, when the party undertook 'a prolonged investigation of the Pembrokeshire coast, under the guidance of Mr. Holcombe's companionship' (Riddelsdell 1905). There is also documentary evidence, in the form of the letter to Cullum, that he was active in 1775.

FIG 16. The nationally scarce galingale, first recorded in Pembrokeshire by John Ray, is now found only at Cors Penally near Tenby. (Stephen Evans)

Three years later, in 1808, Richard Fenton, who was born near St Davids, toured Pembrokeshire with the antiquarian Sir Richard Colt Hoare to collect material for *A Historical Tour through Pembrokeshire*. This was published in 1811 and, despite its title, mentions many items of geological and natural interest. In describing Goodwick Sands, near his house in Fishguard, for instance, Fenton notes that:

The barrier to these sands consists of a beach formed with pebbles, the aggregate of ages, backed by a high mound of sand, consolidated by sedge and the dog-rose, over which the horn poppy luxuriates with its delicate, but perishable, yellow bloom.

TENBY NATURALISTS

When Tenby became a fashionable spa town in the Victorian period, natural history provided a useful focus for those who were in a position to achieve a change of air and scene. Numerous books assured 'botanists, geologists, zoologists and the curious layman, that at every turn Tenby's natural delights were waiting to dazzle and amaze' (Tenby website 2016). One of these, a 'little tract' entitled *Contributions Towards a Catalogue of Plants Indigenous to the Neighbourhood of Tenby*, was published in 1848 by Dr Randle William Falconer. Falconer had founded the Tenby Literary and Scientific Society a year earlier, and his opening address drew attention to 'the need to record local plants and animals, for geological work and for a local museum to exhibit the results and to demonstrate the archaeological interest of the locality' (Davies 1981). The museum was finally opened in 1878 (Fig. 17).

Charles Babington, Professor of Botany at Cambridge, was a friend of Falconer, and mentions that his book was printed 'solely that the materials collected might not be lost, when a change of residence deprived him of his "long-cherished hope" of publishing a perfect catalogue of the native plants to be met with near Tenby' (Babington 1863). In his checklist of the *Plants of*

FIG 17. Tenby Museum and Art Gallery is the oldest independent museum in Wales. Founded in 1878, and overlooking St Catherine's Island, it incorporates one of the remaining parts of the medieval castle. There were many naturalists exploring the natural history of Tenby and its surroundings in the nineteenth century, and the museum holds some of their collections. (Jonathan Mullard)

Pembrokeshire (published in 1970 and still the only county flora), Tommie Warren (T. A. W.) Davis claims that 'no copy is known to have survived', but this is incorrect. I have a copy that was once owned by Professor Albert Howard Trow, Professor of Botany at University College Cardiff. Slightly battered, after being bequeathed to the University Library on Trow's death in 1939, it nevertheless represents one of the earliest attempts to 'draw attention to the natural vegetable productions of the neighbourhood of Tenby, and afford some assistance to its Residents, and numerous Visitors, who take a delight in the interesting study of Botany' (Falconer 1848). Like many early books, Falconer's 54-page *Catalogue* is now available as a free download through the British Library.

In 1863 Babington himself published a list of 550 species found in south Pembrokeshire in the first volume of the *Journal of Botany*. He had been at Cambridge with Charles Darwin, and in 1829 they apparently argued over who should have the pick of beetle specimens from a local dealer.

Ten years before Babington, in 1853, Edwin Lees had published a list of Pembrokeshire plants, observed in May and June of that year, in *The Phytologist*. Even then the flora of the area was under threat, Lees recording that:

> I only intend, in this paper, to note the vegetation that fell under my own view at this particular time. Such notices, by competent botanists, are, I think, advantageous; because changes are always in progress, more or less influencing the continuance, or causing the banishment, of particular species of plants. The Backwater at Tenby is now partially drained, and inclosed by stone-walls, to the detriment of its pristine beauty; and while I was there the northern side of the Burrows themselves was invaded, and columns of smoke daily trailing heavily over the ground, from heaps of smouldering gorse and withering plants; appearing, to a botanical eye, like the destruction and desolation of a battle-field.

Tenby Backwater, sometimes known as Holloway's Water, was an inlet immediately to the west of Tenby that was finally blocked by the construction of the Tenby to Pembroke railway line, which opened in 1863.

Frederick Walker, another enthusiastic amateur botanist, contributed several papers on the flora of Tenby and its environs to learned societies (Fig. 18). In Richard Mason's *Guide to the Town of Tenby and its Neighbourhood*, published in 1875, he wrote that '671 out of the 750 species recorded in the 7th edition of the *London Catalogue of Plants* have been recorded in the area'. His obituary in the *Tenby Observer* stated that 'he helped to bring the district into repute as a place where nature could be studied in some of her most varied and beautiful forms'.

FIG 18. Frederick Walker (1815–99) was one of the founders of Tenby Museum and a Fellow of the Royal Microscopical Society. He contributed several papers on the flora of Tenby and its surroundings to learned societies. (Tenby Museum)

Much of this activity was reinforced by the presence of some of the leading biologists of the day, including Fredrick Daniel Dyster, a surgeon naturalist with strong interests in marine zoology. Dyster was present at the meeting of the Linnean Society in July 1858 when Darwin and Wallace's papers on evolution by natural selection were first read. One of the many visitors to Tenby, Philip Henry Gosse, bought Dyster's microscope for £30 and used it for the rest of his life. In natural history circles, Gosse is best known for his book *Tenby: a Sea-Side Holiday* (1856), which contained details of his excursions along the shore with his family and friends in the summer of 1854. In the preface, he states that:

> *Nearly every day's occupation is set down, just as it occurred; tide pool excavation, cavern searchings, microscopic examinations, scenery huntings, road-side prying – here they all are a faithful narrative of how the Author was engaged for about six weeks at that pleasant little watering place.*

The book sold extremely well, at least in Tenby – perhaps too well. By 1856 Gosse was already complaining that the caves and rocks in the area had been badly damaged and many species, especially sea anemones, had been taken by collectors to display in their new aquariums – a fashion which Gosse himself had developed and popularised in other books.

At the bottom of the windows in St Anne's Chapel, inside St Mary's Church in Tenby, there is a note about William Lyons and his extensive family. Another local naturalist, Lyons lived in the town from about 1796 until his death in 1849 and created one of the most important collections of shells in Wales. This was donated to Tenby Museum by two of his daughters, Jane and Sarah, when the museum opened. The donation formed the basis of the museum's natural history collection, and includes a small bivalve shell identified by him that now bears his name, *Lyonsia norwegica*. A blue plaque, funded by a grant from the Royal Society, was erected outside Tenby Museum to celebrate his life and remember him as a 'local hero'. The museum is currently researching his family and his work to find out how, and when, he collected so many shells.

Tenby has a special place in the history of the study of the seashore, as indeed does Dale Fort on the western side of Milford Haven. The West Wales Field Society bought the fort for £6,000 in 1946 and leased it to the Council for the Promotion of Field Studies, now the Field Studies Council, who subsequently purchased it in 1959. John Barrett, the first warden at Dale Fort, was the author, with C. M. Yonge, of the influential *Collins Pocket Guide to the Sea Shore* (1958). Yonge had already written *The Sea Shore*, number 12 in the New Naturalist series, published in 1949.

ORNITHOLOGISTS

Birdwatchers in Pembrokeshire celebrate the fact that one of the great naturalist vicars of the late Victorian period, and author of *The Birds of Pembrokeshire and its Islands* (1984), the Reverend Murray A. Mathew, lived in the county from 1880 until 1888. Resigning the living of Bishops Lydeard in Somerset because of ill health, he settled in north Pembrokeshire, where he was curate of St Laurence at Wolf's Castle. Living at Stone Hall in the hamlet of Welsh Hook, he immediately began to explore the area. He was a member of the British Ornithologists' Union and a Fellow of the Linnean Society, and was thus well connected with developments in natural history. Locally, he was involved with the Pembrokeshire Naturalists' Field Club (forerunner of the West Wales Field Society), which began in 1879, and later the Pembrokeshire Natural History Society. There is not space here to do justice to these last two organisations and their evolution, but it is hoped that, in due course, a proper appreciation of their activities will be produced. As to Mathew's interests, he records that:

> *Much of our eight years' residence in the county … was devoted to the study of its birds. All the noted bird resorts were visited, as well as the various collections*

of stuffed birds we could hear of within the county; while from numerous sporting friends, and from others with a taste for natural history, whatever information they were able to impart was sought after and noted down.

Photographs of the Reverend Mathew have been hard to track down. For many years David Saunders, a notable Pembrokeshire naturalist, tried to locate one – and then in 2015 an entry in *The Sunday Telegraph* 'In Memoriam' column on Remembrance Sunday caught his eye. It referred to a Murray Mathew who had died on 4 July 1915 of wounds received at Gallipoli a month earlier. David knew this was a nephew of the Reverend, and through the newspaper was able to contact Mrs Rosemary Cadenhead (a great-great-granddaughter of the Reverend Murray's brother, himself a distinguished naturalist). Luckily, she managed to unearth three photographs from deep in the family archives, one of which is reproduced here (Fig. 19) (David Saunders, personal communication).

FIG 19. The Reverend Murray Mathew (1838–1908), one of the great naturalist vicars of the late Victorian period and author of *The Birds of Pembrokeshire and its Islands*. (Rosemary Cadenhead/David Saunders)

Many of the stuffed birds referred to by Mathew were probably prepared by James Tracy, a well-known Pembroke naturalist and taxidermist, and author of a 'Catalogue of Birds taken in Pembrokeshire; with Observations on their Habits, Manners, etc.', published in *The Zoologist* in 1850 and 1851. Unfortunately, or perhaps fortunately, given that those listed had been shot, the catalogue omitted seabirds. Through Tracy's hands passed several of the specimens used to illustrate William Yarrell's *History of British Birds*, which was published in 1839. The specimens included the first yellow-billed cuckoo *Coccyzus americanus* to be recorded in Britain, Tracy noting that:

> *The specimen from which Mr. Yarrell figured his bird was killed by my brother, near Stackpole Court. I first noticed it on the top of an ash-tree, in the act of feeding on some small insects on the wing, very similar to the golden-crests; seeing it appeared a nondescript it was shot immediately, and nothing more observed as to its habits.*

The yellow-billed cuckoo is a vagrant, which until 1994 had been recorded only three times in Wales, all in the nineteenth century. The 1994 record was at Porthclais near St Davids, and – times having changed – this bird, at least, was not killed (Green & Roberts 2004).

Later in the nineteenth century, following Tracy, more detailed accounts of Pembrokeshire's birds were produced by Thomas Dix and published in *The Zoologist* in 1866 and 1869. Dix notes that the magpie *Pica pica* was very common, 'but so readily destroyed that I fear it will soon be a rarity'. Indeed, his lists again include a depressing number of birds that have been killed.

The outcome of Murray Mathew's research, *The Birds of Pembrokeshire and its Islands*, was published in 1894 (like Falconer's botanical guide, this is also now available as a free download). In addition, Mathew wrote several papers and the section on birds in an early edition of Mason's *Guide to Tenby*. Accompanied by Mortimer Propert of St Davids, 'an enthusiastic oologist' (egg collector), Mathew stayed overnight on Skomer, describing, among other sights, the great kittiwake colony on the Wick and his nocturnal rambles among the Manx shearwaters *Puffinus puffinus* when 'the whole island seemed alive and the air vocal with their unearthly wailing cry (rendered as) – come over the wall, come over the wall'. He likewise visited Ramsey with Propert, noting that they 'had a troublesome pull across on our return' against the tide, a brisk east wind raising a choppy sea. Around 1900, in failing health, he retired to Bath, where he died in 1908. He is buried at Buckland Dinham in Somerset, the location of his last church.

TWENTIETH CENTURY AND BEYOND

Bertram 'Wolf' Lloyd, an all-round amateur naturalist, visited Pembrokeshire, often several times a year, between 1925 and 1940. A member of the Humanitarian League, whose aims were to 'enforce the principle that it is iniquitous to inflict avoidable suffering on any sentient being', he remained a close friend of Henry Salt, the organisation's founder, until Salt's death in 1939. Lloyd himself died in 1944 but his widow extracted all his plant records from his notebooks and published them in the *North Western Naturalist* (Lloyd 1948). For many years Secretary of the Hertfordshire Natural History Society, his self-penned epitaph includes the lines:

> *For he worshipped the Earth and all its creatures ...*
> *Ay, even its geological features;*
> *Plants, birds and beasts, whatever he saw,*
> *To know and love them was his chief lore:*
> *He cared not a farthing for Heaven or God,*
> *But valued far more an inch of green sod.*

Many of Lloyd's records were subsequently incorporated into Lillian Rees's very useful *List of Pembrokeshire Plants*, published in 1950, which was compiled at his suggestion. The list mainly covers the area around Tenby and south Pembrokeshire. A few years later Mary Gillham (1956a, 1956b), a well-known botanist and writer in South Wales, produced a number of papers on the botany of the Pembrokeshire islands, concentrating mainly on the effects of birds and mammals on the maritime flora. On International Women's Day in 2012, Gillham was one of thirteen women to be featured in an exhibition called 'Inspirational Botanists – Women of Wales' at the National Botanic Garden of Wales. The exhibition celebrated the contribution of women to the botanical sciences in Wales over the past 200 years.

R. M. Lockley arrived in Pembrokeshire in 1927, having taken a lease on Skokholm. His activities on the island, for which he is best known, and which are related in Chapter 5, ended in 1940 when he had to leave because the island was requisitioned by the army in the Second World War. Although Lockley never returned to live on Skokholm, he was the driving force behind the reopening of the bird observatory there in 1946, and in the same year helped establish a field centre on Skomer with his brother-in-law John Buxton, who later wrote *The Redstart* in the short-lived New Naturalist Monograph series (Buxton 1950). Lockley was also instrumental in the acquisition of Grassholm by the

FIG 20. R. M. Lockley moved to Orielton Manor in 1954. Nine years later it was bought by the Field Studies Council for use as a study centre, and it continues in this role today. (Jonathan Mullard)

RSPB in 1948 and was its honorary warden for many years. The same year, in collaboration with Colonel Harry Morrey Salmon and Geoffrey C. S. Ingram, he produced *The Birds of Pembrokeshire*; the first such publication since Mathew's book, over half a century earlier.

In 1954 Lockley moved to Orielton Manor (Fig. 20), where, among other things, he was involved with restarting the ringing of wildfowl at the duck decoy, one of the few such structures in Wales. He was also involved in the establishment of the Pembrokeshire Coast National Park and the Coast Path and was one of the founders of the Pembrokeshire Bird Protection Society, which later became the West Wales Field Society and is now part of the Wildlife Trust of South and West Wales.

In this latter role Lockley managed David Saunders, the first warden of the Skomer Island National Nature Reserve, who was appointed in 1960 (Fig. 21). This began David's long association with Pembrokeshire, and for over twenty years he was Director of the West Wales Trust for Nature Conservation (the successor to the Field Society) and, following Lockley, honorary warden of Grassholm. He also coordinated Operation Seafarer, a major census conducted in 1969 and

1970, which provided the first comprehensive account of the distribution and abundance of seabirds breeding around the coasts of Britain and Ireland.

For European birdwatchers the ultimate rare bird sighting must be a first record for the Western Palearctic (the Palearctic being one of the eight recognised global ecozones). Saunders was lucky enough to have such an experience when he noticed a strange bird at North Haven on Skomer in October 1961. He had no camera, so could only make detailed field notes and a rough sketch. This was duly submitted to the British Ornithologists' Union Records Committee, and after lengthy consideration it was thought to be a Blackburnian warbler *Setophaga fusca*, a North American bird. At the time the occurrence each autumn of birds from across the Atlantic was not recognised, so it was only considered a 'probable' sighting until 1990, when the record was finally accepted (Saunders & Saunders 1992).

Other notable Pembrokeshire naturalists include the various wardens of Skokholm and Skomer – especially Mike and Rosanne Alexander, who are still the longest-serving Skomer wardens, having spent ten years there from 1976 to 1986 (Fig. 22). Both have written books about the island and their experiences living and working there (Alexander 2017). Many of the island wardens, such as

FIG 21. R. M. Lockley (right) heading for Skokholm on a return visit with David Saunders in 1991. (Jack Donovan)

FIG 22. The author (left) and Mike Alexander, still one of the longest-serving Skomer wardens, on Stackpole National Nature Reserve in May 2019. (Jonathan Mullard)

Peter Conder, Mike Harris and Michael Brooke, went on to occupy key posts in conservation and research organisations. Conder was a prisoner of war between 1940 and 1945 along with John Barrett, John Buxton and George Waterston. He subsequently become Director of the RSPB in 1963, with Waterston as Director of RSPB Scotland (Niemann 2013). Mike Harris is a leading expert on the puffin, while Michael Brooke is now Curator of Ornithology at the Cambridge University Museum of Zoology and still continuing his interests in islands and seabirds. To these luminaries must be added the many people who have worked for the Countryside Council for Wales and its successor organisation, Natural Resources Wales, including Bob Haycock, former Senior Reserves Manager for Pembrokeshire. A number of other birdwatchers, including T. A. W. Davis, Jack Donovan and Graham Rees, whose exploits are recounted in full in *Birding in Pembrokeshire* (Green & Roberts 2004), have also played key roles in developing our understanding of the county's birdlife.

Finally, mention must be made of Sam Bosanquet, who managed to complete two bryophyte floras for southwest Wales within five years, firstly Carmarthenshire in 2005 and then, in 2010, Pembrokeshire. The product of a 'detailed, exhaustive, and at times exhausting, survey', it aimed to look at habitats in as much of the county as possible. Since most of the resident naturalists had previously neglected bryophytes, this was an important publication. The book includes a detailed history of bryophyte recording in the county. Interestingly, Stackpole National Nature Reserve was found to hold the greatest number of nationally and locally rare species.

The presence or absence of bryophytes is, however, strongly influenced by the local climate, which has been commented on by many writers over the centuries. George Owen, for instance, mentions that:

> *Giraldus Cambrensis highly commended the healthy air of this country, saying it is purified by the Irish air blown over the seas from Ireland into this part of the land, accounting their air of Ireland pure and healthy as coming from a country free by nature from venomous worms.*

Today we know that Pembrokeshire lies on the boundary between two climatic zones: the 'hyperoceanic' zone covering the north coast, St Davids peninsula and the Dale and Castlemartin peninsulas, and the 'euoceanic' zone which covers the rest of the county (Bendelow & Hartnup 1980). The hyperoceanic zone is characterised by mild winters and long, cool summers; winter air temperatures here rarely drop below 4°C, while summer temperatures above 20°C are seldom recorded. In contrast, the euoceanic zone is defined by colder winters and shorter, warmer, summers. Within each of these zones, any climatic differences are governed largely by the topography. The cooler and wetter area surrounding Mynydd Preseli, for example, contains more wetlands than the drier lowlands further to the southwest. The topography is, of course, determined by the geology, and this is considered next.

CHAPTER 3

The Underlying Rocks

A walk along the coastal path parades one rock formation after another, delineated by contrasts in colour and texture: here massive yellow or purple sandstones plunge like naked ribs into the churning foam, there a group of contorted dark shales zig-zag up the cliffs like some sort of berserk concertina.

Richard Fortey (2000)

PEMBROKESHIRE IS ONE OF THE classic areas for the study of geology and geomorphology and is said to have a greater variety of scenery than any other similarly sized area of the British Isles. The coast itself has been described, by the National Park Authority, as a 'superlative natural classroom in which large sections of the story of Planet Earth can be unravelled' (Fig. 23). The geology of the county is complex, since it includes rocks from six different eras and igneous as well as sedimentary rocks. Most were formed by the end of the Carboniferous Period, some 299 million years ago, when what is now Pembrokeshire was situated just north of the Equator. The majority of the rocks younger than this have been lost, principally due to weathering and erosion, particularly marine erosion. A small area of Oligocene clays from the Tertiary Period, between 66 and 3 million years ago, occurs near Flimston Chapel but is no longer easily accessible, as it is on the Castlemartin Range. The county's geological importance is recognised by the designation of 50 Geological Conservation Review sites; out of only 400, or so, in the whole of Wales. They are supplemented by a further 65 Regionally Important Geological Sites.

FIG 23. Coastal scenery at the Deer Park, overlooking Jack Sound near Marloes. Nowhere is the geology of Pembrokeshire more exposed than where the rocks are being eroded by the sea. (Blaise Bullimore)

GEOLOGICAL DISCOVERY

These riches have attracted the attention of geologists for centuries. The first person to take an interest in the geology of the county appears to have been George Owen. On a brass tablet in the parish church at Nevern this stalwart of Elizabethan Pembrokeshire is, perhaps confusingly, described as 'the patriarch of English geologists', based on a reference to him in the *Edinburgh Review* (Anon. 1841). It was Owen's desire to improve the agriculture of the county that led him to investigate and describe its geology. The soil, he claimed, needed lime and his purpose in identifying the distribution of limestone was to enable farmers to find where 'it lyeth hid, and save labour to others in seakinge it where there is no possibilitie to finde it'. He accurately identified the outcrops of limestone and demonstrated that they occurred in close parallel with the Coal Measures. Owen likewise recognised and described glacial till, or boulder clay, which he knew as clay marl, but followed the beliefs of his time in assuming that it had been deposited by Noah's Flood. Thomas Jehu, who quoted Owen's work at length in his classic paper on the glacial deposits of north Pembrokeshire (1904), stated that

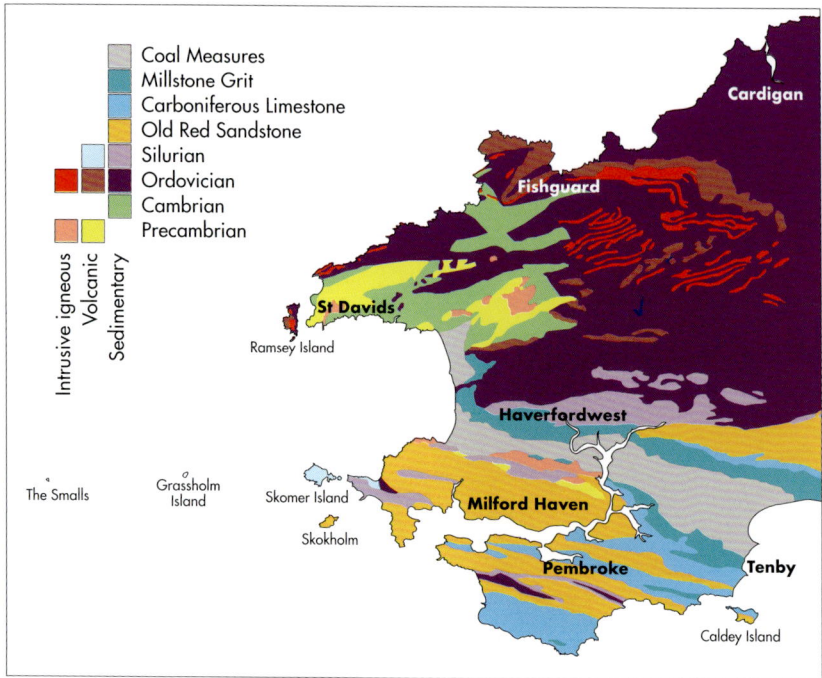

Coal Measures
Millstone Grit
Carboniferous Limestone
Old Red Sandstone
Silurian
Ordovician
Cambrian
Precambrian

Intrusive igneous
Volcanic
Sedimentary

Cardigan
Fishguard
St Davids
Ramsey Island
Haverfordwest
The Smalls
Grassholm Island
Skomer Island
Skokholm
Milford Haven
Pembroke
Tenby
Caldey Island

FIG 24. A simplified map showing the solid geology of Pembrokeshire. The lack of an adequate topographical base for geological maps hindered the progress of the early geologists for many years. (Pembrokeshire Coast National Park Authority)

this was 'perhaps the earliest attempt to give a full description of the boulder-clay'. Referring to the substance as 'claye marle', Owen (1599) wrote:

> *This kind of Marle is digged out of the Earthe where it is found in great quantitie and thought to be in rounde great heapes and lompes of Erthe as bigg as round hills, and is of nature fatt toughe and clamye. This marle is of couler with us most commonlie blwe and in some place red. It is verie hard to digg by reason of the toughness, much like to waxe, and the pickax or mattock beinge stroken into it, is hardlie drawne out againe, so fast is it holden; it is also verie heavie as ledd.*

It is unfortunate that Owen did not record the results of his surveys on a map. By doing so he would have preceded by 140 years the work of Christopher Packe, whose *A New Philosophico-chorographical Chart of East Kent*, published in 1743, was the first regional geological map. He certainly had an interest in cartography, criticising the depiction of Pembrokeshire in Christopher Saxton's *Atlas of England and Wales*, which was published in 1579. One of Owen's scribes, John Browne, also drafted a fine map of Pembrokeshire for inclusion in Camden's *Britannia* (Miles 1994).

Despite Packe's pioneering work, the lack of an adequate topographical base for geological maps continued to hinder progress for many years (Fig. 24). The problem was only solved following the establishment, in 1791, of the Trigonometrical Survey under the Board of Ordnance, which began to publish topographical maps at a scale of one inch to a mile (1:63,360) in 1805. When Henry De la Beche, the first director of the Geological Survey of Great Britain, came to map south Pembrokeshire in 1822 he was therefore able to utilise the newly published Ordnance maps for Pembroke (1818) and Haverfordwest (1819). He records how his good friend, the Reverend William Daniel Conybeare, Dean of Llandaff Cathedral, had earlier been defeated in his attempt to geologically survey the region for want 'of anything deserving the name of a map' (De la Beche 1829). Conybeare (Fig. 25) is probably best known for his ground-breaking work on marine reptile fossils in the 1820s and the first published scientific description of a plesiosaur.

The results of De la Beche's initial mapping were published in the Geological Society's *Transactions* in 1829, and by 1845 the whole of Pembrokeshire had been surveyed in some detail. One of the rocks which he mapped was the Old Red Sandstone, the sequence of which, in Pembrokeshire, differs north and south

FIG 25. The Reverend William Daniel Conybeare (1787–1857) was one of the first people to attempt to map the geology of south Pembrokeshire. (Wikimedia Commons)

of the Ritec Fault. This is one of the major faults in South Wales, running from Tenby through Pembroke Dock and Milford Haven to St George's Channel at Dale. Some geologists have suggested that the fault continues eastward under Carmarthen Bay to Gower, where it separates the peninsula from the mainland, but this remains unproven.

In 1831 Roderick Murchison, a Scottish geologist and geographer, arrived in Wales with the aim of finding out whether the rocks underlying the Old Red Sandstone, formed in the Devonian Period, could be grouped into a definite order of succession. These formations were, at that time, completely unknown. Following years of fieldwork, Murchison established the outlines of the Silurian system under which were grouped a remarkable series of formations, each with its own particular fossil record and distinct from those of the other rocks. These discoveries were published by Murchison in two books, *The Silurian System* in 1839 and *Siluria* in 1854. Unfortunately, he relied heavily on the work of other geologists, without always bothering to mention this in print, although he does acknowledge 'the great aid I derived from being in possession of the previous labours … of Mr. De la Beche, whose maps in fact are able outlines, which I have endeavoured to work up to the existing state of knowledge'. Murchison devotes a whole chapter of his first publication to the geological structure of Pembrokeshire, considering that 'washed on two sides by the sea, and fissured on its southern face by the deep bay of Milford Haven and its accessary inlets, [it] affords extraordinary facilities for the study of its mineral structure' (Murchison 1839).

On 26 May 1837, while Murchison was still exploring Wales, Henry Hicks was born in St Davids, where his father practised as a surgeon. After attending the cathedral school, he studied medicine at Guy's Hospital in London, becoming a member of the Royal College of Surgeons in 1862. Returning home to work, he met John Salter, a palaeontologist with the British Geological Survey, who was surveying Pembrokeshire. Inspired by Salter's enthusiasm for discovery, Hicks started to look for fossils himself. Before long he found a fossil brachiopod, *Lingula*, in the Cambrian rocks near St Davids, which were previously thought to be fossil-free. This discovery received a positive response, and supported by a grant from the British Association, Hicks succeeded the following year in discovering as many as 30 species in the Early Cambrian beds. He subsequently wrote over 60 papers on geological subjects, covering the earlier Palaeozoic strata of Pembrokeshire and the beds underlying conglomerates at St Davids, and in North Wales, which in his opinion marked the base of the Cambrian rocks. Unfortunately for Welsh geology, in 1871 Hicks moved to Hendon in Middlesex to further his medical career. He continued to be active in scientific organisations, however, especially the Geological Society, of which he was president from 1896 to 1898.

Thomas Jehu, mentioned earlier in connection with glacial till, was another Welsh physician and geologist, born in Llanfair Caereinion, Powys, in 1871. Author of many works on the geology of Wales and Scotland, he was appointed to the first lectureship in geology at the University of St Andrews. After his appointment he pursued his interests in glaciation through detailed studies of deposits in Caernarfonshire and Pembrokeshire.

Among the early geologists who worked in Pembrokeshire, Ernest Dixon deserves special mention. He completed the mapping of south Pembrokeshire for the Geological Survey of England and Wales and wrote the accompanying memoir, which is an outstanding record of his meticulous observations. The current British Geological Survey 1:50,000 map covering Pembroke and Linney Head is unchanged from the earliest 1:63,360 map produced from his work. Over a century later this is still an essential source of information for anyone studying the geology of the area. Dixon also made significant contributions to the mapping and description of the adjoining Milford and Haverfordwest areas.

The work of Thomas Neville George (1904–80), Professor of Geology at Glasgow University from 1947 to 1974, was said to have 'left an unmistakeable signature in the geological literature of Wales' (Howells 2007). His main research interests were carboniferous rocks and fossils, evolution, and geomorphology and paleogeography. Born in Swansea, he produced numerous papers on the geology of Wales. He was particularly known for writing a volume in the British Regional Geology series on South Wales, the last revision of which was published in 1970. Gareth T. George (no relation), whose family comes from St Davids and the surrounding farming communities, has similarly produced a very useful guide to the geology of South Wales, which includes the classic Pembrokeshire sites (George 2015). Finally, Sid Howells, the former Regional Earth Scientist for the Countryside Council for Wales and then Natural Resources Wales, has done an enormous amount to explain and promote the geology of Pembrokeshire to a wide variety of audiences.

PRECAMBRIAN AND CAMBRIAN

The oldest rocks in Pembrokeshire date from the late Precambrian, 650 to 541 million years ago. Today rocks from this period can be found at Hayscastle and Johnston, and on the Treginnis peninsula opposite Ramsey Island, where they were created by the intrusion of molten magma into accumulated lava and ash. (The igneous rocks from Bolton Hill quarry at Johnston supply much of the surfacing material for the roads in Pembrokeshire.) There was then a period of uplift and erosion, resulting in the volcanic rocks being folded. In the Cambrian

FIG 26. The west façade of St Davids Cathedral was completely rebuilt by George Gilbert Scott in the late 1860s, using the striking purple stone quarried from the nearby coastline. (Jonathan Mullard)

Period, 541 to 485 million years ago, this volcanic landscape was flooded by the sea, leading to an accumulation of sediments as the depth of water increased. These shallow seas on the coastal shelf persisted throughout much of the period, except during the Middle Cambrian when the water was deeper.

The St Davids peninsula is crossed by a major anticline, an upfold, which bought these Precambrian volcanics and Cambrian sediments, the conglomerates, shales and sandstones, to the surface. Only the sandstones are really suitable for building material, purple-coloured stone being used in the construction of both the cathedral and the nearby Bishop's Palace. Other materials used include an attractive pale green stone found in the same areas as the purple material, a deep red sandstone and brown mudstones (Lloyd *et al.* 2010). Together they form part of what is known as the 'Caerfai Group', which also includes St Non's Sandstone and Caerfai Bay Shales. The cliffs were originally quarried in the twelfth century, the stone probably being levered out with iron bars from several locations along the coast. Further quarrying took place in the 1860s to supply stone for the cathedral's restoration by George Gilbert Scott, a prolific 'Gothic revival' architect (Fig. 26). In 1998 one of the sites used as a source of stone in the nineteenth century was reopened for quarrying, under a special licence from the National Park Authority, for further repairs.

FIG 27. Porth y Rhaw near Solva, the classic location for the trilobite *Paradoxides davidis*. (Jonathan Mullard)

In rocks more recent than the Early Cambrian, fossils are fairly common, the most notable being the 'giant' trilobites which occur in the area around Solva. In 1862, while examining the coastal exposures of St Davids peninsula by boat, John Salter landed in the small inlet of Porth y Rhaw, apparently thinking that it was Solva harbour (Fig. 27). It was a lucky mistake, however, because in the rocks there he discovered the remains of one of the largest trilobites ever found (over 50 cm long), ensuring that the locality became established as a classic, and well-known, source of their fossils. Salter, who subsequently became the leading authority of the time on trilobites, named his find *Paradoxides davidis*, after his close friend David Homfray, an amateur fossil collector from Porthmadog. This species is now one of the best-known trilobites in Britain, and is illustrated in numerous publications. Indeed, it has been suggested that if there were a 'national fossil' for Wales, this species would be the prime contender (Fig. 28).

The trilobites at Porth y Rhaw played a key role in the career of Richard Fortey, who who worked for many years as a palaeontologist at the Natural

FIG 28. *Paradoxides davidis* is one of the best-known trilobites found in Britain and has been suggested as the 'national fossil' for Wales. This specimen was collected from Porth y Rhaw by Michael Lewis in the 1970s, during research for his PhD at Cardiff University on Cambrian stratigraphy and trilobites (Lewis 1988). (National Museum Wales)

History Museum in London. At the age of fourteen, he discovered his first specimen here, sparking a passionate and lifelong interest in these species. In his book *Trilobite!: Eyewitness to Evolution* (2000), he describes how he 'beat the rocks at Nine Wells and Porth-y-rhaw' to extract the fossils. It should be noted though that the procedure now is to collect only from loose material that would otherwise be lost, although the winter storms do, admittedly, erode significant areas of rock.

ORDOVICIAN

Following the Cambrian there was another period of uplift and erosion at the beginning of the Ordovician, 485 to 444 million years ago. This was again followed by the incursion of the sea and the creation of deeper water (known as the Welsh Basin) characterised by the accumulation of various muds, which commonly contain fossils of graptolites. The word 'graptolite' comes from the Greek *graptos* (writing) and *lithos* (stone) and refers to the fact that their fossils resemble pencil marks made on stone. They appear to have been planktonic organisms, like jellyfish, drifting about the sea feeding on algae or microscopic animals. Graptolites are sometimes found arranged in rings, as if they were attached at the time when they fell to the seafloor. The remains are found in rocks from the Cambrian to the Carboniferous, but they are most common in the Ordovician and Silurian.

Graptolites evolved rapidly, resulting in a huge variety of body shapes that had limited time spans. As a result, they are extremely useful for the relative dating of rocks. It was Murchison who first realised that while graptolite fossils often looked very similar at first glance, individual species could be linked to particular geological strata and their occurrence correlated across different localities, making the sequencing clear. Indeed, he even has one species, the tuning-fork graptolite *Didymograptus murchisoni*, named after him. *Didymous* in Greek means 'twin', and this accurately describes the shape of the graptolite, many of which are found in pairs, joined together by a short stem known as the prosicula (Fig. 29).

One of the best places in Pembrokeshire to find the tuning-fork graptolite is in the slates at Abereiddi Bay, to the north of St Davids (Monks 2017). These rocks formed in environments devoid of oxygen and consequently the graptolites did not decay as they would have done elsewhere. Given the importance of the area, the rocks should not be hammered but fossils are easily found in eroded pieces of rock. Apart from graptolites, other fossils recorded from Abereiddi include

FIG 29. Tuning-fork graptolites are extremely common in the shales at Abereiddi. They can sometimes be found still in pairs, joined by a short stem, known as the prosicula, which is said to give them the appearance of a tuning fork. Each tiny saw tooth on the long fossil fragments represents the end of a tube where the 'tuning forks' were attached. (Jonathan Mullard)

planktonic trilobites called agnostids and, in certain strata, *Lingula* brachiopods similar to those found by Henry Hicks.

The tuning-fork graptolite existed for only a relatively short period of time, during the latter part of the Middle Ordovician, about 470 to 464 million years ago. As a result, it is often used as an 'index fossil' for this period, which geologists refer to as the Llanvirn stage, after a cottage on the lane leading to Abereiddi. While graptolites quickly became widely used as index fossils, the nature of the animal itself took a long time to resolve. In life they were complex colonial animals, each tooth a small tube in which individual animals, zooids, lived. Nothing of the zoids remains, but their shared skeleton ultimately became preserved as the fossil known as a graptolite. Eventually, it was realised that they were related to similar species, pterobranchs, that live in the seas today. Like graptolites, pterobranchs are colonial organisms. The individual animals within the colony are small, inhabiting tube-like structures, which form clumps a few centimetres across. Unlike graptolites, however, they form a crust on rocks, rather than being planktonic.

A study of the geology of the north Pembrokeshire coastline between Dinas Island and Cemaes Head by the British Geological Survey in 2004, using graptolites to date the rock sequences, revealed that most of the material, representing more than 1.3 km of rock thickness, lay within one graptolite 'zone'. This thick sequence was deposited between 446 and 444 million years ago, a relatively short period of geological time. Elsewhere in Wales over the same timescale only a few tens of metres of black mudstone was deposited. The explanation seems to be that the area was subsiding rapidly during the Late Ordovician, creating a marine trough which then filled with mudstone and sandstones (Williams *et al.* 2004).

There was widespread marine volcanic activity during this period, which may have resulted in the formation of volcanic islands. The basalt pillow lavas of Strumble Head and spectacular rhyolitic rocks of Ramsey Island provide good examples of the products of these underwater eruptions. Rhyolite is an igneous rock formed during eruptions of granitic magma. Pillow lavas are formed

FIG 30. Pillow lavas on Strumble Head. These are formed when molten basalt lava erupts onto the sea floor and cools to produce formations resembling pillows or sacks. (Jonathan Mullard)

FIG 31. The cross-section of columnar basalt on Pen Anglas headland, one of the classic sites in the British Isles for pillow lavas and the related tuffs and breccias. (Jonathan Mullard)

when molten basalt lava erupts onto the sea floor. If the water is deep enough, the pressure stops the lava from exploding the water into steam. Instead, the liquid rock and liquid water coexist, the lava being extruded into successive oval globules resembling pillows or sacks (Fig. 30). Associated intrusions are marked by the prominent tors along the north Pembrokeshire coast, such as Carn Llidi and Garn Fawr, and Carn Menyn on Mynydd Preseli.

To the east of Strumble Head, Pen Anglas headland, near Goodwick, is one of the classic sites in the British Isles for pillow lavas and the related tuffs and breccias (Fig. 31). On the northeast-facing cliffs, below the Ordnance Survey triangulation pillar, there are some excellent examples of the hexagonal jointing. The inclined rock face consists of sections through the columns, demonstrating the hexagonal cooling joints. On the southwest side of the headland the long axes of the columns can be seen. The whole intrusion therefore appears to be dipping at about 45 degrees towards the northeast.

The geology of Pen Anglas has been celebrated for centuries, Richard Fenton (1811) noting that:

*On the Goodwick side of the projecting headland, called Penainglas, a ledge of
rock sloping to the sea presents a front curiously reticulated by the outbreak of
an almost horizontal stratum of basaltic columns, called by the common people,
torthau ceiniogau, or penny loaves, and uniting with as much regularity as the
cells of a honeycomb. There are two other columnar strata, one on each side of
the estuary of Fishguard, portions of which have been taken out for gateposts,
from six to seven feet high, and about twenty inches in diameter, but more
irregularly sided than these.*

SILURIAN AND DEVONIAN

During the Silurian Period, 444 to 419 million years ago, what is now
Pembrokeshire was covered by warm tropical seas with volcanic islands. There
were also shallow lagoons, indicated by ripple marks in the rocks, which are
thought to have been formed around 439 million years ago. Brachiopods and
corals were especially abundant during this period, exemplified by the fossil-rich
rocks on Marloes Sands. There is a long history of study of the classic exposures
at Marloes Sands, beginning with Murchison in 1839. The near-vertical beds of
Silurian sandstone rocks known as 'The Three Chimneys', created by the differing
degree of weathering between the alternate sandstones and siltstones, provides an
excellent introduction to the past environments of the period (Fig. 32).

While Marloes Sands is undoubtedly one of the most scenic locations in
Pembrokeshire, the same cannot be said for another key locality for Silurian
rocks, the famous Gasworks Lane exposures in Haverfordwest. Here, in a very
urban setting near the railway bridge, is a key locality for the Llandovery Age,
a subdivision of the Silurian Period named after that area of Carmarthenshire.
It has yielded one of the best assemblages of fossils that have so far been found
in rocks formed about 440 million years ago. A wide range of brachiopods,
including *Leptaena haverfordensis*, *Leptostrophia reedi*, *Dolerorthis sowerbyiana* and
Resserella llandoveriana, have been found here as well as solitary corals, crinoid
columnals, tentaculitids, bryozoa and sponges (Potter 2013). As with graptolites,
the age of the sandstone exposed at this site can be determined with reasonable
precision because of known evolutionary changes in some of the fossil groups.

In the late Silurian and Devonian, 419 to 359 million years ago, the Caledonian
Orogeny, a mountain-building event caused by the closure of the Iapetus Ocean,
created the folds and faults seen in north Pembrokeshire, in places such as
Ceibwr Bay. In contrast, the Old Red Sandstone in south Pembrokeshire (Fig. 33)

FIG 32. The Three Chimneys on Marloes Sands, a rock formation of near-vertical beds of Silurian sandstones that stand out in the cliff from the less resistant, and interbedded, sandy mudstones. (Blaise Bullimore)

FIG 33. A striking exposure of the Old Red Sandstone strata in the aptly named Red Berry Bay on Caldey Island. Here vertical beds of the rock, derived from sediments deposited on coastal plains, are being eroded along a fault. (Jonathan Mullard)

represents sediments deposited on coastal plains, where there were braided river channels with a fringe of early land plants, examples of which have been found at Freshwater East. The rivers contained primitive armoured fish, while there is evidence of amphibians burrowing on the floodplains.

CARBONIFEROUS

The Early Carboniferous, 359 to 323 million years ago, was also a time of warm tropical seas with abundant corals, sea lilies and orthocones, squid-like creatures with elongated conical shells. In the Middle and Late Carboniferous, 323 to 290 million years ago, swamps formed in river deltas and there was an accumulation of peat beneath giant clubmosses, ferns and horsetails that provided the source of the anthracite coalfield in Pembrokeshire. Coal mining here dates back at least to the early fourteenth century, with the coal being used locally, or exported by sea. The coalfield stretched across the centre of the county from Saundersfoot in the east to St Brides Bay in the west. It produced anthracite of the very finest quality, having a very high carbon content which meant it burned brightly, while giving off very little smoke. Despite this, the coal seams were usually thin and plagued by complex fault lines. Marine fossils, such as ammonites and scallop-like bivalves, indicate that the delta swamps were periodically inundated by the sea.

As George Owen described, the outcrops of limestone occur close to the Coal Measures. The Carboniferous Limestone, which covers much of south Pembrokeshire, was deposited between 359 and 325 million years ago. It is composed of layers of shelly sand and mud that accumulated in warm seas near the Equator. The total thickness of the layers is approximately 1,500 m, equivalent to the height of Ben Nevis (Howells 2008). By comparison, the modern cliffs at Stack Rocks are only 40 m high.

Around 290 million years ago, towards the end of the Carboniferous Period, the Rheic Ocean, to the south of what is now Wales, began to close as the continent of Gondwana (containing most of modern-day Africa) moved northward. Eventually, it collided with the European landmass to form the

FIG 34. The Ladies Cave anticline at Saundersfoot, often pictured in geology textbooks, may well be the most famous such feature in Britain. It shows the northward-verging folding of the Variscan. (Jonathan Mullard)

FIG 35. A classic example of folding near Den's Door, to the north of Broadhaven, where the 'fold-thrust' structures have inspired generations of structural geologists. (Jonathan Mullard)

Variscan Mountains, now largely eroded away. South Wales was greatly affected by this collision, which created the many folds and faults to be seen today in south Pembrokeshire. Previously horizontal layers were folded and contorted, the best-known of which is the anticline known as the Ladies Cave on the beach at Saundersfoot (Fig. 34). Spectacular, and easily accessible, it is well known to generations of geology students because of its frequent inclusion in textbooks.

Another standard textbook example of the folding that occurred during this period can be seen near Den's Door, a sea arch to the north of Broadhaven (Fig. 35). The classic 'fold-thrust' structures here mark the Variscan Front, and were described by Murchison in 1839 as 'a series of breaks, curvatures, thinnings out, and contortions, which cannot be exceeded even in imagination'. The Sleek Stone nearby is a huge 'whaleback' rock and part of an upfold in the Coal

FIG 36. A large sea stack, with the larger of the two natural arches known as Den's Door. The huge 'whaleback' rock to the right is the Sleek Stone, part of an upfold in the Coal Measures. (Richard Law – Geograph/Creative Commons)

Measures (Fig. 36). Since Pembrokeshire is an important training ground for geology students, the Broadhaven outcrops are often the first place many future earth scientists experience the delights of structural geology.

WEATHERING AND EROSION

Following the Carboniferous Period, during the Permian, Triassic and Jurassic Periods, 299 to 142 million years ago, the older rocks in Pembrokeshire were weathered and eroded. In the Cretaceous Period which followed, 145 to 66 million years ago, the sea levels were extremely high and, while chalk was deposited in southern England, around 70 million years ago the surface of the county was cut by marine erosion (Fig. 37). The resulting plateau is a modified remnant of a marine erosion surface which is similar to, but obviously much more extensive than, the wave-cut rock platforms seen around the present coastline. It was subsequently raised to 60 m above sea level by a combination of tectonic forces and falls in sea level that came about because of the development of continental ice sheets.

Following this uplift, further marine erosion produced the impressive cliffs which extend from Stackpole to Linney Head. The coastline here is extremely exposed and waves around 5–10 m high, and occasionally higher, are often encountered in the area, as it receives the full force of Atlantic weather systems (Howells 2008). The resulting erosion of the seaward face of the limestone has produced a range of spectacular coastal features. There are promontories, arches, stacks, caves, cauldrons, blow-holes and huge crevasses, all within a relatively short distance of each other. Many of these features have developed as the sea has eroded dry watercourses and cave systems produced by drainage from an ancient land surface, or faults. In his Presidential address to the Geologists' Association in 1933, on 'The geology and scenery of Tenby and the South Pembrokeshire Coast', Arthur Leach, Curator of Tenby Museum, noted that 'It is a district worn down to a gently undulating surface without any striking inland features but the coast attains a high level of geological interest and of beauty, even grandeur, of scenery.' Murchison (1839) considered that 'There is no part of the coast of Great Britain where a longer continuous zone of the carboniferous limestone is laid open, nor is it anywhere so much contorted as in this promontory.'

FIG 37. Looking east along the south Pembrokeshire cliffs towards St Govan's Head. The coastal plateau, a relic of marine erosion in the Late Cretaceous Period, is obvious from this viewpoint. (Jonathan Mullard)

The best-known arch is the Green Bridge of Wales, one of the most famous landmarks in the country (Fig. 38). A short walk from a car park on the Castlemartin Range, it attracts thousands of visitors each year. The standard explanation is that the bridge was created as a result of erosion on both sides of a headland, caves being formed and gradually enlarged until they met in the middle, creating an arch. In this case, however, the explanation may be more complex. Whatever the cause, the resulting structure is around 24 m high and has a span of more than 20 m. Its outer support rests on a broad pedestal of rock, which clearly shows the landward dip of the strata. In the autumn of 2017 Storms Ophelia and Brian resulted in the bridge losing a huge section of its seaward leg. Similar events in the future are, unfortunately, likely to lead to its complete collapse.

A short distance to the east of the Green Bridge are two isolated stacks, known as Stack Rocks, or the Elegug Stacks after the local name for guillemots *Uria aalge*, which nest here each year. The stacks represent the final stage of cliff

FIG 38. The spectacular limestone arch known as the Green Bridge of Wales was created by erosion on both sides of a headland. This photograph was taken before the storms in the autumn of 2017, which removed a huge section of its seaward leg (see Fig. 255). (Jonathan Mullard)

FIG 39. Pen-y-holt Stack, on the left of the photograph, is a significant geological feature within the Castlemartin Range. (Jonathan Mullard)

erosion, when an arch has collapsed leaving the supporting pedestals. In time they will be undermined by the sea, becoming merely stumps covered at high tide. Not all stacks develop from arches though, some being formed when the joints in a headland are widened by the sea and rainwater. Another significant stack is located in Pen-y-holt Bay (Fig. 39), but owing to the military training the feature can only been seen a few times a year on guided walks led by the National Park Authority.

Further east again lies the Devil's Cauldron, an enclosed shaft some 45 m deep and up to 55 m across. The shaft has been formed by the growth, coalescence and collapse of a number of blow-holes, or cavities, which developed along 'master joints' in the rock, planes of separation where no detectable displacement has taken place. Blow-holes are formed as a result of water pressure, caused by waves, creating upward erosion. Eventually, this reaches the top of the cliff and the cave is joined by a narrow shaft, the blow-hole, to the cliff top. On the southern side

FIG 40. Huntsman's Leap, south of Bosherston, is a deep and narrow crevasse formed by erosion along a fault in the Carboniferous Limestone. (Jonathan Mullard)

the Cauldron is connected to the open sea by a bridge and, as erosion continues, it will eventually collapse, to form a series of stacks and arches that will dwarf the Green Bridge and Stack Rocks (Goudie & Gardner 1992).

Near the ancient St Govan's Chapel lies Huntsman's Leap, a narrow and shear-sided coastal crevasse created by erosion along a fault (Fig. 40). The name is said to be derived from an occasion when a man on horseback jumped from one side to the other while being pursued. On looking back and seeing the gap that he had jumped, he supposedly died of shock. The nearby Stennis Ford and Rousehole at Giltar have been formed in the same way.

Other cavities along the coast include Lydstep Caverns, which were heavily promoted in the Victorian period as a tourist attraction. *Allen's Guide to Tenby* mentions a Betsy Brinn, who with her child lived in a nearby cottage, and would, for a small fee, show visitors the caves (Gwynne 1868). The 'Wonders of Lydstep Bay' included the 'Cave of Beauty', 'the Tower', 'the Drot' (the local pronunciation of draught) and the 'Smugglers Cave'. The last of these is a large sea cave about 92 m long with an average passage cross-section of 6 m wide and 9 m high. At the rear of the cave is a blow-hole which emerges on the western face of Lydstep Headland. The Victorians installed steps with railings, creating an additional route into the cave, but today it is a dangerous descent which is not recommended. To the west, the Drot has a similarly sized entrance, which opens out into a chamber roughly 18 m square.

QUATERNARY

The Quaternary Period, 3 million to 10,000 years ago, was marked by very cold conditions, with ice covering much of Britain. Material from glaciers in the mountains on the western coast of Britain joined with sea ice to form the Irish Sea ice sheet, which advanced across Cardigan Bay. During an early glaciation (c. 450,000 years ago) the ice completely covered Pembrokeshire and reached as far as the north Devon coast. During the most recent glaciation (c. 18,500 years ago), however, only north Pembrokeshire and St Brides Bay were covered by ice and south Pembrokeshire experienced tundra conditions. The ice carried boulders hundreds of kilometres from their original location. These glacial 'erratics' take their name from the Latin word *errare*, which means to wander. Thomas Jehu called them 'travelled boulders' – and descriptions of them are scattered in the geological literature of the county, as they are themselves scattered across the land.

At Flimston Chapel, Castlemartin, two large boulders have been used as headstones for the graves of Colonel F. W. and Lady Victoria Lambton of Brownslade House, the Colonel having collected many erratics from the surrounding area (Fig. 41) (Chandler 1909). The most spectacular location for an erratic though is St Govan's Head, where a large boulder, probably from North Wales, overhangs the cliff edge.

A particularly good exposure of Quaternary deposits, dating to the Late Pleistocene, can be found at Aber Mawr. Here there is a comprehensive sequence of glacial deposits, consisting of till, angular bedrock fragments and erratics. They represent a complete cycle of advancing and retreating ice formed close to the margin of a glacier. As a result, there is a continuous stratigraphic record here, going back around 125,000 years.

At the maximum extent of glaciation sea levels were around 50 m lower than they are today, and it would have been possible to walk from Tenby across to

FIG 41. Erratic boulders at Flimston Chapel, used as headstones for the graves of Colonel F. W. and Lady Victoria Lambton who lived in the nearby Brownslade House. The remains of the house were demolished around 1980. (Jonathan Mullard)

Gower. Rising sea levels, however, following the melting of the ice at the end of the last glaciation, drowned many river valleys, producing spectacular 'rias'. The Milford Haven Waterway is one of the most spectacular examples. The vast volumes of water created a huge temporary lake, Lake Teifi, and channels such as Cwm Gwaun and the deep gorge of the Afon Teifi between Llechryd and Cardigan were probably originally cut by subglacial meltwater flowing outwards from the Welsh ice cap. The Western Cleddau, in particular, is a 'misfit' river, a relatively small watercourse running through a deep valley formed at the end of the last glaciation. At this time the Afon Teifi, swollen by meltwaters, was unable to reach the Irish Sea along its previous course since it was still blocked by ice. Instead it flowed west, through the Nyfer (Nevern) and Cwm Gwaun, before heading south along the course of the Western Cleddau, scouring out the landscape as it went. The deep wooded gorges at Treffgarne are just one feature created by this massive movement of water, some 12,000 years ago.

The postglacial rise in sea level which occurred as a result is associated with the migration of shingle ridges, such as that at Newgale, and the submergence of previously wooded coastal plains.

LOST LANDS AND SUBMERGED FORESTS

All around the British coast are the remains of lands which have been destroyed by the sea, due to gradual erosion, sudden subsidence or the rising sea levels after the last glaciation, and there are numerous documents and legends describing the submerged areas. For example, Manuscript 3514 in the library of Exeter Cathedral, dated around 1280, records that 'The Kingdom of Tewthi, son of Gwynnon, King of Kaerrihog between Mynwy [St Davids] and Ireland [was submerged by the sea.] No one escaped from it, neither man nor beast, except Teithi Hen and his horse, and for the rest of his life he was sick with fright' (Pennick 1987).

For centuries, storms along the Welsh coast have revealed the remains of inundated forests. Gerald of Wales noted the submerged forest at Newgale in 1188 during his tour:

> We then passed over Niwegal sands, at which place (during the winter that king Henry II spent in Ireland), as well as in almost all the other western ports, a very remarkable circumstance occurred. The sandy shores of South Wales, being laid bare by the extraordinary violence of a storm, the surface of the earth, which had been covered for many ages, re-appeared, and discovered the trunks of trees cut off, standing in the very sea itself, the strokes of the hatchet appearing as if made only yesterday. The soil was very black, and the wood like ebony. By a wonderful revolution, the road for ships became impassable, and looked, not like a shore, but like a grove cut down, perhaps, at the time of the deluge, or not long after, but certainly in very remote ages, being by degrees consumed and swallowed up by the violence and encroachments of the sea.

This theme was repeated by William Camden, the antiquarian, who noted that in the time of Henry II storms exposed 'the trunks of trees which had been cut down standing in the midst of the sea with the strokes of an axe as fresh as if they had been yesterday with very black earth'. Before their true nature was understood, they were believed to be the result of the biblical flood and were called 'Noah's trees'. Later seekers of curiosities also noted the phenomenon, and Jehu in his geological paper of 1904 mentions the remains of a similar forest in Whitesands Bay:

> Again about 16 years ago [c. 1888] a big storm washed away the sand and exposed roots of great trees in Whitesand Bay. Huge logs of oak trees were carried away by the neighbouring farmers, some of which are still stored, and were shown to the writer. Twigs and branches of hazel were found in abundance, although no hazel grows now near St David's. The writer is also informed that horns of deer were picked up.

No serious analysis of the subject was, however, undertaken until 1913 when Clement Reid, another geologist, published a book on the subject. His volume *Submerged Forests* attempted to put these trees into a wider archaeological context and argued conclusively that they were the result of a rise in sea level. Later, in 1957, F. J. North in his book *Sunken Cities* included a photograph of a tree stump and root system at Marros Sands, just over the border in Carmarthenshire. Part of a submerged forest which was regularly exposed at low tide, the stumps can still be seen when low tides and rough seas combine to remove the accumulated mud and sand. Both this and the similar features at Amroth are rooted in peat levels lying below the marine sand ('the very black earth' noted by Gerald and Camden) and preserved by the continuously waterlogged conditions (Fig. 42). There are several beaches where this phenomenon can be seen, but the stumps and roots of oak *Quercus* sp., alder *Alnus glutinosa* and willow *Salix* sp. are better preserved here than anywhere else in Wales.

These tree stumps around the Pembrokeshire coastline, and elsewhere, are the strongest direct evidence we have for the existence of the lost lands of Wales, and the remains of terrestrial animals have been also excavated from the deposits around the tree stumps. In 2014, when a storm scoured the beach and moved the shingle bank at Newgale inland, it revealed more tree stumps, together with human footprints. The devastation was remarkably similar to the twelfth-century account by Gerald of Wales of the earlier storm at Newgale (Brown 2015). The footprints in the exposed peat, which probably date from the Mesolithic, around 10,000 years ago, suggest people may have been tracking animals, perhaps an aurochs *Bos primigenius*. The ancestor of domestic cattle, these huge animals survived in Europe until the last recorded live aurochs, a female, died in 1627 in the Jaktorów Forest, Poland, from natural causes. Aurochs horns, together with

FIG 42. The remains of the submerged forest on the beach at Amroth. This is one of the best examples in Wales of a drowned landscape. (David Evans)

skull fragments, were found at Whitesands by Shaun Thompson, a local resident, after the 2014 storm (Fig. 43). One was almost complete, but only the bottom half of the other was found. This was not the first find of aurochs bones in Pembrokeshire, since some horns were discovered at Wiseman's Bridge, between Saundersfoot and Amroth in 1914 and are now in Tenby Museum. In addition, the 'greater part' of an aurochs skeleton was dug up on the shore about 3 km from the earlier site in the 1930s, and this is now in National Museum Wales in Cardiff (Matheson 1932).

Other finds include those of red deer *Cervus elaphus* (the 'horns of deer' mentioned by Jehu) and brown bear *Ursus arctos* at Whitesands, and the famous Lydstep Pig, a wild boar *Sus scrofa*. The pig was discovered by Arthur Leach in 1917, when the skeleton, now estimated to be 6,300 years old, was found trapped

FIG 43. One of two aurochs horns found at Whitesands in 2014 by Shaun Thompson, a local resident. It is held here by Phil Bennett, former Culture and Heritage Manager for the Pembrokeshire Coast National Park Authority. (Pembrokeshire Coast National Park Authority)

beneath a tree trunk, with two broken flint points in its neck (Leach 1918). It was assumed that the animal escaped from the people that were hunting it, only to die later of its wounds. Although numerous animal bones have been found in association with the submerged forests, it is very rare to find clear signs of hunting such as this. The skeleton of the pig is now stored in the Natural History Museum in London but Tenby Museum and Art Gallery have a display based on images of the bones, a glass slide of the skeleton made at the time of the discovery and original documentation.

Further evidence about the Lydstep Pig emerged in 2010 when a local resident contacted Dyfed Archaeological Trust to report footprints on the beach close to the location where the pig was discovered. On site the Trust found red deer hoofprints preserved in the surface of a solidified peat deposit that once formed the floor of a shallow lagoon. Human footprints were also present, including those of children, their prints deeply embedded into the peat as if they had stood patiently waiting in one place, perhaps as part of a hunting party hiding near the watering hole. Now the immediate environment of the find is clearer, an alternative interpretation of events has been put forward, which is that the pig did not escape but its body was placed in the lagoon as an offering and secured by a tree trunk. We will never know the true story.

These lost lands, however, represent only a comparatively small area of inundation, and further advances by the sea, due to global warming, are now inevitable. The sea contains an enormous variety of species, which will be affected by the changing climate. In order to know what is at risk, therefore, we need to understand the current situation.

CHAPTER 4

Inshore and Offshore Waters

Britain's shallow seas are a mysterious domain. They remain largely unseen and unexplored except by marine scientists and divers who have been documenting their wondrous discoveries over many years.

Keith Hiscock (2018)

T HE WATERS AROUND PEMBROKESHIRE AND its islands are rich in wildlife, though much of this is invisible to visitors and naturalists, unless they are experienced divers. An interactive display about the Skomer Marine Conservation Zone at Martin's Haven provides a glimpse of what can be found in the impressive underwater landscape here, but this represents only a fraction of the marine environment (Fig. 44). Other offshore locations are equally diverse, and this region of the Celtic Sea, covering around 1,500 km², includes subtidal sandbanks, reefs and marine caves, alongside many other features.

The topography of the seabed varies considerably, with some of the deepest inshore waters within Ramsey Sound, where it drops to a depth of 66 m. Elsewhere there are underwater cliffs, or rocky slopes, particularly around the western headlands, islands and offshore rocks, with depths of over 100 m west of the Smalls. From a global perspective, however, these are shallow seas, part of the European continental shelf, and extremely productive in biological terms. Bedrocks of slate, mudstone and sandstone and sediments of sand and gravel add to the variety. The highly variable topography, together with the indented coastline and high tidal ranges, produces strong tidal streams particularly around headlands, through the sounds or channels, and in the inlets.

FIG 44. Martin's Haven and 'Fisherman's Cottage'. Along with the nearby Lockley Lodge run by the Wildlife Trust, the base for the Marine Conservation Zone is the main interpretation centre for the area. Boats from the Haven run through the Zone to Skomer and Skokholm. (Jonathan Mullard)

Although there is still much to learn about their wildlife, these seas, and the waters of Milford Haven, are one of the best-studied marine environments in the United Kingdom. In 1977 the area around Skomer and Middleholm, and most of the adjacent Marloes peninsula, was declared as one of the first two Voluntary Marine Nature Reserves, the other being Lundy in the Bristol Channel. Despite having no legal basis, and being entirely dependent on the cooperation of users, the voluntary approach was surprisingly successful. A few years later, in 1981, the Wildlife and Countryside Act included provisions for the legal establishment of Marine Nature Reserves, and this led to the designation of Skomer Marine Nature Reserve in July 1990. The negotiations were protracted, however, and it was only after four years of discussions, and with a number of the originally proposed byelaws deleted, that the Marine Nature Reserve was finally designated.

Once established, the new reserve was almost immediately effective. Some of this success was a consequence of the foundations laid during the life of the voluntary reserve, but most of the credit must go to the professional staff employed to manage the area. In 2014 the Marine Nature Reserve was replaced by a Marine Conservation Zone, established under the Marine and Coastal Access Act 2009. So far there have been no significant management changes and no apparent loss of protection. The area currently benefits from interim safeguards in the legislation that maintain levels of protection until such time as specific measures are established by the Welsh Government. For 24 years the reserve provided security for the marine and terrestrial wildlife around Skomer, and it can only be hoped that the new Marine Conservation Zone will be as effective.

The purpose of the Marine Conservation Zone is to safeguard the full range of marine wildlife diversity, but in practice the protection provided by the designation is limited, since, except for scallop dredging and beam trawling, fishing continues (Bullimore 2017). While great scallop *Pecten maximus* populations and sedimentary habitats are stable and improving, other species are still under pressure. Sadly, there is still no marine area in Wales that is completely free of commercial fishing pressure and which could be studied to demonstrate the benefits of marine protected areas for the wider marine ecosystem. But surveys of scallops indicate the potential. Populations have recovered dramatically since collection, by dredging or diving, was prohibited in 1988. Their mean density has increased by more than 25 times, though recovery at some sites slowed after 2006 and one or two locations have not responded so well (Bullimore 2014, Burton *et al.* 2017).

Much of the sea around Pembrokeshire falls within the Pembrokeshire Marine Special Area of Conservation. The Wales National Marine Plan commits the Welsh Government to maintaining and enhancing marine ecosystems, but it remains to be seen what resources are made available for this objective.

SKOMER MARINE CONSERVATION ZONE

The Marine Conservation Zone is at the northern limit for many warm-water species and the southern limit for some northern species, as at Skomer the warm waters of the Gulf Stream meet the cold-water currents from the Arctic. These strong tidal currents are rich in plankton, which forms the base of the marine food chain.

In the shallow areas of the Conservation Zone there are dense 'forests' of brown kelp *Laminaria* spp. These light-dependent seaweeds cannot survive here in water much deeper than 15 m. Below this depth diminishing light levels mean that algae give way to animals. The wildlife includes over 100 different sponges,

FIG 45. Pink sea fan, one of the most exotic-looking animals on the seabed, is actually a colony of anemone-like polyps. It is one of the few marine invertebrates specifically identified for protection in the Wildlife and Countryside Act. (Blaise Bullimore)

40 species of anemone and soft coral, and 76 species of sea slug (Jones *et al.* 2016). Steep walls off the north coast of Skomer are covered with soft corals such as dead man's fingers *Alcyonium digitatum* and red sea fingers *A. glomeratum,* pink sea fan *Eunicella verrucosa* and a rich animal turf composed of sea mats, hydroids, sponges and anemones. The intricately branched pink sea fan is one of the most exotic-looking animals on the seabed and one of the few marine invertebrate species to be listed for protection under the Wildlife and Countryside Act (Fig. 45). It is classified as 'vulnerable' on the global International Union for Conservation of Nature (IUCN) Red List, a comprehensive inventory of the global conservation status of plants and animals.

An individual sea fan can grow up to 80 cm high and 100 cm across, but it is not a single animal. It is actually a colony of tiny anemone-like polyps with stinging tentacles, which capture microscopic animals from the passing water. They usually grow at right angles to the prevailing water currents, to catch as much food as they can. The colonies are extremely slow-growing and vulnerable to damage. Numbers in one area of the Marine Conservation Zone declined sharply in 2016, perhaps as a result of the impact of lobster-potting (Newman *et al.* 2018). Elsewhere in Pembrokeshire, threats to the species include beam trawling, scallop dredging and boat anchoring. Between 2002 and 2014 the conservation team reattached seven fans that had become dislodged, either fastening them to the pitons used to mark the survey sites, or gluing them into small rock crevices with epoxy resin. Most of these colonies subsequently re-established their holdfasts on the rock, but without this intervention they would have been lost (Bullimore 2014).

Pink sea fans were one of the first species to be included in the monitoring programme for the Marine Conservation Zone. The number of sites and fans under observation increased from the initial 4 specimens photographed in 1982, to nearly 114 fans in 2017, a slight decrease from the maximum of 129 recorded in 2013 (Newman *et al.* 2018). Each individual fan is photographed, on each side, every year and 'repair' work is carried out by staff and volunteers, as necessary. For instance, a fan that was found to be broken off the rock in 2011 was subsequently reattached with a ring bolt and cable tie. It was a large sea fan and continued to thrive until it was last seen in 2016. Occasionally seen during this exercise is the sea-fan sea slug *Tritonia nilsodhneri* which feeds on the fan and is superbly camouflaged, individuals mimicking the colour of the colony.

On the south side of Skomer large silted boulders and rock surfaces support rarities such as the nationally scarce scarlet and gold star coral *Balanophyllia regia* (Fig. 46). This is a brightly coloured solitary coral, with up to 48 short, stout, evenly tapering tentacles arranged in groups of six around the mouth.

FIG 46. The nationally scarce scarlet and gold star coral occurs on the south side of Skomer. The tentacles are a rich, translucent, golden yellow colour and shade into a yellow, orange or scarlet disc. (Blaise Bullimore)

Skomer is considered to be one of the richest areas in the United Kingdom for marine sponges, and with each survey additional species new to science are discovered. They are, however, notoriously difficult to identify so the species here are being 'barcoded' as part of the Sponge Barcoding Project. This is a worldwide initiative that aims to obtain DNA signature sequences from more than 8,000 described sponge species and the estimated 15,000 unknown species. The barcoding is providing biologists with a simple and rapid method for the identification of samples.

Most sponge species are tolerant of some silt, but since they are filter-feeders, they benefit from being in an area where there is some water movement. Locations subject to very strong tides are, however, not suitable for many species, and the north side of Skomer has noticeably less variety than the south

FIG 47. The sea slug *Polycera faeroensis* is often found feeding on the bryozoan *Bugula flabellata*. It is common on the west and southwest coasts of Scotland, Wales, England and Ireland. (Blaise Bullimore)

side – but instead sponges that thrive in more exposed situations, such as yellow boring sponge *Cliona celata*, are present. This occurs in two distinct forms, a small one that bores into limestone and mollusc shells and a massive wall-shaped sponge as found here. The south side of the Wick, the deep-cut sea cliff on the south side of Skomer, consists of a vertical rock face 50 m high which continues underwater with an overhang, areas of which are covered by mashed-potato sponge *Thymosia guernei*. Other sponges found here, alongside jewel anemones *Corynactis viridis*, include the nationally scarce species *Phorbas dives* and *Stelletta grubbii*.

The 76 species of sea slug recorded in the Marine Conservation Zone represent 70 per cent of the known species in the United Kingdom, in an area of just over 13 km² (Lock *et al.* 2015). As specialised predators, sea slugs are a good indicator of the overall health of an ecosystem, so this again highlights the importance of the site. Some feed on ephemeral prey, such as hydroids, and tend to have several short-lived generations each year, while others feed on perennial prey and live for a year or more. Notable species include animals with mainly northern distributions such as *Doto hystrix*, *Cuthona pustulata*, *Okenia aspersa* and *Doto eireana* and those with southern distributions like *Doris sticta* and *Tritonia nilsodhneri*. Other more widespread species include *Facelina annulicornis* and *Polycera faeroensis* (Fig. 47). The latter was originally described from the Faroes, hence its specific name. In 2002 it was celebrated on a stamp issued by *Posta*, the postal service of the Faroe Islands.

The common sea urchin *Echinus esculentus* plays a key role in the structure of these subtidal communities. An omnivorous grazer, it clears space by removing the hydroid, bryozoan and algal turf on which it feeds, making the rock surface available for colonisation by other species. In low numbers this grazing effect is extremely beneficial, producing a rich mixed habitat; high densities of urchins, however, can be highly destructive, even affecting whole kelp forests. Large numbers of sea urchins were removed from the seas around Skomer during the 1970s, when divers targeted the population for the 'curio' trade. In 1982, for

FIG 48. Surveying sea urchins in the Skomer Marine Conservation Zone. They are now widespread again, despite large numbers being removed in the 1970s for the 'curio' trade. (Blaise Bullimore)

example, despite the creation of the voluntary reserve, the population was 'sparse, ageing and had probably not successfully recruited larvae' in the previous six years (Bishop & Earll 1984). A number of subsequent surveys have shown that numbers have now recovered (Fig. 48). Particularly high densities were recorded at sites along the north of Skomer, with mean densities of 14.04 urchins per 100 m² at Rye Rocks and 10.26 urchins per 100 m² at North Wall. Both these sites are exposed to moderate tidal current and sheltered from the prevailing southwesterly swell and wave action (Burton *et al.* 2016).

A small number of sea urchins in the area are affected by 'bald urchin disease', a bacterial infection known to affect several species of sea urchin. Infection generally occurs at the site of an existing physical injury; the affected area changes colour and the spines are lost (Jangoux 1987). If, however, the lesion remains shallow and covers less than 30 per cent of the animal's surface, the animal usually survives and eventually regenerates any lost tissue. More extensive damage results in death.

Other echinoderms found here include the common starfish *Asterias rubens*, which is locally abundant, spiny starfish *Marthasterias glacialis*, seven-armed starfish *Luidia ciliaris*, bloody Henry *Henricia sanguinolenta* and the common sun-star *Crossaster papposus*, together with a variety of feather stars, brittle stars and sea cucumbers.

There are a number of sea caves in the Skomer cliffs. A survey of a small ball-shaped cave, Tennis Ball Cave on the south coast of the Neck, revealed that, away from the scouring action of boulders on the floor, the overhanging upper walls and roof of the cave supported patches of the jewel anemone *Corynactis viridis*, scattered cup-corals such as Devonshire cup-coral *Caryophyllia smithii* and colonial sea squirts such as *Polyclinum aurantium* (Bunker & Holt 2003). Other nearby caves include Little Growler Cave on the northwest coast of Middleholm. This is exceptional in that it is one of the few caves that contains unscoured habitats and communities. On the southern wall of the largest chamber there are, for instance, large growths of elephant's-hide sponge *Pachymatisma johnstonia*, together with sheets of white purse sponge *Grantia compressa*. The former is a massive sponge which may reach 40 cm in diameter and 10 cm in height. Below these two species large areas of the cave wall are covered by black cave sponge *Dercitus bucklandi* and colonies of *Stryphnus ponderosus*, another enormous encrusting sponge, with a variety of sea squirts.

Most of the work carried out by the Natural Resources Wales staff team in the Marine Conservation Zone is concerned with the biological monitoring of seabed organisms, so it is not surprising that the area is well recorded. In addition, they provide and maintain marker buoys and visitor moorings to ensure that recreational use of the area does not damage its special features. Outside the Zone though there are few comprehensive accounts, material being scattered in numerous survey and monitoring reports. It is beyond the scope of this book to produce a complete list of these and the information they contain, but the following sections summarise what is known of the main habitats. It is hoped that one day all these data will be brought together and made easily available.

ROCKY REEFS

Offshore from Skomer and the other islands, and indeed the mainland coast, there are extensive areas of rocky reefs. The shallower, and southwest-facing, rocky reefs are exposed to severe wave action, while many others are extremely sheltered. Offshore there are extensive areas of tide-swept kelp and red algae and, across the large areas of deeper rock reef, a wide range of invertebrate animal communities. The more sheltered reefs, including those in areas of lower salinity and higher turbidity, typically support communities dominated by sponges and sea squirts.

Spiny lobsters *Palinurus elephas* are found in the subtidal area of rocky reefs (Fig. 49). In Wales they occur mainly around the Pembrokeshire coast and the Llŷn peninsula. Also known in Britain as the crawfish, crayfish or rock lobster, and as *langouste rouge* (red spiny lobster), *langouste commune* or *langouste royale* (royal spiny lobster) in France, it is a large crustacean, capable of growing to 60 cm in length. It has antennae that are longer than its body and it is heavily built, bearing large spines; unlike lobsters, however, its claws are small and hook-like. Spiny lobsters feed on echinoderms, snails and bivalves, shrimp larvae and worms. Egg-bearing females migrate to deeper water while the eggs are developing, and then return to mate. The eggs develop into planktonic larvae, which settle on the seabed as young spiny lobsters about five months later (Picton & Morrow 2007).

There has been a dramatic decline in the number of spiny lobsters around Pembrokeshire since the late 1970s, due to extensive commercial fishing by potting, scuba diving and tangle netting. As a result, it has been identified as a species in need of protection (Jones & Lock 2014), but populations are now slowly starting to recover (Newman *et al.* 2015). The season for spiny lobster runs from July to December, the animals being caught in baited pots. All undersize animals are returned unharmed to the sea, but even so fishing is clearly still having an effect. Numbers are now so small that divers should not take them, even where it is legal to do so, and urgent action is required to save the populations that remain.

FIG 49. A juvenile spiny lobster on Thorn Rock, Skomer. Numbers of this crustacean are now so low, as a result of commercial fishing, that the species is critically endangered in Wales. (Blaise Bullimore)

SUBTIDAL SANDBANKS

Subtidal sandbanks are typically associated with headlands, islands, islets or sublittoral reefs which cause tidal streams to deposit sediment. The major subtidal sandbanks include Bais Bank, located to the north of North Bishop, Turbot Bank to the west of Castlemartin, Wild Goose Race and the Knoll in the vicinity of Skokholm, and the sandbanks associated with Grassholm. There are deep-water sandbanks near the Bishops and Clerks, Hats and Barrels and St Govan's Shoals, and in St Brides Bay. The sediments themselves vary greatly, depending on the topography of the seabed, ranging from well-sorted medium sand to sandy gravel and gravelly sand. The sandbanks are mainly elongated or ovoid structures that lie along the axis of the tidal streams, having an exposed southern or western side and a more sheltered northern or eastern side (Countryside Council for Wales 2009).

Few, if any, large-scale offshore subtidal surveys of soft sediments have been conducted, but a detailed survey of the epifauna and fish of sandbanks off the Welsh coastline was undertaken in 2001 (Kaiser *et al.* 2004). Species recorded from Bais Bank and Turbot Bank included common hermit crab *Pagurus bernhardus*, a shrimp *Philocheras trispinosus* and lesser weever fish *Echiichthys vipera*, while Turbot Bank supported a number of additional species such as flying crab *Liocarcinus holsatus*, great sand eel *Hyperoplus lanceolatus* and lesser sand eel *Ammodytes tobianus*. All are able to cope with the mobile substrate, the flying crab having flattened legs which enable it both to swim and to burrow into the seabed.

STABLE SEDIMENTS

In contrast to the mobile sandbanks, there are also large areas of wave-sheltered stable sediments off the Pembrokeshire coast. The main sediment type is, like much of the Irish Sea, sandy gravel, with sand and muddy sand in St Brides Bay. North of Strumble Head, however, there are large quantities of gravelly sand. The character of the seabed is determined by the nature of the source material, the degree of wave action and the strength of tidal currents. The relatively sheltered Carmarthen Bay forms a sink for fine sand, and this substrate is present offshore along much of south Pembrokeshire. It produces a predominantly smooth seabed of fine to medium sand with some small ripples (Mackie *et al.* 2006, James 2008).

Areas of mixed sediments support a wide variety of species, both buried within and living on the surface of the sediment. Within the sediment there are large populations of long-lived species including bivalve molluscs, such as the sword razor clam *Ensis ensis* and ocean quahog *Arctica islandica*, sea anemones such as policeman's helmet anemone *Mesacmea mitchellii* and the uncommon purplish imperial anemone *Aureliania heterocera*, which uses its broad base as an anchor. Animals on the sediment surface include a relatively isolated population of great scallop and species such as ross coral *Pentapora folicea*.

ABEREIDDI QUARRY

In November 1969 a group of lecturers and students from the University of London spent a weekend diving in Pembrokeshire. When a force eight gale made diving at sites on the open coast impossible, they sought the shelter of Abereiddi Quarry, which was flooded in the 1930s to create a harbour (Fig. 50). The group discovered that, protected from waves and tidal currents, the quarry supported a rich variety of wildlife, although a thermocline develops in the summer which isolates waters below a depth of about 12 m. The site was subsequently studied by Keith Hiscock in the course of his PhD research (Hiscock 2018). Shaded surfaces in the quarry are characterised by the double spiral worm *Bispira volutacornis* and the attached stage of the moon jellyfish *Aurelia aurita*, together with a variety of sponges characteristic of these locations. Algae typically include various filamentous red algae, together with stringy red weed *Polysiphonia elongata*, sugar kelp *Saccharina latissima* and bootlace weed *Chorda filum*. Known today as the 'Blue Lagoon', it is a favourite area for coasteering and often used by divers when the weather is too bad to go elsewhere. 'Coasteering', derived from 'coast' and 'mountaineering', involves traversing the intertidal zone of the rocky coast on foot, or by swimming, without the aid of boats, surfboards or other craft. It is an increasingly popular recreational activity in Pembrokeshire.

MARINE FISH

Moving across all these habitats, and providing a rich source of food for a variety of other wildlife, are marine fish. The first reference we have to them is again from George Owen (1603), who wrote that:

Another great blessing has of late years fallen to this county in greater abundance than heretofore has been seen in this shire, which is the great abundance of her-rings, taken all along the coast round about the whole shire, as if the same were

FIG 50. Abereiddi Quarry, the 'Blue Lagoon', was flooded in the 1930s to create a harbour and now supports a rich variety of marine life. (Pembrokeshire Coast National Park Authority)

enclosed in with a hedge of herrings, which being held in great store and sold to parts beyond the sea procure also some store of money.

During the main spawning period, between August and October, huge shoals of herring *Clupea harengus*, up to 2 miles (3.2 km) wide and 4 miles (6.4 km) long, could be found. In the daytime, herring shoals remain close to the seabed, so most of the catches occurred at night when the fish were feeding close to the surface, or early in the morning when they came inshore. The fishing season usually began in September and lasted for three or four months. It was recorded that after one night's fishing, on 5 October 1745, 47 boats landed over 1.3 million fish (Perry 2015). Although the basic techniques for catching herring did not change over the centuries, the boats and equipment used developed substantially, enabling the fishermen to catch and transport much larger numbers of fish. Herring boats in the nineteenth century, for example, were 25–30 feet (7.5–9 m) long and able to carry a load of up to 20 tons, with a crew of around six or eight people. Lower Fishguard developed as a herring port, trading with Liverpool, Bristol and Ireland. The fish were sorted at the warehouse that

FIG 51. A sculpture of a herring shoal in Lower Fishguard, by John Cleal, recalls the large fishery that was once the main element of the town's economy. (Jonathan Mullard)

is now used by the Sea Cadets. With around 50 vessels registered, it is claimed that so many herring were being landed in Lower Fishguard that the fish were spread across the fields, as a form of fertiliser. There are records, dating back to 1550, of the Bishop of St Davids complaining about the scarcity of herring due to the greed of fishermen, who took too many herring when they were plentiful (Perry 2015). On the harbour wall a sculpture of a shoal of herrings by John Cleal, a renowned local artist and sculptor, commemorates this important part of the area's history (Fig. 51).

The fishing industry in South Wales underwent significant changes following the opening of a new dock in Milford Haven in 1888. This provided opportunities for large steam-powered trawlers, and within weeks of the dock opening around 55 steam trawlers and 200 sailing smacks were based there. The enormous catches of herring that resulted inevitably had an effect on the population in the surrounding seas. Numbers remained relatively high though until the early 1970s, when the overfishing finally caused a major collapse in numbers. The management of the stocks has since improved, and herring is now harvested sustainably throughout much of its range. Today the larger boats, fishing offshore, target fish such as bass *Dicentrarchus labrax*, turbot *Scophthalmus maximus*, brill *S. rhombus* and sole *Solea solea*, while the smaller inshore vessels operating in the coastal waters, up to 6 nautical miles offshore, target a wide range of species, including lobster *Homarus gammarus*, spider crab *Maja brachydactyla* and brown, or edible, crab *Cancer pagurus*. Many of these fishing boats attract a number of different species of birds, including gannet *Morus bassanus*, which feed on the discarded part of the catch, and can be seen passing the coast at locations such as Strumble Head.

At least 21 species of shark are resident in British seas, and several of them occur in the waters around Pembrokeshire. To date there have been no detailed surveys but there are often reports in the press about notable catches. In June 2018, for example, a 324-pound (147 kg) porbeagle shark *Lamna nasus* was caught by an angler off the coast of Milford Haven. Thankfully, it was released, relatively unscathed, back into the sea (*ITV News*, 14 June 2018). The porbeagle is classed as vulnerable worldwide, and as either endangered or critically endangered in different parts of its northern range (Shark Trust, www.sharktrust.org). Similarly, a 4-m thresher shark *Alopias vulpinus* was caught off Dale the following month (*BBC News*, 4 July 2018). This too was released, but unfortunately sightings of sharks are often sensationalised in the media to generate news, causing unnecessary concern. The porbeagle, for instance, was described in the title of another article as 'a relative of the deadly great white' (*Daily Mirror*, 14 July 2019). Only at the very end was it stated that porbeagles are not a danger to humans.

Smaller sharks also occur, and in 2018 a dead angelshark *Squatina squatina* was left on the sand by a receding tide at Saundersfoot. With its flat body, broad trunk and large, very high pectoral fins, it looks more like a large ray than a shark. They lie on the seabed covered in sand to ambush unsuspecting prey such as fish, crustaceans and molluscs. After the species suffered widespread decline across its range over the last century, there have been increasingly frequent sightings of this once common shark along the Welsh coast in recent years. The angelshark's only established stronghold is the Canary Islands, and it is not known whether the individuals seen around Wales are moving between Wales and the Canary Islands, or whether they are two distinct populations. The Angel Shark Project is working with a range of organisations in four project areas in Wales, including the seas between Fishguard and Milford Haven, to research the shark's ecology and develop conservation measures (angelsharknetwork.com/wales).

STRUMBLE HEAD

Strumble Head is regarded by birdwatchers as the best place in Wales to see migrating seabirds (Fig. 52). Well over 200 species of birds having been recorded from this location in north Pembrokeshire. Specialities include four species of shearwater – great shearwater *Ardenna gravis*, Cory's shearwater *Calonectris*

FIG 52. The lighthouse and the Lookout at Strumble Head, a key area for sea watching in Wales and the source of many rare bird records. Before the Strumble Head lighthouse was automated birdwatchers had access to Ynys Meicel, the island on which it stands. (Jonathan Mullard)

borealis, Balearic shearwater *Puffinus mauretanicus* and sooty shearwater *A. grisea* – three species of skua – great skua *Stercorarius skua*, pomarine skua *S. pomarinus* and long-tailed skua *S. longicaudus* – along with Leach's petrel *Oceanodroma leucorhoa* and Sabine's gull *Xema sabini*. Rarities seen over the years include little shearwater *Puffinus assimilis*, soft-plumaged petrel *Pterodroma mollis* and Wilson's petrel *Oceanites oceanicus*. The best time is in autumn when south or southwesterly winds have been followed by winds from the west or northwest. These conditions move seabirds into the Irish Sea and then bring them close inshore as they leave, sometimes in a day-long procession involving a great variety of species and thousands of birds. It was this type of weather pattern that resulted in the setting of record totals on 3 September 1983. Over 100 storm petrels *Hydrobates pelagicus*, 103 arctic skuas *S. parasiticus*, 198 great skuas and 397 sooty shearwaters were logged, all against a backdrop of a dawn-to-dusk procession of at least 40,000 Manx shearwaters. As Graham Rees, one of the people present, later recounted, this was 'not just a day of statistics but of sheer gripping spectacle' (Rees 2005).

The area's potential was first recognised by David Saunders, who, with friends, conducted a series of sea watches there between 1965 and 1975. Previously, David had been sea watching at St Davids Head, but Strumble could be reached directly by road and it saved the long walk from Whitesands Bay to the Head (David Saunders, personal communication). Recording many species that had previously only been associated with the islands, they soon realised that it was worthwhile to watch from the mainland. The most memorable sighting was on 8 September 1974 when they counted 45 arctic skuas passing in three hours. Up until that time the average annual total for the whole of Pembrokeshire had been 10 birds. This record completely changed the perception of the species' abundance in Pembrokeshire and, over the following years, set in motion a reappraisal of the status of other seabirds.

In the beginning David's group had access to Ynys Meicel, the island on which the Strumble Head lighthouse stands, so they sometimes sat in the shelter of the lighthouse wall. Sea watching from exposed headlands can, however, be very cold and it is difficult to protect telescopes and binoculars from rain and salt spray. When the light was automated in 1980, and access to the island was no longer possible, birdwatchers began to use a derelict building on the nearby mainland. Built in the 1940s, it was one of two identical buildings that housed equipment used to develop air-to-sea radar systems, which played a large part in the battle against German submarines during the Second World War. Conditions for sea watchers were still fairly basic until 1987, when the National Park Authority purchased the headland and renovated the building. The 'Lookout', as the building is now known, means that long sea watches can be carried out even in poor weather conditions. In the early days the Lookout was often crowded with birdwatchers. On two notable occasions people turning up at dawn were too late to get inside, as the building had filled up with people while it was still dark. A total of 120 birdwatchers assembled on one occasion, most travelling from South Wales, but some from Bristol and Swindon. The Lookout can still become crowded at times, but thankfully not to the same degree as previously.

The seabirds recorded at Strumble Head originate from many different locations. Birds like great and arctic skuas, for example, come from Iceland and Scotland, pomarine skuas from Russia, black-winged terns *Chlidonias niger* and little gulls *Hydrocoloeus minutus* from northern Europe and Mediterranean gulls *Ichthyaetus melanocephalus* from Belgium and France. On the other hand, great and sooty shearwaters are on their way to their breeding grounds in the South Atlantic, while Sabine's gulls have flown from Canada. More locally, roseate terns *Sterna dougallii* have made the short sea crossing from Ireland (Rees 2005).

Sea watching here though is about more than seabirds, as there can be large 'passages' of waders. The majority of knots *Calidris canutus* recorded in Pembrokeshire, for example, were not seen in the estuaries but logged passing the head. Ringing recoveries have shown that knots migrate from North America and cross to this side of the Atlantic before travelling on to Africa. Those seen at Strumble have almost certainly recently departed from staging areas such as the Dee estuary and do not stop again until they reach the Bay of Biscay, or even further south. Other waders frequently seen from the Lookout include dunlin *Calidris alpina*, oystercatcher *Haematopus ostralegus*, bar-tailed godwit *Limosa lapponica* and whimbrel *Numenius phaeopus*. Sometimes these movements occur suddenly and finish quickly. For instance, in a period of about an hour and a half on 5 August 1998 several large groups of whimbrels went by, a total of 370 birds. Nowhere else in Pembrokeshire have numbers of this magnitude been seen at this time of year, most large concentrations having been recorded in spring. Other species observed on migration include common scoter *Melanitta nigra*. Like many birds, scoters migrate at night, as well as during the day, so care is needed in interpreting the sightings. There seems, for instance, to be little correlation with the bird's movements elsewhere in Wales. The majority of scoters passing Strumble are seen in the early morning, and it may be that they have migrated at high altitude during the night, dropping down to sea level at daybreak.

CETACEANS

The seas around Strumble Head are rich in cetaceans, especially harbour porpoise, and a local group, Sea Trust, undertakes several surveys here each year. Indeed, the abundance of porpoises at this site was largely responsible for the Trust being formed. Although Welsh seas do not support a great variety of porpoises, dolphins and whales, a number of species are either present throughout the year, or are regular seasonal visitors (Baines & Evans 2012). The most common offshore species is the short-beaked common dolphin *Delphinus delphis* (Fig. 53). These animals are particularly abundant over the Celtic Deep, west of the Smalls, but there are some sightings off Skokholm, Skomer and Grassholm all year round, with the greatest numbers being recorded from July to September. They often occur in large pods, which can include anything from forty to a hundred individuals or more. In areas with a high abundance of food, these pods can merge temporarily, forming a 'superpod' consisting of well over a thousand dolphins (*BBC News*, 9 January 2013). Such a gathering was filmed off the Pembrokeshire coast in September 2014, and the resulting video makes for spectacular viewing. Despite the United Kingdom's biggest pod of bottlenose dolphins *Tursiops truncatus* living relatively close by, in Cardigan Bay, they are rare visitors to Pembrokeshire.

There are often very large numbers of harbour porpoise, including females with calves, all year round. Counts have recorded well over a hundred individuals in the summer and more than fifty during the winter. Ramsey Sound is a good place to see harbour porpoise, and since they are tidal feeders it is relatively easy to predict where they will occur. Animals can often be observed at the north end of the sound during flood tides, and also at the beginning and end of the ebb tide. During the ebb tide itself, however, they tend to congregate at the south end (Barradell 2009). These movements are probably related to the availability of fish. When the tide reaches the shallower waters at the ends of the sound it scours the seabed, bringing nutrients to the surface which attract fish, which, in turn, attract porpoises. Porpoises are present every month of the year, but the actual number fluctuates quite dramatically with records ranging from single

FIG 53. The short-beaked common dolphin is the most frequently encountered cetacean off the Pembrokeshire coast. It is distinctively coloured, being dark grey to black on top, with a lighter underside, yellowish-brown and light grey patches on the sides and a black 'mask' connecting the eye to the snout. (Richard Crossen)

individuals to groups of eighteen or more. From April to October gannets can often be seen flying above groups of porpoises and, at times, diving among them. It seems likely therefore that the gannets follow the porpoises in order to locate prey (Pierpoint 2008).

The porpoises in Ramsey Sound are an important species for wildlife tour operators in this part of Pembrokeshire, with boats present throughout the day between March and October. There is an actively promoted boat-use code in 'cetacean sensitive areas' all around the Pembrokeshire coast, including Ramsey Sound, but the area has become a focal point because porpoises and a busy tourist boat operation happen to coincide here. Further offshore there are at times enormous numbers of mackerel *Scomber scombrus*, and the first indication of something interesting is usually the sight of gannets plunge-diving into the shoal. Any tourist boats in the vicinity head straight for the area in the hope that whales and dolphins are feeding on the mackerel as well.

During the summer, weather and sea conditions combine to draw plankton, fish and squid into the Irish Sea in large numbers. This abundant food source attracts marine mammals that would normally only be seen in the open ocean. In addition to short-beaked common dolphin, species include minke whale *Balaenoptera acutorostrata*, fin whale *B. physalus*, long-finned pilot whale *Globicephala melas* and Risso's dolphin *Grampus griseus*. Recent sightings of fin whales have occurred west of Ramsey Island and near the Smalls. There are occasional sightings of orca, or killer whale *Orcinus orca*. In 2015 an orca was seen eating a seal pup by observers at Strumble Head, where the Sea Trust carry out regular cetacean watches. Very little is known about their movements on the west coast of Britain so this was an important sighting.

Other species recorded around Wales include humpback whale *Megaptera novaeangliae*, pygmy sperm whale *Kogia breviceps*, Blainville's beaked whale *Mesoplodon densirostris*, striped dolphin *Stenella coeruleoalba*, Atlantic white-sided dolphin *Lagenorhynchus acutus* and white-beaked dolphin *L. albirostris*. It is therefore possible that these species may be encountered occasionally off Pembrokeshire, especially around the islands to the west.

As well as sightings of live animals, dead cetaceans and other marine species are often washed up on the Pembrokeshire coastline. Ancient laws give the Crown rights to stranded cetaceans, which were known as 'Royal Fish', but it was not until 1913 that the Natural History Museum began to collate information on these. In 1988 an outbreak of phocine distemper resulted in the deaths of many thousands of seals throughout Europe and as a result, in 1990, the then Department of the Environment initiated the funding of a long-term monitoring programme involving the systematic post-mortem examination of United

TABLE 1. Cetacean strandings in Pembrokeshire in 2018. Adapted from Penrose & Gander 2019.

Date	Species	Condition	Location
1 February	Harbour porpoise	Dead	Broad Haven
6 February	Common dolphin	Dead	Goodwick
13 February	Harbour porpoise	Dead	Poppit Sands
16 February	Harbour porpoise	Dead	Wiseman's Bridge
29 June	Harbour porpoise	Alive – refloated	Whitesands Bay
18 July	Harbour porpoise	Dead	Ramsey
29 July	Common dolphin	Dead	Manorbier
4 August	Striped dolphin	Dead	Pop-pit Sands
7 August	Common dolphin	Alive – but died on beach	Newgale
18 August	Common dolphin	Dead	Saundersfoot
1 October	Harbour porpoise	Dead	Poppit Sands
13 December	Harbour porpoise	Dead	Freshwater East

Kingdom-stranded marine mammals, to investigate the cause of death. In Wales the Cetacean Strandings Investigation Programme is coordinated by Marine Environmental Monitoring, which provides annual summaries of the species involved (Penrose 2019). In 2018, as in most years, the most commonly recorded cetaceans were harbour porpoise and common dolphin (Table 1).

MARINE TURTLES

Only one marine turtle, the leatherback turtle *Dermochelys coriacea* is reported annually and considered a normal part of our fauna. They do, however, travel vast distances. A female turtle found in Carmarthen Bay in September 1997 had previously nested, and been tagged, in French Guiana. Loggerhead turtles *Caretta caretta* and Kemp's ridley turtles *Lepidochelys kempii* occur less frequently, with most specimens being carried north from their usual habitats by ocean currents. Sightings of two other vagrant species, the hawksbill turtle *Eretmochelys imbricate* and the green turtle *Chelonia mydas*, are very rare. The leatherback, the world's largest marine turtle, is unique amongst reptiles in that it is able to raise its body temperature above that of its immediate environment, allowing it to survive in colder waters. The other four species have hard shells and usually only occur as 'cold-stunned' juveniles, which should not be placed back in the sea if found.

There are 13 records of leatherback turtle in Pembrokeshire. The oldest, which dates back to June 1960, was an individual at Little Haven. The latest was an animal seen around 25 m off the beach at Newport on 19 September 2014, where it stayed for 20 minutes basking at the surface until it was disturbed by a boat. It then retreated to around 800 m offshore and stayed for 45 minutes (Penrose & Gander 2018). Loggerhead turtle sightings include a juvenile that was found on the beach at Freshwater West on 13 February 2014. Named Stormy, it was taken to Bristol Aquarium the following day. That was a better fate than the one suffered by another individual on 11 December 1938, which had been observed swimming along the shoreline the day before. It was taken alive to the Norton Hotel to be used to prepare soup. Apparently, this specimen had one flipper missing and appeared to have been in this condition for some time. Similarly, in 2015 a loggerhead turtle was found alive on Caldey but both front flippers had been cut off and the animal later died. Since 2015 eleven loggerheads have been recorded in the United Kingdom, both alive and dead, with single, partial or both front flippers missing. It seems that these individuals, like the one recorded in 1938, have unfortunately been caught in fishing nets as by-catch, have had their flippers cut off, and have then been thrown back into the sea to die. Becoming entangled in fishing gear is now a major cause of death for turtles of all species.

Only two records of Kemp's ridley turtle are known, the first at Broadhaven on 30 November 1999. This was taken to the former Oceanarium at St Davids for rehabilitation and then flown to Seaworld in Florida, for release at Cape Canaveral. The second specimen, which had a satellite tag, was found stranded on Poppit Beach on 15 January 2016. Turtles have also been recorded further offshore. A loggerhead, for instance, was seen off Skokholm in June 1992 and later that month a leatherback was also sighted. A leatherback was recorded again in July 2001. Like the seas around them, the islands themselves are also rich in species, and these are covered in the next four chapters.

CHAPTER 5

Skokholm and Skomer

The broad Atlantic, and ships sailing the sea. I see the Islands, and of
lighthouses three.

W. H. 'Clockwinder' Williams of Marloes

THERE ARE SAID TO BE a thousand islands lying off the Pembrokeshire coast, but of course it depends on how you define an island. The vast majority are simply sea-washed rocks. Of the eight main island groups, five – Skokholm (Fig. 54), Skomer and Middleholm, Ramsey, Bishops and Clerks, Caldey and St Margaret's – are situated relatively near the mainland, while the remaining three – Grassholm, the Smalls, and the Hats and Barrels – lie far out to the west in the Celtic Sea. The lighthouse at Strumble Head, constructed to guide vessels entering Fishguard Harbour, stands on Ynys Meicel, a tidal island, which, together with Ynys Onnen and Carreg Onnen, is close to the north coast of the mainland.

The larger tidal islands, or half-islands as George Owen called them, include St Catherine's Island at the eastern end of Tenby's South Beach, and Gateholm at the western end of Marloes Sands. Dinas Island, on the coast between Fishguard and Newport, despite its name, is not actually an island, since it is connected to the mainland by a valley which opens to the sea at both ends. A relatively small rise in sea level, however, would soon create a new island. Reviews and surveys of islands often exclude the tidal islands, since the presence of brown rats *Rattus norvegicus*, or hedgehogs *Erinaceus europaeus*, can have severe effects on populations of ground-nesting birds, and their vegetation differs little from that of the mainland to which they are attached.

FIG 54. An aerial view of Skokholm from the southwest. The island was publicised by R. M. Lockley, who established the first bird observatory in the United Kingdom there, in 1933, and wrote numerous books about his experiences. (Janet Baxter)

ORIGIN OF THE ISLANDS

Despite their relative proximity the Pembrokeshire islands owe their origin to a number of quite different geological processes, which took place over vast periods of time in widely separated locations (Fig. 55). Ramsey, in fact, has the most complex geology of all the Welsh islands, being composed of both Cambrian and Ordovician sedimentary rocks and Ordovician igneous rocks. The latter including lava and ash derived from underwater eruptions around 470 million years ago. To the west of Ramsey, the geology of the southern Bishops and Clerks can be matched with the Ordovician rocks on the island, while the northern group consists of the same igneous rocks that can be found on the mainland at St Davids Head (Bates *et al.* 1969). On Skomer there is a thick sequence of Silurian lavas, which accumulated from eruptions from fissures on a low-lying volcanic island. Lavas similar to these also formed Grassholm and the series of rocky reefs which extend for 14 km west from Grassholm towards the Smalls. In contrast, Skokholm is composed of Old Red Sandstone, which was derived from sediments deposited by ephemeral rivers on an arid coastal plain, as is the southern half of Caldey. The northern half of Caldey consists of Carboniferous Limestone, formed, as described in Chapter 2, in a shallow tropical sea between 360 and 330 million years ago.

Like the adjacent areas of mainland, most of the islands look relatively flat when viewed from a distance. These flat areas are the result of marine erosion, similar to the intertidal and subtidal rock platforms of the modern coast, but formed more than 50 million years ago. At that time, the three isolated hills on Ramsey (Carnysgubor, Carnllundain and Foel Fawr) would have been small islands, being composed of very hard igneous rocks which are only slowly eroded. The present coastline reflects their resistance to marine erosion, with Carnysgubor and Carnllundain forming headlands on either side of Aber Mawr Bay, which has been created by the loss of softer sedimentary and volcanic rocks. Foel Fawr forms the headland at the southern end of the island.

Following the uplift of the Earth's crust and falling sea levels around 70 million years ago, the old marine erosion surfaces were exposed as coastal plateaus which were then gradually altered by terrestrial erosion, and more rapidly by marine processes at the new sea level. This erosion was concentrated along fault planes, the present-day islands being formed when narrow necks were breached. This process is continuing today – so, for example, on Skomer the erosion of softer sedimentary rocks at South Haven will eventually mean that it joins up with North Haven. This will isolate the eastern part of Skomer, forming another separate island, which will, in turn, be dissected into smaller islets by

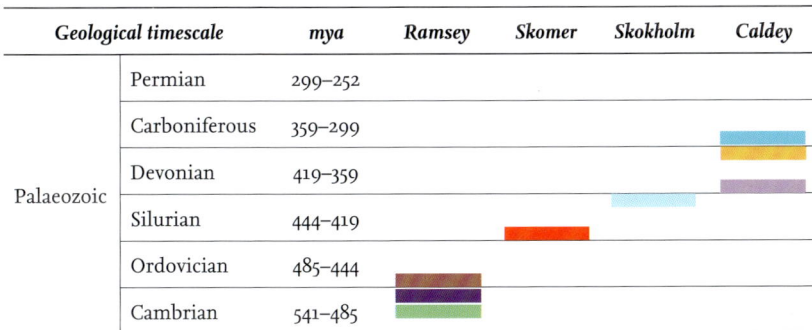

Geological timescale		mya	Ramsey	Skomer	Skokholm	Caldey
Palaeozoic	Permian	299–252				
	Carboniferous	359–299				▬
	Devonian	419–359				▬
	Silurian	444–419		▬	▬	
	Ordovician	485–444	▬			
	Cambrian	541–485	▬			

FIG 55. The stratigraphic range of rock sequences on the largest Pembrokeshire islands stretches from the Cambrian to the Carboniferous Period, covering over 280 million years of the Earth's history. (Adapted from Howells 1997)

further erosion along faults. This process has already started, since Matthews Wick on the south side of the Neck is being enlarged along a significant fault.

It is likely that Ramsey was also separated from the mainland by erosion along a fault, which is now concealed beneath Ramsey Sound. This was later considerably deepened and enlarged by meltwater which flowed south under the ice sheets that covered the Irish Sea on several occasions during the last glaciation. The movement of the ice rounded the hills on Ramsey and planed down the northern part of Skokholm. At this time, because the sea level was at least 40 m lower than it is today, the islands would have been isolated hills surrounded by a wide coastal plain. During the relatively short interglacial stages, when sea levels rose again due to the melting ice, they became islands once more. In the most recent of these interglacials the sea level was 5–6 m higher than it is now. On Caldey, for example, consolidated beach sediments are found several metres above the level of high tide.

Because of the significance of their geological features, a number of Geological Conservation Review sites have been designated on Ramsey, Skomer and Gateholm. These are sites of national and international importance that have been selected to show all the key scientific elements of the United Kingdom's geological heritage.

PEOPLE AND ISLANDS

The Pembrokeshire islands were exploited in various ways for centuries. George Owen refers to the birds found there in terms which clearly suggest that they were an 'article of commerce' (Matheson 1932). For example, he mentions two rocks adjacent to Ramsey as yielding 'small proffitte, saving some gulls', while the Bishops and the Clerks 'all yield store of gulls in the time of the yeare'. Ninety years later, an edition of Camden's *Britannia* stated, with reference to the seabirds of Ramsey:

> *The Harry-birds [shearwaters] are never seen on land but when taken; and the manner of taking these and the Puffins, is commonly by planting nets before their berries, wherein they soon entangle themselves. These four sorts [including guillemots and razorbills] cannot raise themselves upon the wing, from the land; but if at any distance from the cliffs, wadle (for they cannot be well said to go, their legs being too infirm for that use, and placed much more backward than a Duck's, so that they seem to stand upright) to some precipice, and thence cast themselves off, and take wing: but from the water they will raise to any height.* (Gibson 1695)

Richard Fenton tells us that the eggs of the seabirds on the Bishops Rock, by which he probably meant South Bishop, were 'often eaten, and by some esteemed a luxury; but they are now princially taken to be sent to Bristol for the purpose of fining wines' (Fenton 1811). The puffins themselves also suffered, Fenton mentioning that a very profitable voyage used to be made to the Bishops every summer from Ramsey 'for the purpose of taking several dozen Puffins, from whose breast and backs, being the only parts they stripped of feathers, they filled an immense sack with the soft plumage not inferior to eiderdown'.

It was not until the nineteenth century that the islands' importance for breeding seabirds was properly recognised by ornithologists. In 1894 the Reverend Murray Mathew wrote:

> Were it not for the islands off the coast there would be little to write about the Birds of Pembrokeshire, but these are, in the summer time, when the various cliff birds resort to them to nest, so thronged with countless birds, that they serve to redeem the county from the charge we have had elsewhere to bring against it of being, comparatively, uninteresting to the ornithologist, and also afford a justification for our book, which, without them, we should have felt no incentive to compile.

Mathew, like many others of his time, however, was seemingly not concerned about bird conservation, and even though the wanton killing of seabirds and the plunder of their eggs did not result in their extinction, the populations were severely affected. For example, in 1867, no fewer than 1,400 seabirds are stated to have been slaughtered *in one week* on Ramsey (Morris 1897). The first person to take substantive action to protect the wildlife of the islands was Mathew's contemporary Joshua John (J. J.) Neale.[*] An owner of fishing trawlers, and a key member of the Cardiff Naturalists' Society, he leased Skomer and Grassholm from 1890 to 1915, after witnessing a company of seamen slaughtering gannets on Grassholm by using them for target practice. Occasionally, he invited fellow members to join him for long weekends on the islands, and it was probably on one of these trips, in 1896, that Robert Drane, a Cardiff pharmacist and 'founding father' of the Cardiff Naturalists' Society, discovered the Skomer vole *Myodes glareolus skomerensis* (Fig. 56). Drane had a wide range of interests, including natural history, porcelain and antiquities, but was probably best known by the public as an antiquary.

---

[*] Neale bought the first Cardiff steam trawler in 1888, eventually owning a fleet of 17 vessels, industrial fishing only starting in 1880. The latter is an important date because it is the first time fossil energy was used to go after fish. Unwittingly, Neale, while being keen to conserve seabirds, helped kick-start a massive reduction in fish stocks, which is now affecting their very survival.

FIG 56. Robert Drane, the discoverer of the Skomer vole (left), and his close friend J. J. Neale, deep in conversation at an unknown locality in April 1913. (Cardiff Naturalists' Society Archive)

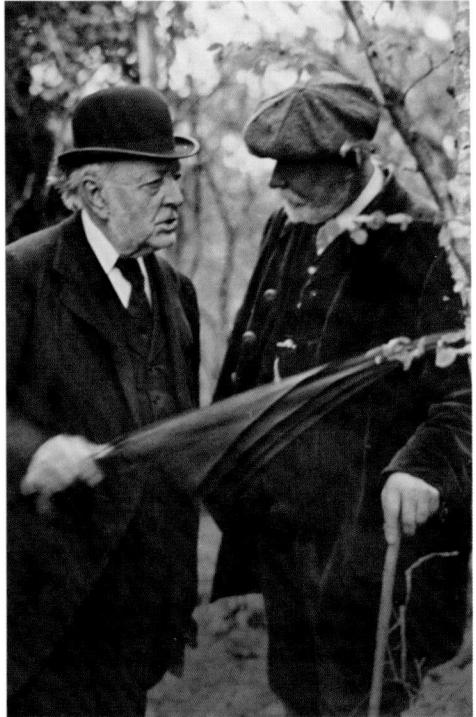

In those early days the Society had a close relationship with the Pembrokeshire islands, and a few decades later its members, such as Morrey Salmon, worked with Lockley on a number of projects. In 1934, for example, Salmon was the first person to record the nocturnal habits of Manx shearwaters and storm petrels on Skokholm, using flash photography. Together with his friend Geoffrey Ingram, he often visited the islands, and they spent long periods there.

Skokholm had become the property of the Phillips family of Sandy Haven and Haythog in 1713 and, through inheritance and marriage, part of the Dale Castle estate in 1740. The last man to make a living solely from farming on Skokholm was Captain Henry Edward Harrison, who was resident there in the 1860s with his wife, three daughters, one son and two servant girls. He died on the island in 1881, aged 64. From 1905 to 1916 John 'Bulldog' Edwards held the lease, and while he is credited as being the last man to farm Skokholm, he was based mainly on the mainland. Lockley, however, was the last farmer on the island, even if he soon moved on to other activities. By the time Lockley negotiated his lease the farmhouse was badly damaged and the other buildings were in ruins, but he repaired them using timbers from the wreck of the *Alice Williams*, a two-masted schooner which ran aground on the island in 1928 and was a total loss.

In 1933 Lockley established the first bird observatory in Britain on Skokholm, with help from Salmon and Ingram, who assisted in building the first Heligoland trap (Fig. 57). The Heligoland trap was so named because it was originally developed from the *troosel-goards* or 'thrush-bushes' once used by the Heligoland islanders, who live on a small archipelago in the southeast corner of the North Sea, to catch migrant birds for food (Brownlow 1952). Lockley had previously visited Heligoland to see the traps in action. They consist of a tapering netting enclosure, open at the wide end and closed at the narrow end by a collecting box with a transparent back, which appears as an escape route to birds that enter. To finance these activities Lockley initiated a public appeal for funds, for which Ingram and Salmon acted as treasurers. The first birds trapped were five willow warblers *Phylloscopus trochilus* on 7 August 1933 (Lockley 1936).

In 1922 Major Sturt, a retired dentist, and his wife, Violet, bought Skomer. Reuben Codd, often remembered as the last man who tried to farm Skomer, eventually married Sturt's daughter, Betty. Around this time the members of the Bird Protection Society were challenged by their President, Dr Julian Huxley, to be more adventurous and enterprising in creating nature reserves

FIG 57. A modern Heligoland trap near the farm buildings on Skokholm. It is in the same location as one of Lockley's original traps. (Jonathan Mullard)

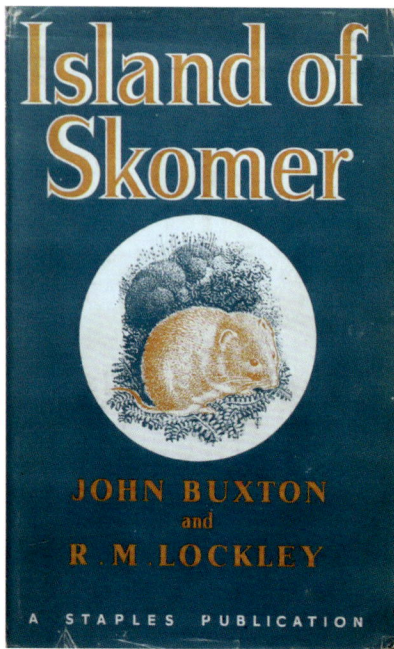

FIG 58. The front cover of *Island of Skomer*, featuring a Skomer vole drawn by Charles Tunnicliffe. (Jonathan Mullard)

on the Pembrokeshire islands. As a result, the Society had rights to protect wildlife on Skokholm after 1948, and St Margaret's Island after 1950. It was also active on Ramsey and Caldey, but Skomer at that time was still owned by the Sturts. The Society responded by establishing a temporary base on Skomer in 1946 to conduct a field survey, leasing the island for eight months. This research resulted in a book, *Island of Skomer*, edited by John Buxton and Lockley (1950) (Fig. 58). Their original intention was to acquire a longer lease, or even purchase the island, but their offers were not accepted. It seems though that the Society's time on Skomer was instrumental in establishing formal government interest in the island, for in 1947 Skomer was recognised as a potential nature reserve by the Wildlife Conservation Special Committee, of the National Parks Committee – which, helpfully, was chaired by Julian Huxley. The role of this subgroup was to advise the National Parks Committee, which was developing proposals for National Parks in England and Wales, on modifications that could be made to enhance wildlife conservation in these areas.

Farming on Skomer had been largely abandoned after the outbreak of the First World War in 1914, although there was a revival in the 1920s and 1930s, due to the activities of Reuben Codd. Agricultural activities finally ended in 1950 when the island was put up for sale by Major Sturt's executor. Ronald Staples, of the Staples Press, who had published Buxton and Lockley's book, had promised what was now the Field Society £8,000 as an outright gift to purchase the island, but when they offered Betty Codd £9,000 for Skomer she refused to sell. Later the same year, Skomer, Middleholm and Gateholm were sold privately to Leonard Lee, an industrialist from the Midlands, for £12,500. The Society, unable to maintain their operation on Skomer, had already purchased Dale Fort on the mainland instead, in order to establish a nature study centre and a base for the Skokholm bird observatory. Interestingly, Bill Condry, later author of *The Natural History of Wales*, number 66 in the New Naturalist series, and his wife, Penny, helped in the restoration work at the fort during the very cold winter of 1947. Subsequently, Dale Fort and Skokholm were sublet to the Field Studies Council, and the first Dale Fort warden, John Barrett, took on responsibility for the management of the island. By 1954, as a result of the research undertaken, most of the larger islands in Pembrokeshire had been notified as Sites of Special Scientific Interest.

In April 1958 Lee finally agreed to sell Skomer to the Field Society and gave them permission to occupy the island even before the purchase was completed. After a long period of negotiation between the Society and the Nature Conservancy, a forerunner of Natural Resources Wales, the latter agreed to finance the purchase. The Society purchased the island from Lee, as the result of an appeal for funds, for £10,000 and immediately sold it on to the Nature Conservancy for £6,000, this being the maximum amount the Conservancy was allowed to spend by the Treasury. In return for the £4,000 shortfall, the Conservancy gave the Field Society a lease for 21 years, with an option to renew for a similar period. It imposed a nominal rent of £5 per annum, and accepted responsibility, as landlord, for all maintenance and repairs. In addition, the Conservancy committed to building a warden's house at North Haven, at an estimated cost of £5,000. The purchase was completed at the beginning of April 1959, and on 4 April the Nature Conservancy granted the lease of Skomer to the Trustees of the West Wales Field Society.

The RSPB purchased Grassholm from Malcolm Stewart in 1948, making it their first nature reserve in Wales. Stewart's father was Sir Percy Malcolm Stewart (1872–1951), an English industrialist and philanthropist. As an undergraduate at Oxford, reading geology, thanks to his father chartering the steam yacht *Golden Eagle* for cruising among the Western Isles, he managed in 1930 and 1931 to land on North Rona, a remote Scottish island in the North Atlantic which lies about 70 km north of the Butt of Lewis. A few years later he wrote a book, *Ronay*, a description of the islands of North Rona, Sula Sgeir and the Flannan Islands (Stewart 1933). Shortly after the Second World War Stewart bought about 4,000 ha of land on Hoy in Orkney, which included the Old Man of Hoy (137 m). Like Grassholm, this is now an RSPB reserve. The previous owner of Grassholm, Martin Coles Harman, who was also the owner of Lundy, had originally bought it in 1937 for his son John, but John was killed in April 1944 during the siege of Kohima in India (and was posthumously awarded the Victoria Cross).

Ramsey was owned by the Church until the early 1900s, when it was sold into private ownership and farmed, initially as a general farm, until the 1970s. After that it became a deer farm; red deer were introduced in 1976 and are still present today. The RSPB was, however, involved in the management of Ramsey for many years, before acquiring the island in 1992. Finally, the National

Trust bought Middleholm and Gateholm, as part of their wider estates on the Marloes peninsula, in 1981. This completed the hold of the various conservation organisations on the islands, apart from Caldey.

James Taylor Hawksley (1839–91) became the owner of Caldey when the island was bought for him by his father, James Wilson Hawksley of Middlewood Hall, Yorkshire, for £15,950 on 5 November 1867. He had previously studied farming science, and he pioneered farm management and agricultural techniques on the island – the effects of which are still visible today. Hawksley was appointed High Sheriff of Pembrokeshire in 1884 and died on 3 August 1891, being buried in Tenby. After his death, his wife sold Caldey Island in 1894 to Thomas Dick Smith-Cuninghame from Lanarkshire for £12,750, who then resold the island on 21 December 1897 for £12,200 to the Reverend William Done Bushell. In 1928 the island was again sold, this time to a Cistercian order, and it is still a working monastic community.

SKOKHOLM

Situated just 3 km off the mainland coast, Skokholm is a small island, only 1.6 km long and 0.6 km wide. The first reference to it appears in a charter, now in the British Library, in which William Marshal the Younger, Earl of Pembroke (1190–1231), grants to Gilbert de Vale land in Ireland in exchange for lands in Pembrokeshire, including the island of 'Scoghholm' (Lockley 1938). The first Skokholm lighthouse was built in 1776, and the present building was constructed between 1910 and 1915 and officially opened in 1916. Along with the lighthouses at South Bishop and the Smalls, it is still important for guiding ships into Milford Haven, despite the widespread use of modern navigational aids. Automated in 1983, it is now monitored and controlled from the Trinity House Operations Control Centre at Harwich. The attached building is utilised as accommodation for the island's wardens, while the original farm cottage, restored by Lockley, and the surrounding farm buildings are used for visitors (Fig. 59).

Skokholm was made famous by Lockley. He was a prolific author who wrote over fifty books on natural history and island life, including *Dream Island* (1930), *Island Days* (1934) and *I Know an Island* (1938). According to his daughter, Ann, Lockley never really lived the simple island life described in his books, as he had too many commitments both onshore and offshore, relying on the Sturley family of Dale to transport him to and from the island (Fig. 60; Lockley 2013). He was, however, one of the first people to study the breeding biology of storm petrels, Manx shearwaters and puffins. His monograph on shearwaters was the result of

FIG 59. The original farm cottage on Skokholm, restored by Lockley, is now a Grade II listed building. (Jonathan Mullard)

twelve years' study. These activities inevitably brought him to the attention of a wider circle of conservationists and naturalists, including Peter Scott, the famous ornithologist and conservationist.

The Eighth International Ornithological Congress, which met in Oxford in 1934, ended with an excursion for 145 participants to the Pembrokeshire islands, including Skokholm to view the bird observatory (Anon. 1934). The excursion was aided by the Royal Navy, who provided two destroyers, HMS *Windsor* and HMS *Wolfhound*, for transport (Fig. 61). The guests consisted of a large proportion of the ornithological world at that time, including Max Nicholson,[*] who reported on his 1928 national heron survey, Konrad Lorenz, later famed for his work on animal behaviour, and the ex-king of Bulgaria, Ferdinand the First.

As a result of Lockley's work, and that of his successors, Skokholm is recognised as internationally important for its breeding seabirds, several thousand birds

---

[*] A pioneering environmentalist, ornithologist and internationalist and a founder of the World Wildlife Fund – and an early mentor of the author.

FIG 60. Lockley and Edgar Sturley off Skokholm in 1939. In this period visitors to the island were usually transported from Dale by the Sturley family, with Edgar, the skipper, hardly moving from his place at the tiller. (John Fursden/David Saunders)

nesting here each year. Together Skokholm and Skomer form one of five United Kingdom core seabird monitoring sites under the national Seabird Monitoring Programme, providing essential demographic data on numbers, productivity and survival rates. They are especially important for storm petrels and Manx shearwaters, which also breed on Ramsey. Over 50 per cent of the world's Manx shearwaters now depend on the Pembrokeshire islands to raise their young. A whole-island shearwater census kept the Wildlife Trust's wardens, Richard Brown and Giselle Eagle, and volunteers busy during the 2018 nesting season and revealed that there were an estimated 89,000 breeding pairs on Skokholm. This was an increase on a 2012/13 census which resulted in a figure of approximately 63,980 pairs. Alongside these two species can be found fulmars *Fulmarus glacialis*, great black-backed gulls *Larus marinus*, herring gulls *L. argentatus*, lesser black-backed gulls *L. fuscus*, ravens *Corvus corax*, guillemots, razorbills and puffins.

The great black-backed gulls are the island's apex predators, the vast majority of Manx shearwater kills being due to them – although in 2018 a raven was observed extracting a shearwater from its burrow and killing it. Fulmars first bred on the island in 1967 but since then there has been a steady increase in numbers, which is starting to impact negatively on other species. Recent years have seen both adult and young herring gulls 'oiled' by nesting fulmars. Fulmars store a foul-smelling oil in a section of their stomachs called the proventriculus, which they can spray out as a defensive measure. This oil can be deadly to other birds, as it can cause matting of the feathers leading to the loss of waterproofing and even render the bird unable to fly. The name 'fulmar' comes from two Old Norse words, *fúll* (foul) and *már* (gull) – though in fact the 'foul gulls' are related to petrels and albatrosses. Razorbills, adults and chicks, have also been evicted from ledges by prospecting fulmars and a juvenile peregrine *Falco peregrinus* oiled. These interactions are, however, a normal part of a healthy seabird colony.

Other breeding birds include choughs *Pyrrhocorax pyrrhocorax*, oystercatchers, lapwings *Vanellus vanellus*, skylarks *Alauda arvensis* and wheatears *Oenanthe oenanthe*. While he was warden on the island in the 1940s and 1950s, Peter Conder studied the wheatears, eventually publishing a monograph on the species in 1990.

FIG 61. Delegates from the Eighth International Ornithological Congress, 1934, arriving on Skokholm to view the bird observatory set up by Lockley. (Harry Morrey Salmon/ David Saunders)

Storm petrels

Dark summer nights on Skokholm are filled with noises. In amongst the raucous cries of the Manx shearwaters can be heard the eerie calls of storm petrels, which wait for the relative safety of darkness before returning to feed their young in the nest. It is an unforgettable experience for visitors as they stand there with the birds arriving all around them. The smallest oceanic seabird in the world, the storm petrel has a wingspan of only 36–39 cm. It is square-tailed and has black plumage except for a snow-white rump that extends to the sides of the tail base and a broad white band on the underwings. Storm petrels only return to land to breed, nesting in holes and crevices and emerging at night in an attempt to avoid predators (Fig. 62). Like other petrels, the storm petrel cannot walk properly, but shuffles along, making it extremely vulnerable until it reaches the safety of its nest. Short-eared owls *Asio flammeus* have been seen hunting petrels in the Quarry, an inlet adjacent to the lighthouse. Their main predator used to be the introduced little owl *Athene noctua*, which breeds nearby on Skomer. The Skokholm Bird Observatory Report for 1936 recorded a little owl nest containing the corpses of nearly 200 petrels, but until 2014 they had not been seen on the island since 1995. Data from the United Kingdom Breeding Bird Survey suggest that the number of little owls is decreasing, with the population estimated to have reduced by 24 per cent between 1995 and 2008. There are probably now fewer than 20 pairs breeding in the county (Green & Roberts 2004).

The storm petrel's behaviour means that obtaining accurate estimates of breeding numbers has been extremely difficult. *Birds in Wales*, the official journal of the Welsh Ornithological Society, mentions around 6,000 pairs on Skokholm in 1989 (Powell 1990). Recent surveys appear to suggest that the population has decreased since then, although much of this may be a result of errors in earlier counts. More recently, the method of playing recordings of storm petrel calls to elicit a response from adults occupying burrows during the day, while they are incubating eggs, has been used. Unfortunately, not all adults present at a colony will respond to recordings, and counts therefore underestimate the number of occupied nests and have to be adjusted by a response rate measured on site. Despite these problems, current estimates are that a total of 1,910 pairs of storm petrels breed on Skokholm. A further 220 breed on Skomer, and including non-breeding individuals, it is estimated that the total population in this part of Wales is just over 5,000 birds, representing around 10 per cent of the United Kingdom breeding population (Mitchell *et al.* 2004).

Storm petrels breed in rock crevices in the majority of bays and inlets around the coast and in the stone walls around the farmstead, which are sometimes

FIG 62. A storm petrel returning to its nest on Skokholm under cover of darkness. The characteristic white band on the underwing is clearly visible. (Bart Vercruysse and Pol Dewulf)

FIG 63. A storm petrel chick, temporarily removed from a crevice in a wall near Skokholm farm, being weighed by a researcher. (Jonathan Mullard)

shared with other seabirds, or rabbits *Oryctolagus cuniculus*. Storm petrels usually mate for life and do not breed until they are around four or five years old. The birds use the same hole every year, the female laying a single white egg on the bare soil. They share the lengthy incubation, of up to 50 days, and both feed the chick, which is not normally brooded after the first week (Fig. 63). Very few nesting sites, however, are accessible to researchers, which makes the birds very difficult to study. In the late 1970s wooden nest boxes were installed in the walls around the farm, to be replaced in the 1990s with plastic ones. There is a general consensus though that the number of pairs utilising the eighteenth-century herringbone walls has declined, perhaps due to a loss of suitable nest sites as vegetation and soil fills gaps in the collapsing walls. So, in 2016, a new wall was built near the Quarry. Although it looks like one of the existing herringbone walls, it contains over 100 storm petrel nest boxes with access hatches. The wall, inevitably named 'the Petrel Station', will hopefully replace some of the nesting sites lost from the original drystone walls. There were four boxes containing eggs in 2018 and more birds in 2019, so the initiative appears to be working.

Only small numbers of chicks are ringed each year, but over the seven years from 2012 to 2018 well over 4,000 full-grown storm petrels were ringed, many of which have subsequently been retrapped away from Skokholm. Over 150 birds were also recorded on the island after being ringed elsewhere in the United Kingdom, Ireland, France and Portugal. The information generated from this exercise suggests that there is a relatively discrete Irish Sea population. The non-breeding birds which enter this area generally stay here until they next journey south for the winter, although a few have entered the North Sea and skirted Ireland. There had never been a recovery from anywhere north of the Farne Islands in Northumberland, despite a significant amount of ringing going on in Scotland, until 2018 when a bird ringed on Skokholm on 5 August reached Sandoy in the Faroe Islands just nine days later.

Migratory birds

Skokholm is excellently placed to attract passage migrants, and during spring and autumn thousands of birds pass through the island. In 1946, after a break due to the Second World War, the observatory set up by Lockley was reopened, but ringing on the island ceased in 1976. Following two trial seasons Skokholm successfully applied to the Bird Observatories Council for reaccreditation in December 2013, joining the 18 other observatories that now exist in Britain, Ireland and the Isle of Man. Once again at the forefront of ornithological research, it is contributing valuable data on migratory species, with a total of 8,353 birds caught and ringed in 2018. There are three Heligoland traps on Skokholm, constructed on the same sites as those originally used by Lockley.

As in Lockley's time, the daily census data are still collected at the 'evening log' by the fire in Lockley's cottage. All the handwritten logs have recently been digitised and checked by volunteers, resulting in a total to date of 13,711,286 records of 296 species. Peak day counts include 12,979 swallows *Hirundo rustica*, 710 house martins *Delichon urbicum*, 3,000 willow warblers, 500 whitethroats *Sylvia communis*, 164 blackcaps *S. atricapilla*, 133 chiffchaffs *Phylloscopus collybita*, 3,200 chaffinches *Fringilla coelebs*, 800 bramblings *F. montifringilla* and 1,200 siskins *Spinus spinus* (Richard Brown & Giselle Eagle, personal communication).

The key months for migrants are May, August and September – and you never know what is going to turn up. Over the years the island has hosted many special rarities, as well as being a hotspot for the commoner European breeding warblers, flycatchers, pipits and thrushes. Goldcrests *Regulus regulus* are also often recorded in large numbers on autumn passage (Fig. 64). Scarcer visitors include birds such as wryneck *Jynx torquilla*, hoopoe *Upupa epops*, melodious warbler *Hippolais polyglotta* and red-breasted flycatcher *Ficedula parva*. Exceptional rarities

FIG 64. A goldcrest captured in a Heligoland trap on Skokholm in September 2015. Large numbers move through the county during the autumn migration, with a breeding population of 5,000 pairs. (Jonathan Mullard)

seen on the island include scops owl *Otus scops*, rose-breasted grosbeak *Pheucticus ludovicianus*, Baltimore oriole *Icterus galbula*, bobolink *Dolichonyx oryzivorus* and Swainson's thrush *Catharus ustulatus*. The scops owl is a migratory bird, normally wintering in east and west Africa, while the rose-breasted grosbeak is North American, as are the Baltimore oriole, bobolink and Swainson's thrush. Unlike David Saunders' 1961 record of a Blackburnian warbler, these regular influxes of birds from across the Atlantic are now considered normal.

House mice

While the Skomer vole is well known among naturalists, it is the house mice *Mus musculus* on Skokholm that have been the subject of one of the longest genetic studies of the adaptation of mammals to island life. The research by R. J. (Sam) Berry, the author of *Islands*, number 109 in the New Naturalist series, shows natural selection in action (Berry & Jakobson 1975, Berry 2009). Islands are ideal 'laboratories' for this sort of study, because a discrete population can be observed without the danger of confusion caused by individuals migrating to the study site from elsewhere. Unusually for an island population, the origins of the house mice seem clear. There were apparently no house mice on the island in 1881, when it was last intensively farmed, but Lockley recorded that when the current lighthouse was being built, between 1913 and 1915, they were so abundant that special precautions had to be taken to ensure the building was mouse-proof. Trinity House cannot, however, confirm this statement. It has been speculated that the mice reached the island in the 1890s in sacks brought across by rabbit-catchers for transporting their harvest back to the mainland.

The mice on Skokholm are peculiar in that, in more than half of them, the vertebral neural arches have failed to fuse – producing a very mild form of spina bifida. This feature is very rarely found in wild mice but is present in just under 10 per cent of the mice on the mainland opposite Skokholm. It seems to be a very clear case of a 'founder effect' in the island mice. Studies show that the gene frequencies change one way during the summer breeding period and the opposite way during the winter survival phase, when the chief problem for the mice is finding enough food to survive. For more details the reader is recommended to consult one of Berry's other books in the series, *Inheritance and Natural History* (Berry 1977).

Vegetation

Skokholm means 'wooded island' in Norse, but no trees survive today. Woodland plants such as lesser celandine *Ranunculus ficaria* and bluebell *Hyacinthoides non-scripta* provide the only clues to the former vegetation cover. When the trees were cleared is difficult to tell, but prehistoric settlement is indicated by finds of scattered flints, and a laser survey carried out from the air in 2012 for the Royal Commission on the Ancient and Historical Monuments of Wales revealed details of Iron Age and medieval enclosures and fields beneath the field pattern created by nineteenth-century farming. Although it is one of the windiest places in the British Isles in winter, Skokholm is not nearly so exposed in summer, and for this reason it has been possible to grow crops successfully in the past. Along with natural processes such as exposure to wind, salt spray, waves and drought, traditional farming has played a key role in the development of the current flora. The many thousands of seabirds present have also contributed to the vegetation types and species present, principally through disturbance, or deposition of guano, which has resulted in nutrient enrichment.

Rabbits were introduced to nearly all the Welsh islands from about 1180 onwards, at a time when their fur and meat were of considerable economic importance. Black fur and felt were particularly in demand, and the Bishop of

FIG 65. An extensive band of thrift hummocks occurs on the southern and western coasts of Skokholm. (Jonathan Mullard)

St Davids deliberately put black rabbits on Ramsey as they were worth five times the usual price. It is not known when rabbits were first introduced to Skokholm, but the introduction of these then rare and expensive animals, probably by the Earl of Pembroke, would not have been easy. The island had previously only been used for sheep grazing. In 1997 around 6,000 rabbits, descendants of the original stock, were estimated to graze the 'lawns' around the farm buildings. Lockley tried introducing commercial breeds, such as Chinchilla, Skewbald and Long-haired but they did not survive the harsh conditions. A few of their genes still linger on though in odd-looking rabbits. Outbreaks of myxomatosis, particularly in 1988, and viral haemorrhagic disease, in 1995, have threatened the population but there are no rabbit fleas *Spilopsyllus cuniculi* on the island to spread them.

The abundance of rabbits, together with the influence of seabird guano, has resulted in extensive modifications to the maritime vegetation that have been much studied by botanists. The island is of special interest for its maritime grassland, an extensive band of thrift *Armeria maritima* hummocks occurring on the southern and western coasts (Fig. 65), along with extensive areas of sea campion *Silene uniflora*. The zonation of the constituent plant communities depends, in general, on exposure to the prevailing winds and spray (Gillham 1955).

Further inland and along the eastern coast, red fescue *Festuca rubra* characterises the short sward of the old field systems, along with Yorkshire-fog *Holcus lanatus*, common bent *Agrostis capillaris*, sea plantain *Plantago maritima* and English stonecrop *Sedum anglicum*. Bracken *Pteridium aquilinum* dominates sections of the more sheltered eastern coast, but is much sparser elsewhere and has an understorey consisting of sheep's sorrel *Rumex acetosella* and wood sage *Teucrium scorodonia*. Springs and seasonally inundated depressions in the boulder clay that caps the centre of the island have a marsh flora of rushes *Juncus* spp. and sedges, particularly sand sedge *Carex arenaria*, with a lot of marsh pennywort *Hydrocotyle vulgaris*. The island also supports a large population of sea stork's-bill *Erodium maritimum* and wild pansy *Viola tricolor* ssp. *tricolor*, and on the cliffs, out of reach of the rabbits, small colonies of tree mallow and sea-purslane *Atriplex portulacoides* persist. Plants like thrift, sea campion and goldenrod *Solidago virgaurea*, however, are unpalatable, and provide colour in the spring and summer (Fig. 66).

FIG 66. Large stands of goldenrod in front of the former cowsheds, which now house the library and visitor accommodation on Skokholm. The plant is avoided by rabbits and thrives as a result. (Jonathan Mullard)

FIG 67. One of the best specimens of golden hair lichen on Skokholm is located on the north side of the Neck. (Jonathan Mullard)

An exciting new record for the county in 2018 was the discovery of small adder's-tongue *Ophioglossum azoricum* on the island. The fern was found by Fiona Gomersall, Wildlife Sites Officer with the Shropshire Wildlife Trust, while she was re-recording permanent quadrats in May. A plant of well-drained maritime turf, being widespread on cliff tops and in damp sandy hollows in coastal dunes, it had been tentatively identified from Skomer in the 1960s but never confirmed. Like other ferns it dies down in mid-summer and survives over the winter as dormant root-buds, the leaves regrowing the following spring.

Besides the nesting birds, one of the other glories of Skokholm are the lichens, which include many nationally scarce species.[*] Almost every square

---

[*] Controversy surrounds the subject of English names for lichens, and many members of the British Lichen Society view them as undesirable, while others think they help to engage the general public. While I am of the latter opinion, not all lichens have been given common names, so the terminology in this book is necessarily inconsistent.

centimetre of exposed rock is covered by sea ivory *Ramalina siliquosa*, along with *Lecanora rupicola*, *Ochrolechia parella* and the brown *Anaptytycha runicata*, among many others. The real star of the show, however, is the rare golden hair lichen *Teloschistes flavicans* (Fig. 67). This grows mainly on other lichens, or bryophytes, in coastal situations. In Pembrokeshire it is recorded from 14 other locations, including Skomer and Ramsey. Where it does occur, it can be surprisingly abundant, a survey on Skokholm in 2013 revealing 600 individual plants. The severe and prolonged storms which hit the United Kingdom the following January and February took their toll though and the following summer only 302 plants could be found (John Jones, personal communication). Other rarities include *Rocella fuciformis*, a grey strap-shaped lichen with hanging branches, *R. phycopsis*, a tufted and densely branched species, and *Rinodina confragosa*, a grey crustose lichen.

The larger fungi of Skokholm were extensively studied in the 1940s and 1950s by Arthur Frederick Parker-Rhodes, a Cambridge polymath, who published 13 papers based on his research on the island between 1949 and 1955 in the *New Phytologist* and the *Transactions of the British Mycological Society*. For nearly thirty years he taught a course on fungi at Flatford Mill Field Studies Centre in Suffolk, and in 1950 he published a popular book, *Fungi, Friends and Foes*. Parker-Rhodes identified several taxa new to science, including the crust fungus now known as *Trechispora clanculare*, which he found in a puffin burrow on the island.

Slow worms

Skokholm is home to what are said to be the largest wild slow worms *Anguis fragilis* in Britain. This is often attributed to the fact that they lack one of the gut parasites prevalent in mainland populations, but this seems unlikely. Many species on islands increase in size in comparison to those on the mainland, and island gigantism is a well-known biological phenomenon. A lack of predators may also be a factor, since there are no foxes, stoats, weasels, shrews, hedgehogs or rats on the island. Jackdaws *Coloeus monedula*, however, eat a lot of slow worms, so they are not entirely free from threats (Giselle Eagle, personal communication). One of the best places to find slow worms on Skokholm is under corrugated iron sheets laid out for this purpose by the reserve wardens (Fig. 68). Being burrowing lizards, they spend most of their time hiding underneath objects, feeding at night on slugs and worms. Males are pale coloured and sometimes have blue spots, while the females are larger, with dark sides and a dark stripe down the back.

FIG 68. The slow worms on Skokholm are said to be the largest examples of this species in Britain. The corrugated iron sheets to the left have been removed temporarily, in order to count and photograph the animals, which hide under them. (Mervyn Greening)

Invertebrates

Until comparatively recently, many of the records of invertebrates on the Welsh islands originated from single visits by specialists. These include the substantial list of springtails for Skokholm produced by H. J. Gough, who collected 34 species between 27 August and 5 September 1969, and the visits by William Syer Bristowe, a renowned spider expert and island lover (Bristowe 1958, Gough 1971). Bristowe's book *The World of Spiders* (1958) was number 38 in the New Naturalist series. He considered that the 'diminution in fauna with the dwindling size of the islands' was exemplified by the number of spider species he collected on the

Pembrokeshire islands: 77 on Skomer, 53 on Ramsey, 51 on Skokholm and 11 on Grassholm (Bristowe 1935).

Today, however, we know far more about the range of species present and their relative distributions. Moths recorded on Skokholm, for instance, include nationally scarce species dependent, for the most part, on a coastal location. These include Barrett's marbled coronet *Hadena luteago barrettii*, whose larvae feed on the roots of sea campion, black-banded moth *Polymixis xanthomista statices*, whose larvae feed mainly on thrift, and marbled green *Cryphia muralis*, which is dependent on saxicolous lichens, especially *Diploica canescens*. The adults of the marbled green are on the wing during July and August, and rest during the day on rocks and walls covered with lichen, being superbly camouflaged. The darkling beetle *Cylindrinotus laevioctostriatus*, which feeds on algae growing on dead wood, or lichens on rocks, has been recorded from both Skokholm and Skomer.

SKOMER

Skomer is the third-largest island in Wales, after Anglesey and Holy Island (Fig. 69). Although only separated from the mainland by some 600 m of sea, the rock-strewn Jack Sound with its fierce tidal streams, along with the island's high cliffs and awkward landing places, can make access difficult. The resistant

FIG 69. Skomer is the third-largest island in Wales and supports huge numbers of breeding seabirds each year, including Manx shearwaters. (Janet Baxter)

volcanic rocks, with some interbedded sediments, form a plateau bordered by steep sea cliffs up to 60 m high. On old maps, Skomer is sometimes labelled Skalmey, the name deriving from two Norse words, *skalm*, meaning a short sword, or cleft or cut, and *ey* denoting an island – therefore Skalmey, or 'cleft island'. This probably refers to the fact that two of the narrow inlets, or 'wicks', North Haven and South Haven, as mentioned earlier, almost cut the island in two. On the south side of the island is another deep inlet, known as the Wick, and it is here that the main seabird breeding colonies can be found. There are a number of small islets around Skomer, the main ones being the Mew Stone at the southern tip of the island, the Garland Stone and its outliers in the north and the Pig Stone and Littlewill Bench to the west. To the east of the Neck and separated from it by Little Sound lies a more substantial island, Middleholm, or Midland Isle. Unlike Skomer, Middleholm is not open to the public, but it supports a similar range of species.

Breeding seabirds

Together Skomer and Middleholm support the greatest concentration of breeding seabirds in England and Wales, surpassing even the Farne Islands. Thousands of razorbills nest on the cliffs, as do guillemots, while in burrows on the open ground there are similar numbers of puffins and a few hundred storm petrels. In addition, there are large colonies of lesser black-backed gulls, guillemots, kittiwakes and fulmars, together with herring gulls, great black-backed gulls and shags *Phalacrocorax aristotelis*.

Manx shearwater

Skomer is particularly noted for having the largest Manx shearwater breeding colony in the world, a census in 2011 providing an estimate of 316,000 breeding pairs. The census, with some improvements to the methodology, was repeated in 2018 and this indicated a population of at least 350,000 pairs, with a further 16,000 on Middleholm and the 89,000 on Skokholm. So, in total, during the summer on the three Pembrokeshire islands there are over 455,000 breeding pairs of shearwaters present (Fig. 70). Around 100,000 pairs also nest on Rum, in the Inner Hebrides, and there are 40,000 pairs on the islands off County Kerry in Ireland (Cabot 1999).

When Robert Drane visited Skomer in June 1897 he found a farm boy collecting Manx shearwaters to sell for bait to the lobster fishermen. 'The boy got thirty and buried them, waiting for the fishermen and pay; but they came not. Meanwhile the carrion crow did.' Manx shearwaters arrive on Skomer to breed towards the end of March. They spend the winter off the coast of South America, undertaking

FIG 70. A Manx shearwater returning to its nest on Skokholm in the dark, to avoid predators. There are over 455,000 breeding pairs on the Pembrokeshire islands during the summer. (Bart Vercruysse and Pol Dewulf)

a 22,000-km round journey each year. Like storm petrels, shearwaters only come onto the island under the cover of darkness and nest in burrows to avoid predators such as great black-backed gulls. Similarly adapted to life at sea, with long, narrow wings and feet placed far back on the body for efficient swimming, they are clumsy on land and during daylight hours very vulnerable to predation. The number of carcasses strewn around the island highlights the danger they face. Occasionally, the birds can get blown off course. In September 2017, for example, hundreds of Manx shearwaters were stranded on Newgale beach after being blown ashore by high winds. The RSPCA mounted a rescue operation and around 144 birds were saved, but a further 100 were found dead.

The nests of breeding birds frequently contain the shearwater flea *Ceratophyllus fionnus*, and where their burrows are near those of puffins the tick *Ixodes uriae* is commonly found. The mite *Neotrombicula autumnalis* is also often present and has been implicated in spreading puffinosis, a viral disease that can kill up to 70 per cent of infected birds (Brooke 2010). The disease, first recorded

FIG 71. A young Manx shearwater, suffering from puffinosis, sitting by the edge of a path on Skomer in daylight. The bird was subsequently killed by a great black-backed gull. (Jonathan Mullard)

in 1948, is only known to occur among fledglings and is characterised by blisters on both surfaces of the feet, together with conjunctivitis and stiffening of the legs (Harris 1965). In severe cases, paralysis on one side of the body results in the individuals being unable to remain upright. Young birds become lethargic, sitting exposed in the open, and can sometimes be seen alongside a path (Fig. 71). Individuals are often covered in flies, an indication that their body temperature is dropping and death is near. Despite the mite being implicated, it is still not clear what actually causes the disease. At one stage a researcher from the University of Gloucester was monitoring changes in the body temperature of affected birds on Skokholm, using a thermal imaging camera, to investigate how temperature changes were related to the physical symptoms. Investigations are coming to a halt though, since despite many years of study it has not been possible to progress our understanding of this disease. It does not, however, have a significant impact on the population, as evidenced by the increase in breeding pairs referred to earlier.

Counting birds

In 1962 David Saunders, appointed as Skomer warden two years earlier, initiated the first long-term censuses of seabirds in the United Kingdom. Motivated by concern over the decline of puffins and guillemots on the island, he was a pioneer, since at that time no one knew the best way to survey seabirds. Numbers of guillemots had been higher, as shown by the series of photographs Lockley took at the Wick, in 1934. Unfortunately, although the resolution on these photographs is quite good, the photographs are just not detailed enough to be able to count individual birds. It is clear though that enormous numbers of birds were present, with the guillemots forming a continuous line along the main ledge. A drastic decline in the number of both guillemots and puffins took place between 1934 and 1946, probably because of oil pollution from ships sunk during the Second World War that continued to release oil as they broke up on the seabed (Birkhead 2016). Buxton and Lockley noted the extent of this oiling in *Island of Skomer* (1950):

> When we had last seen these ledges in the summer of 1939 there had been no gaps in that long line of breeding guillemots; but in 1946 there were three distinct gaps of unused ledge, marked plainly by a green growth of some sort, probably the nitrogen loving Stellaria media [chickweed]. For want of a better explanation we were inclined to blame these reductions in the ranks of the guillemots to losses at sea through waste bilge oil from oil-burning and oil-carrying ships. The beaches along the whole coast of south west Wales were in 1946 (and each year after, to the date of this book going to press) smirched with this tarry residue; one could not navigate Skomer in a small boat without seeing numbers of seabirds dead, dying, or badly contaminated by this horrible filth.

David Saunders counted individuals rather than pairs, or nests, because guillemots breed closely together and do not construct a nest (Fig. 72). The guillemot population was subsequently the subject of one of the United Kingdom's longest and most detailed seabird studies. Professor Tim Birkhead of the University of Sheffield has followed the fate of individually marked birds since 1972 to examine the role of survival, breeding success and recruitment in driving population changes (Birkhead 2016). When the study started the number of guillemots on Skomer was at the lowest level ever recorded, with only about 3,500 individuals present. From 1980 onwards, however, the number of birds has increased steadily, and 24,788 individuals were recorded in 2017. Judging by the earlier evidence there is still some way to go before the population fully recovers – and the possibility of oil pollution remains. Oiling incidents, even as far away

FIG 72. Guillemots on a small section of the main breeding ledge at the Wick, Skomer. (Mike Alexander)

as northern Spain, continue to have a negative effect on the population, due to the fact that guillemots overwinter across such a wide area.

Birkhead's study is designed to understand the processes responsible for long-term changes in the guillemot population on Skomer. They are one of our most abundant seabirds, and living for at least twenty years they are reliable indicators of the quality of the marine environment. Every summer, for nearly fifty years, Tim, often accompanied by his colleague Ben Hatchwell, has visited the island together with one or more researchers, to observe and record the birds. Tasks have included monitoring adult and immature survival, age at first breeding, reproductive success, timing of breeding, and the diet and feeding rate of young birds.

No other bird breeds in such close proximity, but there is safety in numbers, the closely packed colony providing a defence against gulls and ravens that might otherwise steal eggs or chicks. Each guillemot pair defends its territory, which consists of a small area in which the female lays a single large egg straight onto the rock. The egg hatches after about 32 days, and 21 days after that the chick leaves the colony at dusk, launching itself off the cliff ledge onto the sea. Here it is met by the male bird, which cares for it out at sea for a few more weeks. The young guillemots are then on their own and they disperse widely, mainly in a southerly direction as far as the Bay of Biscay. When they are two years old a few of the young birds return to Skomer, but they tend to remain on the water or tidal rocks below the breeding ledges. From their third year on the young birds are more confident and visit the ledges close to where they were born. Some exchange between colonies does occur though and a few chicks reared on Skomer have moved on to other Welsh islands to breed. Occasionally, birds arrive from further away. In 1997, for instance, the researchers captured a breeding guillemot that had been ringed as a chick at Stora Karlsö, in the Baltic.

This long-term monitoring has shown that, for long-lived birds such as the guillemot, the single most important factors affecting population size are the life span of individual birds and oil spills. Climate change is also taking its toll, and recent results from the study show that the more extreme the North Atlantic Oscillation, the lower the survival rate. The oscillation results from fluctuations in the difference of atmospheric pressure at sea level between the Icelandic low and the Azores high. These changes influence the strength and direction of westerly winds and storms across the North Atlantic. Fluctuations that result in warmer, wetter and windier winters not only reduce survival but also delay the start of the breeding season. The birds, however, now breed two weeks earlier than they did in the 1970s.

Until 2013 the guillemot monitoring had been funded for many years by the Countryside Council for Wales. Disappointingly, in 2014 Natural Resources Wales, the successor body to the Countryside Council, decided to end funding for the scheme, one of the longest-running animal studies on record. Birkhead therefore turned to crowdfunding in order to secure sufficient money to cover the cost of the 2015 season. In just two weeks, he reached his target of £12,000 and, at the time of writing, a second appeal had reached nearly £100,000, ensuring that this important research can continue for the next fifteen years.

Puffins

The study of the Skomer guillemots highlights the need for this type of long-term monitoring and, luckily, they are not the only seabird under observation. As a result of other monitoring schemes, the Atlantic puffin, the bird with which we are familiar in the United Kingdom, was added to the IUCN Red List of endangered species in 2015. While numbers at colonies in Wales remain steady, or are even growing, in Scotland, Norway, Iceland and the Faroes, which together hold 80 per cent of the European population, numbers have plummeted (Fig. 73). Puffin numbers on Shetland, for example, have fallen from 33,000 in 2000 to only 570 in 2017 (McKie 2018). Research has shown that the birds are particularly susceptible

FIG 73. Puffin amongst the sea campion at the Wick on Skomer. Numbers at colonies in Wales remain steady, or are even growing, in contrast to other locations in Europe. (Jonathan Mullard)

to changes in sea temperature, thermal mixing and extreme weather, all of which affect the sand eels, sprats *Sprattus sprattus* and other small fish they feed on.

Puffins are common offshore around Pembrokeshire between late March and late July but rarely seen outside this period. In the spring and summer they congregate in large rafts close inshore near their breeding colonies (Rees *et al.* 2009). In his 1984 monograph on the species, Mike Harris suggested that the population of puffins on Skomer had declined to between 5,000 and 7,000 pairs by 1963 and perhaps to 6,500 pairs in 1982. In contrast, there has been a very dramatic increase since the 1980s, when a new census method was adopted. This involved counting all the puffins during their peak attendance in spring and provides the population trend, but not the total breeding population, which remains uncertain. The spring count for 1990 was 8,500, and by 2018 the count had increased by almost 300 per cent to 25,227 birds.

Since puffins nest underground in burrows, they compete with Manx shearwaters, as well as with their own species. They prefer burrows close to the cliff edge since they can access these quickly, avoiding predatory gulls. Herring and lesser black-backed gulls often chase puffins that are bringing food back to their chick and try to steal it, but the great-backed gull will kill and eat the adult bird. Towards the end of July, when the chicks are ready to leave the island, the birds are still not fully grown, being only about 70 per cent of the adult's weight. Unlike the young of some other seabirds, however, they are capable of flight and, to avoid being attacked by gulls, they leave at night. Out of the sight of land by dawn, they are on their own, remaining at sea for almost two years before returning to the colony. Less than 20 per cent of these birds survive to breeding age. Once they have begun breeding, however, they are long-lived, with an average life expectancy of around 25 years. Some individuals, of course, live much longer, and the current record on Skomer is 38 years. Puffins are not prolific breeders, however, each pair on average rearing one chick every two years.

Other breeding birds

Apart from the seabirds, Skomer also supports a number of other breeding birds, including peregrine, oystercatcher, skylark, wheatear, linnet *Linaria cannabina*, and reed bunting *Emberiza schoeniclus*. Between one and three pairs of choughs breed on the cliffs, feeding on insects in the short maritime grassland. There

FIG 74. North Pond on Skomer, one of a number of small ponds on the island. In 1999 the first confirmed breeding gadwall in Pembrokeshire were recorded here. (Mike Alexander)

are a number of small ponds on the island, and the largest of these, North Pond, was, in 1999, the location of the first confirmed breeding by wild gadwall *Mareca strepera* in Pembrokeshire (Fig. 74). Although six eggs hatched, only one young gadwall eventually fledged, its siblings almost certainly having fallen prey to gulls. In 2000 a second breeding pair was seen at Marloes Mere.

Gadwall used to be infrequent visitors to Pembrokeshire, and in the nineteenth century the Reverend Murray Mathew (1894) noted only three occurrences, at Pembroke, Orielton and Stackpole. The species began to appear with greater frequency from about 1964, some individuals overwintering. This change was probably the result of widespread introductions by wildfowlers. Today gadwall are regular visitors to Pembrokeshire, with small numbers being recorded, mainly on areas of freshwater near the coast, such as Marloes Mere. The maximum number seen to date is 22 birds at the Gann in Milford Haven during the winter of 1987, the county total at that time being about 40 birds (Green & Roberts 2004).

Skomer vole

Rodents found on the island include wood mice *Apodemus sylvaticus*, pygmy shrew *Sorex minutus* and common shrew *S. araneus*, but the most interesting is the Skomer vole, a subspecies of the bank vole discovered, as mentioned earlier, by Robert Drane (Fig. 75). He recounted his observations on Skomer in a presentation to the Cardiff Naturalists' Society in 1898, titled 'A Pilgrimage to Golgotha':

> We caught many voles which, as far as I understand, do not agree with either of the two, possible, ones found in this country. Of these two, it cannot be the common field-vole with which everybody is familiar and from which it differs widely in appearance, colour and size. It is very much like the remaining one, the bank vole (Microtus glareolus) but is much larger, and I cannot make its teeth agree with Lydekker.

'Lydekker' was a reference to Richard Lydekker's *Hand-book to the British Mammalia*, published only two years earlier, in 1896. Coincidentally, Lydekker, who spent most of his working life in northern India, first attracted attention through his letters to *The Times* in 1913. Writing on 6 February that he had heard a cuckoo *Cuculus canorus*, contrary to the view that the bird never arrived before April, a few days later he had to confess that 'the note was uttered by a bricklayer's labourer' (Gregory 1976). Letters to the newspaper about the first cuckoo subsequently became a tradition, although the custom of welcoming the bird to these shores goes back much earlier (see Chapter 17).

FIG 75. The Skomer vole is an endemic subspecies of bank vole. This is a rare view of the animal under the bracken, since people usually photograph them in the hand after they have been trapped during survey work. During the summer, there are approximately 20,000 voles on the island but, since they remain hidden, most visitors are completely unaware of their presence. (Mike Alexander)

During his visit to Skomer Drane collected a number of specimens of the Skomer vole which, on his return to Cardiff, he sent to the Linnean Society and the British Museum of Natural History (now the Natural History Museum). Some years later, in 1903, Barrett Hamilton of the museum confirmed that the voles were indeed a different race. As a result, in April 1905 Drane again visited Skomer and brought back live voles so that he could study them in his house. In his book *A History of British Mammals*, G. E. H. Barrett-Hamilton (1914) recalled that:

> *Mr Robert Drane always took it [the vole] about or inside farm buildings and Dr Y. H. Mills, in the heaps of swedes in which it was feeding. Whether they have since spread out over the island, or were merely not found in other places by these early investigators, is a matter for speculation.*

In *Island of Skomer* (1950), Buxton and Lockley wrote that:

The vole seems particularly fond of Skomer's many stone walls, especially at points where in summer there is a nearby profuse covering vegetation of bracken and ragwort, and areas near freshwater ponds and even along the banks of streams. It is also to be found on grassy slopes leading down towards the sea and in open land where some plant cover is provided.

Much of the research on the Skomer vole has been carried out by Dr Tim Healing, who has been working on the animals for over forty years, marking and recapturing them to estimate the population size. Recently, this work has been continued by Alice Brooke, a long-term volunteer with the Wildlife Trust. Like other species of vole, the Skomer race has, by our standards at least, a brief life, only surviving for a maximum of eighteen months. The breeding season lasts from May to September, and voles born early in the season may reach sexual maturity in the same year (Coutts & Rowlands 1969). They feed on bracken, bluebells, ground-ivy *Glechoma hederacea* and other vegetation. During the summer months the voles make use of the cover provided by bracken and brambles *Rubus* agg., and it is in these locations that the highest densities occur. The animal's habitat also extends below the surface, and they make use of the numerous rabbit and Manx shearwater burrows, but interactions with these species are unclear.

The Skomer vole is a distinct island race that has evolved after being accidentally introduced to the island at an unknown date, perhaps by a boat delivering supplies from the mainland (Corbet 1964, Hare 2009). An argument has also been put forward that they could be a relict population, resident on the island when it became separated from the mainland thousands of years ago, but this seems unlikely. Despite being geographically isolated from the mainland, the vole is not genetically separate and is able to interbreed with the bank vole *Myodes glareolus*, producing fertile hybrids (Fullagar *et al.* 1963). The Skomer vole is said to be larger, however, than voles found on the mainland, with a distinct coat – and the shape of its nasal passages is apparently unique (Corbet 1964). Most of the island races of voles, however, differ from mainland forms far less than was believed when they were first described as subspecies. It seems that many of the differences were exaggerated by comparing small numbers of specimens in museums (Yalden 1999). Voles from Skomer, for example, while much larger than the voles from southeastern Britain, with which they were initially compared, do not seem so large when compared with Scottish bank voles.

As with the slow worms on Skokholm described earlier, living on islands affects the body size of animals in these situations. There are two main factors:

'competitive release', which favours an increase in body size, and 'resource limitation', which drives a decrease in size (Lomolino 1985). As there appears to be no limit to their food resources, a lack of competition has therefore allowed the Skomer vole to grow larger than some other bank voles. At their largest they are around 12 cm long and weigh a maximum of 40 g. There is little evidence though to support the assumption that small mammals increase in size as the risk from predators decreases. On Skomer, although there are no ground predators, the voles are still at risk from birds, such as barn owls *Tyto alba*, short-eared owls and kestrels *Falco tinnunculus*.

The importance of other characteristics, such as teeth, has also been frequently exaggerated. As immediately recognised by Drane, voles from Skomer, like those on Jersey and elsewhere, do have a different first upper molar to those from mainland Britain. The difference, however, is relatively minor, taking the form of an extra 'loop' on the inner side of the molar. The voles with the more 'complex' tooth come from islands that lack field voles *Microtus agrestis*, and it is therefore likely that they have adapted to a 'harder' diet, perhaps eating more grass than voles on the mainland. As mentioned in relation to the Skokholm slow worms, island races tend to be at the extreme end of the range of variation seen in mainland populations. Given the likelihood of only a few original animals, no ground predators, severe winter food shortages and exposure, the changes could have arisen in only a few generations (Berry & Jakobson 1975).

Vegetation

Bracken, with an understorey of bluebell and red campion *Silene dioica*, covers most of the interior of the island (Fig. 76). In contrast, the maritime grassland in the west and south is characterised by hummocks of thrift and areas of sea campion. As on Skokholm, this is heavily grazed by rabbits and further modified by the manuring and trampling of seabirds. On the more sheltered northern and eastern slopes there are small areas of scrub containing blackthorn *Prunus spinosa*, elder *Sambucus nigra*, bramble, common gorse *Ulex europaeus*, grey willow *Salix cinerea* and common osier *S. viminalis*.

Other plants on Skomer include three-lobed crowfoot *Ranunculus tripartitus*, Portland spurge *Euphorbia portlandica* and the nationally scarce lanceolate spleenwort *Asplenium obovatum* ssp. *billotii*. This perennial, evergreen, calcifuge fern is mainly a plant of sheltered, shady crevices and ledges on maritime cliffs, requiring a frost-free environment. It rarely grows far from the sea but is unable to tolerate dense vegetation.

There is a rich community of lichens, including a large number of specialist species, such as granular bush lichen *Ramalina polymorpha* and *Aspicilia epiglypta*,

FIG 76. During May and June Skomer is carpeted with bluebells and red campion, shown here, which flourishes in the more sheltered areas. (Mike Alexander)

both of which are associated with the nutrient enrichment of rock and soil surfaces by seabirds. Other notable species occurring on rocks are golden hair lichen, the nationally rare *Xanthoparmelia tinctina, Caloplaca britannica, Arthonia atlantica, Porina curnowii* and *Rinodina orculariopsis*. Additional species occurring on the ground include *R. conradii*, found on soil, heather stems or rabbit pellets, and *Trapeliopsis wallrothii*, found on turf and soil.

Beetles and moths

Like Skokholm, Skomer is of importance for a number of scarce beetles and moths. These include the ground beetle *Masoreus wetterhallii* found on rabbit-grazed turf, the tortoise beetle *Cassida hemisphaerica*, which feeds on sea campion, and the golden keyhole weevil *Sibinia sodalis*, which feeds on thrift. Water beetles associated with the ponds include the nationally rare *Graptodytes flavipes* and the scarce water scavenger beetle *Helophorus griseus*. The latter is a characteristic species of impermanent ponds, especially ephemeral pools on grassland. Feeding on thrift are the black-banded moth and thrift clearwing *Bembecia muscaeformis*. Other moths include Barrett's marbled coronet and the Devonshire wainscot *Mythimna putrescens*, both of which have a distinctly coastal distribution in Britain.

ISLAND BATS

Some of these moths may provide food for bats, but, until recently, there were only a few records of bats from Skokholm, Skomer and Ramsey and very little was known about the diversity, abundance and seasonality of the species present. A number were recorded by visitors to Skomer in September 2013, including Nathusius' pipistrelle *Pipistrellus nathusii*, a bat rarely recorded in Britain and Ireland, common pipistrelle *Pipistrellus pipistrellus* and noctule *Nyctalus noctula*. A noctule bat was also caught in a mist net in 1963 on Skokholm, where greater horseshoe bat *Rhinolophus ferrumequinum* droppings were found in a sea cave in 1993 and a soprano pipistrelle *Pipistrellus pygmaeus* recorded in September 2013. In the same year, a greater horseshoe bat was found dead on Ramsey, and other records from the island include pipistrelles and noctules. These chance records did not, however, provide a clear picture of how bats used the islands.

It was extremely helpful therefore that in 2014, as part of ongoing studies of bat migration, a consultancy firm, BSG Ecology, installed a fixed-point automated bat detector to remotely monitor activity on each of the three islands. The intention was to investigate whether bats were migrating between Ireland and Wales (Taylor 2014). Between May and October on Skomer and Skokholm, and until 13 November on Ramsey, the detectors monitored the number of 'bat passes' per night. The number of times a bat passed a detector is a measure of activity, not of the number of bats present. The data from this study suggested that, despite originally being considered to be relatively unsuitable for bats, the islands are regularly utilised by a number of different species.

As can be seen from Table 2, the diversity of bat species recorded on each island was similar, with the same nine species present on Ramsey and Skomer, and seven of these recorded on Skokholm. The lower diversity on Skokholm is probably due to the smaller size of the island and the fact that it is further from the mainland than Skomer or Ramsey. This pattern, with the number of bat passes on each island directly proportional to its distance from the mainland, suggests that most of the bats recorded are visitors.

TABLE 2. Number of 'bat passes' recorded on Skomer, Skokholm and Ramsey in 2014 using an automated bat detector. Identification to species level, especially with *Myotis* species, can be difficult, hence the overlapping columns. (Adapted from Taylor 2014)

Species	Skomer	Skokholm	Ramsey	Total
Nathusius' pipistrelle *Pipistrellus nathusii*	1	1	3	5
Common pipistrelle *Pipistrellus pipistrellus* /Nathusius' pipistrelle	10	0	7	17
Soprano pipistrelle *Pipistrellus pygmaeus*	861	2	119	982
Common pipistrelle / soprano pipistrelle	158	0	195	353
Common pipistrelle	3,676	2	13,612	17,290
Myotis species	8	1	101	110
Myotis sp./ long-eared bat *Plecotus* sp.	4	2	17	23
Long-eared bat species	108	0	73	181
Serotine *Eptesicus serotinus* / Leisler's bat *Nyctalus leisleri*	0	2	0	2
Leisler's bat	38	3	4	45
Noctule *Nyctalus noctula* / Leisler's bat	215	397	13	625
Noctule	831	621	98	1,550
Greater horseshoe bat *Rhinolophus ferrumequinum*	19	3	58	80
Barbastelle *Barbastella barbastellus*	1	0	2	3
Total	**5,930**	**1,034**	**14,302**	**21,266**

A rise in noctule activity observed in September on Skokholm may indicate that noctules are flying past the island during migration, coinciding as it does with the autumn migration period for this species observed in continental Europe (Hutterer *et al.* 2005). Most of the noctule passes were, however, recorded at least an hour after sunset, and no passes were recorded in the 90 minutes before sunrise. If noctules were migrating from the mainland, bat passes would have been expected close to sunset as bats were beginning their migration; or, if bats were migrating to the mainland from across the sea, a more scattered distribution of passes would be expected. Instead it seems likely that at least some noctules are commuting to the island from the mainland in the early evening and returning to the mainland well before sunrise. The increase in activity in September may suggest instead that the bats were dispersing along the coast, or migrating a shorter distance within the United Kingdom, such as south across the Bristol Channel.

Overall the data do not prove that bats are migrating between Ireland and Wales, but there is some evidence that this may occur. On 21 September 2008 naturalists out on a pelagic trip looking for seabirds and dolphins in the Irish Sea were surprised to see a bat flying past in the middle of the afternoon. The bat was flying eastwards, about 30 km off the Pembrokeshire coast towards the Smalls lighthouse. One of the people on board, John Stewart-Smith, managed to get a photo of the animal, and it seems to have been either a Leisler's bat *Nyctalus leisleri* or a noctule, but it was difficult to tell since Leisler's bat is similar to the noctule but is slightly smaller, with longer fur. Leisler's bats are found in Ireland and most of Britain but they have only occasionally been recorded in Pembrokeshire. Noctules, on the other hand, are relatively common in Britain, including Pembrokeshire, but are apparently absent from Ireland.

One of the more unexpected results was the use of the islands by greater horseshoe and barbastelle *Barbastella barbastellus*, two species that prefer wooded countryside and have not been regularly recorded crossing the open sea before, although they are resident on islands such as the Isle of Wight (Altringham 2003). Barbastelles breed in woodland and, although there is a colony in Pengelli Forest, there is no woodland on the islands, so their presence here is difficult to explain. Greater horseshoe bat passes were recorded during three separate nights on Skokholm in September. The previous record of horseshoe bat droppings found in a cave on the island, along with the timing of recorded bat passes, suggests that these bats are flying over to the island in the autumn and potentially hibernating in suitable sea caves over winter. This may also be the case on Skomer. The highest number of bat passes, however, were from Ramsey, and greater horseshoe bats were recorded in every month except November. This suggests that they are either resident on the island, or regularly commuting to the island from the mainland.

CHAPTER 6

Ramsey and the Bishops and Clerks

Taking the islands in order from the north, we have first to describe the one that, in our opinion, is the most picturesque in its rocky scenery, and the brightest in its summer garb of flowers, beautiful Ramsey.

Murray Mathew (1894)

SOME OF THE HIGHEST SEA cliffs in Wales, rising up to 120 m, can be found on Ramsey (Fig. 77). Originally a place of pilgrimage, Ramsey (Ynys Dewi) was named after Saint David, since it was the home of his confessor, Saint Justinian. The embarkation point on the mainland near the lifeboat stations is still known as St Justinian's. Only a short boat ride across Ramsey Sound the island welcomes, like Skomer to the south, day visitors during the spring and summer. There are only about 4,000 visitors annually though, compared with the 17,000 that visit Skomer, mainly to see the puffins.

As mentioned in Chapter 2, one of the early references to Ramsey, and the Bishops and Clerks, can be found in a letter from the Reverend Nicholas Roberts to Edward Lhuyd:

To this Island, and some rocks adjoyning, call'd by the sea-men The Bishop and his Clerks, do yearly resort about the beginning of April such a number of birds of several sorts, that none but such as have been eye-witnesses can be prevail'd upon to believe it; all which after breeding here, leave us before August. They come to these rocks, and also leave them, constantly in the night-time: for in the evening the rocks shall be cover'd with them, and the next morning not a bird to be seen;

FIG 77. An aerial view of Ramsey from the northwest highlights Carnysgubor and Carnllundain either side of Aber Mawr Bay, which was created by the erosion of the softer sedimentary and volcanic rocks. (Janet Baxter)

so, in the evening not a bird shall appear, and the next morning the rocks shall be full. They also visit us commonly about Christmas, and stay a week or more, and then take their leave till breeding-time. Three sorts of these migratory birds are call'd in Welsh, Mora, Poeth-wy, and Pâl; in English, Eligug, Razorbil, and Puffin; to which we may also add the Harry-bird; tho' I cannot at present assure you, whether this bird comes and goes off with the rest.

RAMSEY

Puffins

Like Skomer, Ramsey used to support an enormous puffin colony, but brown rats, which arrived in the nineteenth century from shipwrecks, severely reduced the number of birds breeding on the island. In 1999 and 2000, therefore, the RSPB undertook an ambitious, and successful, rat eradication programme. The project paved the way for more recent eradications on Lundy, Scilly and, in 2016, the Shiants, off the east coast of Lewis. In 1998, the year before the rats were eradicated, there were just 897 pairs of Manx shearwaters breeding on Ramsey, with very few chicks successfully fledging. Numbers increased substantially following the eradication of the rats, with 4,796 pairs now breeding. While storm petrels have bred for the first time, the puffins have been slow to return to their ancestral haunts. In an attempt to lure them back the RSPB installed 200 plastic puffin decoys – a method that has been very successful in Scotland. Welsh puffins seem to be more discriminating, however, and they ignored the strangely immobile birds until 2013, when a sound system was added to broadcast puffin calls. This additional attraction seems to have done the trick and birds subsequently arrived on the island, although they have not yet bred.

Choughs

Ramsey, along with the Castlemartin area, is one of the best sites for choughs in Pembrokeshire, with eight to ten pairs regularly breeding each year (Fig. 78). In fact, the presence of choughs was one of the main reasons for the RSPB acquiring the island. They are often seen feeding on invertebrates in the close-cropped turf

FIG 78. The chough is the rarest member of the crow family in Britain, and can be identified by its blue-black plumage and its red bill, legs and feet. Their Welsh name, *brân goesgoch*, means red-legged crow. Three-quarters of the UK's population can be found in Wales. (Lisa Morgan)

FIG 79. The arms of Rhys ap Thomas are three ravens and a black chevron, but an unknown painter has turned the birds on the roof boss in St Davids Cathedral into choughs. (Jonathan Mullard)

of maritime vegetation and acid grassland. Breeding is so successful that some birds are forced to look for territories on the mainland, although the majority, colour-ringed as chicks in the nest, tend to stay on the island.

The chough is a far rarer bird today than it once was, and it was probably frequent across Wales before its decline (Lovegrove *et al.* 1994). The confiding nature of the bird made it an easy target for trigger-happy gunmen, and it was often the target of egg collectors. In the sixteenth and seventeenth centuries the chough was listed in the Tudor Vermin Acts and bounties were offered for killing it. Gin traps and rabbit traps were a regular cause of death. Indeed, several birds, now in National Museum Wales, were found among a consignment of rabbits from the Pembrokeshire islands in the early twentieth century. At the end of the previous century Murray Mathew (1894) mourned the demise of a bird which he claimed had once been common 'from Tenby to Cardiganshire', and blamed persistent egg collectors. He was not averse to collecting them himself, however, noting that 'We ourselves possess some beautiful eggs of the various Ramsey birds, including those of the Chough.' He recorded that:

In former days, Mr. Phelps says, according to tradition, the Choughs nested in the ruins of the Bishop's Palace at St. David's, until they were driven out by Jackdaws, but as the nests there could have been easily robbed, he suspects they were 'human Jackdaws'.

Choughs still linger in St Davids, in a pictorial form at least, in the nearby cathedral. The roof of the cross passage, towards the east end of the building, has three bosses with coats of arms dating to the early sixteenth century. One of these depicts the arms of Rhys ap Thomas (1449–1525), which are three ravens and a black chevron, but at some stage an enthusiastic painter has coloured the beaks and legs red, turning them into choughs (Fig. 79)! Three choughs are also illustrated on a stained-glass window, in the east wall of the Lady Chapel, dedicated to Thomas Becket (Fig. 80). According to legend, following his murder, 'a crow did stray into the [Canterbury] Cathedral and peck about in the sanguine chaos, staining legs and beak'. They are apparently Cornish choughs, although the Heraldry Society considers that there is 'little heraldry associated with the murder of St Thomas of Canterbury and what there is, is confused and doubtful' (Humphery-Smith 1971).

FIG 80. Three choughs are illustrated on a stained-glass window in the east wall of the Lady Chapel in St Davids Cathedral, dedicated to Thomas Becket. These are allegedly Cornish choughs though, not Pembrokeshire birds. (Jonathan Mullard)

In addition to the Bishop's Palace, other ruined buildings once supported choughs. Samuel Gurney, writing in *The Zoologist* of 1857, describes the ruins of Manorbier Castle as being frequented by choughs, 'which are bred there in great abundance'. Gurney was told by the village schoolmaster that in the breeding season and in the winter the choughs were very tame, 'collecting in numbers around the school-room door at the time the school broke up in order to pick up pieces of bread thrown to them by the children'. A chough that had been raised by some children who lived nearby apparently followed them to school. 'Whenever they left home to go to school the bird would precede them, and arrive there a few minutes after they had started, and some twenty minutes before them. This it did so regularly that the master knew when the children might be expected.'

Other birds

It was not only chough eggs that were collected on Ramsey, however, for Murray Mathew notes that:

> *Mr Wimbush ... went egging oftentimes on Ramsey with ... the Rev. Sydney and Mr Mortimer Propert, when all three used to risk their lives, when boys, in dangling by a rope over its dangerous cliffs while collecting the eggs of the Peregrine, Buzzard, and Raven, and those of the Guillemot and other cliff-birds upon their ledges.*

Dangling on ropes over the Ramsey cliffs was certainly a risky way of collecting birds' eggs, as noted by Browne Willis (1717), quoting from the unnamed manuscript mentioned in Chapter 2:

> *The vast Number of Eggs laid in these Rocks, are, in the Time when they are in Sason, the great Subsistence of the poorer Sort of the Inhabitants of St David's; they are delicately speckled with a charming Variety of Colours. The People hazard their Lives to get them; for whilst one stands upon a steep and high Rock, another is let down by a Rope ty'd to his Middle, one end of which is held by him that stands above; and sometimes they both fall down, and are dash'd to Pieces among the Rocks. That is the Way by which they take Samphire over all this Coast, and with the same hazard. They look upon these Sea-Fowls Eggs to be sweeter than the common Eggs of the Land-Fowl.*

The seabirds found on Ramsey include guillemots, razorbills, lesser black-backed gulls, kittiwakes and shags. Record numbers of guillemots now breed on Ramsey, with over 4,000 individuals counted in 2012. Kittiwakes though are declining nationally and Ramsey is no exception. Only 139 pairs bred on the island in

2013, down from a total of 450 in the 1990s. Kittiwakes used to nest in their millions around the United Kingdom but now only around 300,000 breeding pairs remain. Increasing water temperatures have affected zooplankton with the result that sand eels, a critically important source of food for many birds, have disappeared from large areas of the Atlantic and the North Sea. As a result, seabirds such kittiwakes, which take sand eels from the sea surface, are affected more than species such as guillemots, which utilise fish from greater depths.

Ramsey provides ideal breeding habitat for wheatear, one of the earliest migrants to reach Pembrokeshire in the spring (Rees *et al.* 2009). Most arrive in the second week of March. Up to 90 pairs breed on Ramsey, with some 40 pairs on Skomer and 20 on Skokholm (Green & Roberts 2004). In the last breeding bird survey in the county, between 2003 and 2007 a decrease in the number of pairs breeding on Skomer was balanced by an increase in those on Ramsey. Pairs also breed on the mainland coast and on Mynydd Preseli. In total, the summer population in Pembrokeshire stands at an estimated 400 pairs. Their preferred habitat is open ground with a short sward, with rock crevices or drystone walls nearby, for nesting.

A number of other bird species regularly breed on the island, including raven, little owl, kestrel, buzzard *Buteo buteo,* meadow pipit *Anthus pratensis,* stonechat *Saxicola rubicola,* skylark and linnet. The wardens regularly plant just over a hectare of a seed mixture containing barley *Hordeum vulgare,* triticale (a hybrid of wheat *Triticum* spp. and rye *Secale cereale*), radish *Raphanus sativus,* white millet *Panicum miliaceum* and quinoa *Chenopodium quinoa* as a source of food for flocks of finches throughout the autumn and winter. In 2015, for example, flocks of up to 200 linnets made use of the arable plot. Linnet populations have dropped substantially over the past few decades, mainly due to the reduced availability of weed seeds. Overall the United Kingdom population is estimated to have declined by 57 per cent between 1970 and 2008, so this supplementary feeding is helpful. In hard winters the seed also acts as a supplementary source of food for choughs.

Usually, three pairs of peregrines nest on Ramsey each year, preying mostly on feral pigeons *Columba livia* which live in the caves. The earliest record of peregrines on the island dates from October 1171. In that month King Henry II, waiting in Pembrokeshire to cross to Ireland with an army, found an eyrie of peregrines on the island that got the better of his 'Norway hawks', which were probably goshawks *Accipiter gentilis.* The story goes that he set his bird on a peregrine, but the local bird knocked his larger adversary out of the sky at the monarch's feet. Following this incident, he always procured his birds from Ramsey (Wight 1948). On a visit to Ramsey with a shooting party, Richard Fenton (1811) 'saw here two falcons of that breed this island has been so celebrated for, who being disturbed had deserted their nest, yet often dropped on the wing to flutter round it, testifying their fierceness and parental anxiety by most horrible screams'. In the nineteenth century, shooting parties also caused a great deal of disturbance to the seabird colonies. Visiting the Choir, 'an amphitheatre of rocks, precipitous and of stupendous height', Fenton (1811) noted that 'at the firing of a gun hundreds flew off, succeeded by fresh hundreds in uninterrupted succession, while the sea beneath was covered with other hundreds darkening its surface'.

In 1915 two golden eagles *Aquila chrysaetos*, which had been kept in captivity on Skomer, were released, by J. J. Neale as he gave up the lease on the island that year. Soon afterwards one was shot near Marloes but the remaining bird lived on Skomer for thirteen years before moving to Ramsey in 1928. A 'mate' for this bird, a female, was provided by London Zoo in 1929 but was found drowned on the coast opposite Ramsey later the same year. The remaining bird survived until 1932, when it was shot by the farmer, who maintained it was killing his lambs. Only when the corpse was inspected was it found that this too was female, so it was no wonder that the birds had not bred. There is, however, place-name evidence for the former existence of golden eagles elsewhere in Pembrokeshire, as described in Chapter 13.

Invertebrates

Scarce coastal invertebrates can also be found on Ramsey, including two moths, the dew moth *Setina irrorella* and the Devonshire wainscot, and a scarab beetle *Bodilopsis sordida* whose larvae feed on dung. The dew moth (Fig. 81) is thought to have acquired its name from the male's habit of hanging from a blade of grass during the day, when the yellow, black-spotted, insect appears transparent. The female in contrast is largely nocturnal. It occurs in only a few widely separated sites in Britain, stretching from Kent to Scotland. There is a strong colony on the exposed, western, side of Ramsey, and a few individuals have been recorded between Solva and Newgale. Otherwise it seems to be very localised, possibly constrained by the occurrence of the lichens on which the larvae feed. The distinctive larvae, with their orange diamond-shaped dorsal markings, can be seen sunning themselves on the rocks in April, sometimes in their hundreds (Robin Taylor, personal communication).

Devonshire wainscot was first noted at Torquay in 1859, in Carmarthenshire around 1871 and in Pembrokeshire in 1954 (South 1980). A local species, occurring mainly coastally in southwest England and South Wales, it seems to have gradually spread west after possibly arriving from the continent. The adults fly in late July and August and are attracted to various flowering plants, including wild clary *Salvia verbenaca*. Two other notable moths occurring on the island,

FIG 81. The dew moth is one of a number of rare moths to be found on Ramsey. It is very localised, possibly constrained by the occurrence of the lichens on which the larvae feed. (Lisa Morgan)

Barrett's marbled coronet and hoary footman *Eilema caniola*, are also restricted to southwest Britain. As with the dew moth, the larvae of the hoary footman feed mainly on lichens growing on rocks.

The smallest species of clearwing, thrift clearwing *Synanspecia muscaeformis*, has been found on the exposed headlands at the southern end of the island. The caterpillars feed inside the roots and stems of the foodplant from August to the following May, overwintering as part-grown larvae. Because of its small size and fast flight, it is best located with the aid of a special pheromone lure. Coastal cliffs and rocks are excellent habitats for moths, since the climate is relatively constant and they are generally inaccessible to grazing animals and people. There are also numerous caves and crevices where eggs, small larvae, pupae and hibernating adults can shelter. In addition, natural erosion produces areas of bare ground, which are colonised by pioneer plant communities and provide opportunities for these scarce moth species to spread to new areas (Whitehouse 2011).

Ramsey has a rich spider fauna, including several rarities, such as *Aelurillus v-insignitus*, *Lathys stigmatisata* and *Clubiona genevensis*. The last of these, recorded on the island for the first time in 1933 by Bristowe, is one of the United Kingdom's smallest and rarest spiders (Fig. 82). Orange-yellow in colour, it measures just 3 mm in length. All three species are mainly coastal heathland specialists, usually attaching their egg sacs underneath loose stones among the sparse vegetation. *Clubiona genevensis* has been recorded in Wales only on Ramsey, Skokholm and the Llŷn peninsula. Elsewhere it is equally uncommon, noted only from four sites on the Lizard, the Isles of Scilly, and Lulworth and Ringstead in Dorset. The species is clearly well established on Ramsey, with 27 females, or cells with eggs, being found in 1999 and 24 in 2006. In contrast, *Lathys stigmatisata* was only found on Ramsey in 1999 (Dawson 2000). Similarly, there is only one record for *Aelurillus v-insignitus*, which otherwise is confined to England, south of a line from the Wash to the Severn, and two isolated localities in Scotland.

FIG 82. One of the United Kingdom's smallest spiders, *Clubiona genevensis*, was first discovered on Ramsey by W. S. Bristowe, author of the New Naturalist volume on spiders. This female was photographed on the Llŷn peninsula. (Richard Gallon)

Farming and plants

From the thirteenth century onwards, Ramsey was a valuable part of the Episcopal estate of St Davids. In 1326, for example, the island was said to support 10 horses, 100 cattle and 300 sheep and the lord of the manor could take 500 rabbits a year without appreciably reducing their numbers (Wight 1948). Apart from a short break, farming has continued to the present day. The red deer (Fig. 83) now compete with a resident flock of 100 sheep, 4 ponies and the rabbits for grazing. The original deer herd consisted of 29 individuals, and by 1994 it had increased to 45, including the 14 calves born that year. Since then the population has been limited to around 20 animals, with two mature stags. Trampling and wallowing by the deer in and around a number of shallow pools and the surrounding wet areas create ideal conditions for the island's rare aquatic plants. Grazing on a small island does, however, require a fine balance to be struck, since if the stocking levels are too high the vegetation suffers.

Like Skokholm and Skomer, Ramsey is an island noted for its rare plants and lichens, the nationally scarce granular bush lichen and golden hair lichen

FIG 83. When the RSPB bought Ramsey in 1992, a small herd of red deer was already present, having been introduced in 1976 by the previous owners. (Mike Alexander)

occurring here as well. There are three patches of golden hair lichen on or around the western side of Carnllundain, with over 1,000 thalli known to be present, while granular bush lichen is locally abundant in just a few localities on the tops and sides of coastal rocks in the vicinity of the seabird colonies. Ramsey also has the only remaining specimens of common juniper *Juniperus communis* in Pembrokeshire. Apparently, it was once plentiful here, Browne Willis describing the flora of the island in 1715 as follows: 'The Fern, which grows here in great abundance, is very tall and it produces Thorns of a good Bigness, and great store of Juniper' (Willis 1717). More recently, there are records of only eight bushes. A survey of *Juniper in the British Uplands* published by Plantlife in 2007 noted that:

> *In addition, we received one survey report for a population of 8 plants of Juniperus communis ssp. hemisphaerica, on Ramsey Island, Pembrokeshire. The plants were on steep south facing cliffs, and comprised 7 male and one female mature plants, all of which were green and vigorous. (Long & Williams 2007)*

Only four of these plants survive today, clinging to the cliff edge on the eastern side of the island. They are all that remains of an ancient population, which bears similarities to the coastal subspecies *hemisphaerica* found on the Lizard in Cornwall. But they are not the same subspecies, as incorrectly described in the Plantlife report and some other publications (Long & Williams 2007). Their survival is threatened by the low number of surviving plants, two of which are in poor condition, and their precarious location on the cliff top. A further site was previously lost as a result of a rock fall. Two of the bushes have, however, regenerated through layering, the stems producing roots where they touch the ground. As all but one of the plants are male, this causes problems for their conservation. To achieve regeneration from seed, populations need good numbers of male and female bushes, including productive female plants with viable seed. Where plants are not producing seeds, cuttings can be taken and propagated, and this may now be the only option if the Ramsey population is to be maintained. Juniper also faces a threat from *Phytophthora austrocedrae*, a fungal infection, which has been confirmed in the wild in northern England and Scotland, but not yet in Wales. It is usually fatal.

In the centre of the island are six small ponds, dug to provide drinking water for livestock and obtain culm (low-grade anthracite), which was added to coal dust for fuel. A further five ponds were formed by damming streams. All these ponds contain an interesting variety of plants, including the nationally scarce pillwort *Pilularia globulifera*, floating water-plantain *Luronium natans*

FIG 84. Ramsey supports one of the best populations of three-lobed crowfoot in Pembrokeshire, with over 300 flowering plants visible in early spring. Trampling and wallowing by deer create ideal conditions for this plant. (Lisa Morgan)

and three-lobed crowfoot, which grows in the muddy margins of the two of the ponds. This is one of the best populations of three-lobed crowfoot in Pembrokeshire, and indeed Britain, and most years over 300 plants can be seen flowering in early spring (Fig. 84). Wavy St John's-wort *Hypericum undulatum* occurs in a wet flush, while yellow centaury *Cicendia filiformis* is again associated with damp, open places.

Finally, a new use for the flora of Ramsey has been developed by a local business in St Davids which is producing 'Ramsey Island Gin', varieties of which will include 'botanicals carefully collected from the island under supervision of RSPB staff'. A small percentage of the profits will be donated to RSPB Cymru for general conservation work and special projects on Ramsey and Grassholm. The spring variety of gin will apparently incorporate gorse flowers and thyme *Thymus drucei*, summer gins will include sheep's sorrel and wood sage, while in autumn and winter heather and mint *Mentha* sp. will be used.

Grey seals

Ramsey hosts the largest grey seal colony in Wales, between 500 and 700 seal pups being born on the island's beaches each year, with another 200 pups on Skomer and smaller numbers elsewhere (Baines *et al.* 1995). In total, there are around 5,000 grey seals in the seas around Pembrokeshire and twice a year, when the animals are moulting or breeding, the numbers around Ramsey increase dramatically. All seals need to come ashore once a year to moult their old skin and hair, and from January to April the beaches around the island are filled with large and noisy groups of animals (Fig. 85). Over 300 individuals use them as a refuge during their annual post-breeding moult (Lisa Morgan, personal communication). The juveniles usually moult first, followed by the females and then the adult males, although there is often some overlap. The moult can take as long as six weeks, during which the seals are lethargic and often irritable, since they do not eat during this period. Once their new hair has grown, they head out to sea to resume feeding.

Given their abundance in Pembrokeshire, it is sometimes easy to forget that grey seals are actually one of the rarest seals in the world. In 1914 the British population was estimated to be 700 animals. This was the first time that anything

FIG 85. Grey seals at a haul-out on a Ramsey beach in July 2015. Between 500 and 700 seal pups are born on the island each year. (Jonathan Mullard)

had been published on the seal population, and it is now believed that it was a significant underestimate. By 1970, when the Conservation of Seals Act was passed, the British population was around 70,000. By 2008 it was estimated to have risen to 118,700 individuals, and to about 140,000 in 2015, a figure that seems to be relatively stable (Sea Mammal Research Unit 2017). The United Kingdom now holds around 50 per cent of the world's grey seals, and 95 per cent of the European population. The pupping sites, or rookeries, vary greatly in size, with the largest located in the Hebrides, Orkney, Isle of May, Farne Islands and Donna Nook on the Lincolnshire coast. The population in Pembrokeshire and the surrounding counties is the most southerly in Europe of any significant size, and is relatively isolated from those elsewhere.

During the breeding season, from August through to November, just under 1,000 seals come ashore on Ramsey to breed. The males arrive first and fight among themselves to establish territories on the beach. During this period, they go without food, since if they returned to the sea they would lose the ground they had fought for. Each male may mate with numerous females. Once the egg is fertilised, there is a delay of about three months before the embryo attaches to the wall of the uterus and starts to develop. Following this delayed implantation there is a gestation period of eight months, ensuring that the pups arrive in the next breeding season. Many are born on the small beaches at the back of the island's caves. Grey seals feed their pups with milk for 16–21 days, during which time the pup gains an average of 30 kg, while the mother might lose 65 kg of her own body weight. When she is forced to return to sea to feed, the pup usually remains at the breeding ground for another 14 days or so before heading out to sea, to forage for itself.

At this time of year, the pups are vulnerable to storms. In the autumn of 2017 it was estimated that around 75 per cent of the seal pups on Ramsey were killed by Storms Brian and Ophelia, which hit Ireland and the west coast of Britain with gusts of up to 130 km per hour. The RSPB recorded 120 pups on nine monitored beaches before the storm, but only 31 were found afterwards. At this time the pups were all less than three weeks old and had not been weaned (Fig. 86). Many appear to have been abandoned, the females deciding to save themselves. There was a similar picture on Skomer, where the Wildlife Trust estimated that 50 pups were killed by the storm, battered against the rocks or struck by debris. The few animals that managed to survive sustained a variety of injuries, with others being separated from their mothers by the weather and washing up along the coast. Fourteen dead pups and one live one were found, for example, on Marloes beach. The storms were the strongest to hit Pembrokeshire since 1987 – but with violent storms becoming more common, due to climate change, there could be

FIG 86. During the worst of the weather in autumn 2017 a grey seal pup and its mother shelter in Martin's Haven. This pup was one of the lucky ones, many being separated from their mothers and killed by the storms. (Blaise Bullimore)

long-term impacts on the seal population. Thankfully, in 2018 at least one of the storms coincided with low neap tides, which meant that the pups had somewhere to retreat to, away from the waves.

While today storms represent one of the main threats to seal populations, at one time grey seals were hunted for their oil, which was used as fuel for lamps, and for their skins. Browne Willis (1717) described seal hunting in Pembrokeshire as follows:

And they were hunted out from their caves and on the beaches and killed with sticks by means of a sharp blow on the tip of the nose which is how the inoffensive badger also has to be struck by those misguided enough to want to kill him. In caves about these islands, seals, or sea-calves, bred, which are taken about Michaelmas by the sea-men who go out for them in boats, and kill them in their caves with clubs. They observe when they strike at these sea-calves, that if they do not hit them upon the snout, the blow goes for nothing; and when they run away, they throw great stones behind them with their hind feet, which are very dangerous, if those that hunt them do not take heed.

Towards the end of the nineteenth century, the demand for seal products declined but seals began to be regarded as a competitor for fish. An early written account of seals around Skomer is included in a note written by J. J. Neale on his visit to the island in May 1896:

> *On two or three occasions we saw seals. Once a very large one came up astern of the punt and gazed at us with its human-looking face. No wonder it gave rise to the fiction of there being mermaids in the sea. The so-called gentlemen take a delight in shooting them although they know full well that if shot while in the water the bodies at once sink.*

Despite the 1970 Act, in the decade that followed nearly 2,000 seal pups were killed in Scotland, and the National Trust killed 804 pups, 486 adult cows and 158 bulls on the Farne Islands. By 1978 though, the adverse public reaction that these activities generated meant that the killing ceased. Since then no licences have been issued in the United Kingdom for large-scale 'control measures' and the population has increased significantly. Whether this represents the natural numbers to be expected in the absence of human predation remains to be seen. As with many other species, grey seal populations have been suppressed for so long that estimating their expected abundance is almost impossible, and we will never have enough information on which baselines to use, given the poor early records.

The congregations of seals on Ramsey during the autumn and winter provide an opportunity for the RSPB staff to photograph large numbers of animals, allowing individuals to be identified year after year. This will at least enable us to monitor population trends. For many years the animals were identified by eye, but this is a time-consuming process involving many hours at the computer, and 'patch-recognition' is now being used to match pelage images in females. In the future there are proposals to use face recognition software instead. This should enable many more animals to be identified using their natural markings, which provide a unique 'fingerprint'. Grey seals forage widely, frequently travelling over 100 km between haul-out sites. It is therefore no surprise that seals from the Ramsey 'catalogue' have been matched with sightings made by the Cornwall Seal Group on the Cornish coast. Many animals have distinctive scars, some of which are inflicted during fights, but others, unfortunately, have constriction wounds caused by plastic rope and netting becoming wrapped around the neck and upper back, as do some gannets on Grassholm.

THE BISHOPS AND CLERKS

Immediately to the west of Ramsey is the group of islands and rocks known as the Bishops and Clerks. Probably the first written record of the name dates back to 1595 when George Owen described them in *A Pamphelett conteigninge the description of Milford havon*:

> *The Bishop and these his Clerkes preache deadly doctrine to their winter audience, such poor seafaring men as are forcyd thether by tempest, onlie in one thing they are to be commended, they keepe residence better than the rest of the canons of that see are wont to do.*

The Bishop's entourage of clerks, or choristers, comprising more than 20 islets and rocks, protrude above the surface at low water, but at high tide many of them are hidden beneath the surface, creating hazards to shipping. This 'deadly doctrine', together with the strong tidal currents in the area, has caused hundreds of wrecks over the centuries, and they were long feared by mariners. The unknown writer of the manuscript mentioned by Browne Willis (1717) similarly records their potential effect on shipping:

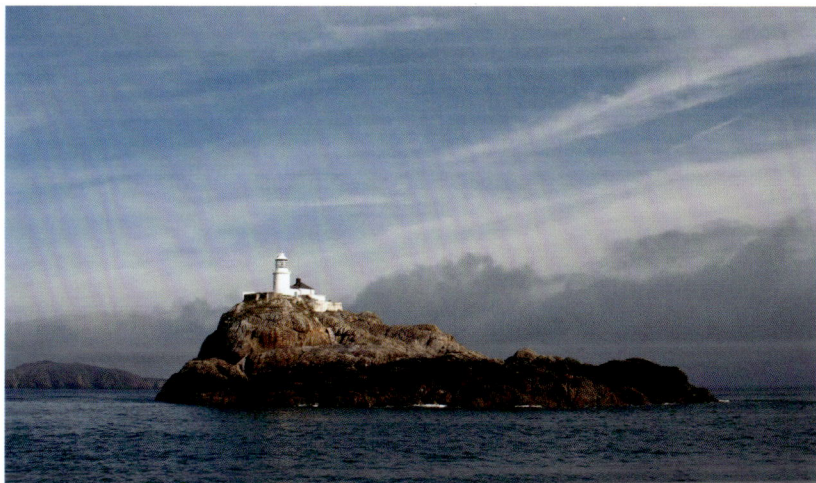

FIG 87. As its name suggests, the island of South Bishop with its lighthouse marks the southernmost extent of the Bishops and Clerks. (Richard Crossen)

It were good for the Mariners to keep aloof from these Fellows; for they are very stout, and will not budge a Foot. I think they (I mean the Bishop and his Clerks) are able to resist the King of Spain's great Navy, and put here Majesty to no change at all.

Extending about 6 km from north to south, the widely scattered Bishops and Clerks can be separated into four distinct groups: North Bishop, Carreg Rhoson and Carreg Rhoson East (with Maen Rhoson nearby), Daufraich (with the three outlying rocks of Maen Daufraich, Moelyn and Cribog) and South Bishop. The last was formerly known as Emsger, the name deriving from the Old Norse *sker*, or skerry, meaning an isolated rock in the sea. Like Ramsey, all of these islands are owned by the RSPB, except for South Bishop, which belongs to Trinity House. The lighthouse here, built in 1839, marks the southern extent of the archipelago (Fig. 87). Unfortunately, since it is located under a major bird migration route, on dark nights the lighthouse lured many birds to their deaths until Trinity House and the RSPB constructed bird perches on the side of the lantern.

The islands were once much richer in birdlife than they are today, Richard Fenton in 1811 noting that:

On these rocks an infinite number of sea birds breed, whose eggs are so thickly deposited all over the surface of them, that if one egg on the summit be stirred in its irregular rotation, it is known to carry hundreds with it … The eggs are very large in proportion to the birds they belong to, all beautifully marked and endlessly varying.

Curiously, although it was obviously only a temporary disturbance caused by the presence of people, Fenton thought that the birds were 'perpetually hovering around the rock' and 'no regular incubation is performed'. Instead he thought that 'the eggs are chiefly hatched by the sun, here felt at the season of their breeding in an almost tropical degree'. There are few other wildlife records from the islands, but North Bishop and Carreg Rhoson between them today support around 150 pairs of storm petrels and a few guillemots and razorbills. There is also a small puffin colony, comprising around 60 individuals, on North Bishop.

As seabird islands, North Bishop and Carreg Rhoson, like Grassholm, support a well-developed mattress of red fescue, with stands of tree mallow in slightly more sheltered, or bird-influenced, areas, together with sea plantain. Other plant species present include sea beet *Beta vulgaris* ssp. *maritima*, sea campion, sea mouse-ear *Cerastium diffusum* and spear-leaved orache *Atriplex prostrata*.

CHAPTER 7

Caldey and the Half-Islands

Caldey Island is one of the lesser-known delights of Wales, although only a short boat-ride from Tenby.

Thomas Lloyd et al. (2010)

CALDEY ISLAND

C ALDEY ISLAND HAS A REMARKABLE history of religious occupation, from the early Christian period to the establishment of a Benedictine abbey in the early twentieth century. There is a well-preserved medieval priory (Fig. 88), and the abbey, dating from 1910, is one of the largest and most complete Arts and Crafts complexes in Wales. It may be one of the 'lesser-known delights', but nonetheless thousands of visitors make the journey across to the island each year, with a fleet of boats running from Tenby Harbour between Easter and October.

In the late nineteenth century, as a result of James Hawksley's activities, Mathew (1894) considered Caldey 'a well-cultivated and fertile farm', and it is still farmed today, over 60 per cent of the island consisting of improved or semi-improved grassland. Extensive areas of conifers, sycamore *Acer pseudoplatanus* and various other non-native species have also been planted by the monks, and this makes it the most wooded of the Pembrokeshire islands. Because of these activities, however, it is not generally considered to be of conservation importance. It is, for example, the only major island in the county not to be

FIG 88. The well-preserved medieval priory complex on Caldey Island consists of the church and three ranges around a small courtyard. (Jonathan Mullard)

designated as a Site of Special Scientific Interest. The adjoining tidal island of St Margaret's nevertheless has long been recognised as important for nesting birds, and the two small sand-dune systems in Priory Bay and Sandtop Bay are interesting areas to explore (Fig. 89).

Despite being intensively farmed in parts, Caldey has, perhaps unexpectedly, one of the richest floras of the Pembrokeshire islands, after Skomer and Ramsey. In contrast to the other main islands, where plant communities characteristic of acid and neutral soils dominate, the combination of Old Red Sandstone and Carboniferous Limestone here produces a more varied flora. Notable species include dotted sedge *Carex punctata*, Portland spurge, the nationally scarce golden-samphire *Inula crithmoides* and ivy broomrape *Orobanche hederae*. In addition, the Welsh monastic islands, such as Caldey and Bardsey, all still support a range of 'herbal relics', including henbane *Hyoscyamus niger*, which is toxic in large doses. Some species like vervain *Verbena officinalis*, deadly nightshade *Atropa belladonna* and white horehound *Marrubium vulgare* are considered native, especially in southern England, but in Wales they are rare and closely tied to ancient human settlements, probably as a result of a tradition of

FIG 89. Herring gull on the pier in Priory Bay. The bay contains one of the two small sand-dune systems on Caldey, while the island itself supports Pembrokeshire's largest nesting colony of herring gulls. (Jonathan Mullard)

keeping herbal remedies nearby. Other notable species include black horehound *Ballota nigra*, common mallow *Malva sylvestris*, hemlock *Conium maculatum*, Alexanders and white dead-nettle *Lamium album*. It may seem odd to include white dead-nettle in this list, but it is comparatively rare in the far west of Britain, and in Wales it is never found far from habitation and is characteristic of castles and abbeys (Conolly 1994).

Approximately 94 species of invertebrates have been recorded from the island but the true count is likely to be much higher (Loxton 1997). Unlike the other islands, however, there are no resident conservation staff and Caldey is therefore under-recorded. The main records to date have come from amateur naturalists on day visits. Between 1996 and 2015, for example, there were only 57 butterfly records of 17 different species (David Redhead, personal communication). All the butterflies listed are common and widespread in Pembrokeshire, except for the wall *Lasiommata megera* and small blue *Cupido minimus*. The wall occurs all around the mainland coast but currently seems to be disappearing from inland areas, whereas the small blue is only found along the southern coast. There

FIG 90. The small blue butterfly is declining across the United Kingdom but can still be found on Caldey and along the southern coast of the mainland. The sexes are similar in appearance, although the upper side of the male is almost black with a dusting of blue scales, whereas the female is a darker brown. (Annie Haycock)

used to be a few coastal colonies in the north of the county, but unfortunately they now seem to be extinct. The small blue relies totally on kidney vetch *Anthyllis vulneraria*, the larvae living only in the flowerheads, where they feed on the developing anthers and seed. Although the kidney vetch is still present the butterfly in these locations has disappeared. Local extinction events are probably common, given its reliance on early successional habitat, and it is declining across the United Kingdom (Fig. 90). In the past these losses would have been balanced by colonisation from other areas but given the increasing rarity of the species this is no longer possible.

Moths on Caldey are poorly recorded, with only 85 species on the current list, when other similar coastal areas in the county support around 400–500 species (Robin Taylor, personal communication). One interesting species on the list is a micromoth *Mompha miscella*. The distribution of this moth follows that of its foodplants, rock-roses *Helianthemum* spp., which are confined to limestone and chalk in southern Britain. There is some doubt about the authenticity of the record, however, as rock-roses are not found in Pembrokeshire, except for a small colony of hoary rock-rose *Helianthemum canum* near St Govan's Head. Other moths recorded on Caldey include pretty chalk carpet *Scotopteryx bipunctaria*, which feeds on bird's-foot trefoil *Lotus corniculatus* and other trefoils, clovers and vetches, and the bordered sallow *Pyrrhia umbra*, the larvae of which feeds mainly on restharrow *Ononis repens*.

There will be more comprehensive records in the future, however, as Caldey Island is now working closely with Butterfly Conservation Wales to improve habitats on the island for butterflies and moths. They will also be undertaking surveys to count and identify the resident moths and butterflies, as well as migrants such as the hummingbird hawk-moths *Macroglossum stellatarum* and convolvulus hawk-moths *Agrius convolvuli* recorded in 2018. The last comprehensive moth count here occurred in 1920, so this will provide an invaluable set of comparative statistics 100 years later.

Until recently, brown rats were present on Caldey, so ground-nesting birds were absent, but an eradication programme is now in place. Hedgehogs, which will similarly eat birds' eggs, still breed here but are being left undisturbed. Following the reduction in the rat population, red-legged partridges *Alectoris rufa* and red squirrels *Sciurus vulgaris*, from captive breeding programmes, have been introduced to the island (Fig. 91). The main red squirrel populations in Wales are on Anglesey and the Clocaenog Forest, in North Wales, and around the Tywi Valley in Mid-Wales, but a commercial attraction in Surrey is breeding them to introduce them to islands free of grey squirrels *Sciurus carolinensis*. Three red squirrels were introduced to Caldey in 2016 after an extensive programme of rat

FIG 91. Red squirrels have been introduced to Caldey following a captive breeding programme to introduce the species to islands free of grey squirrels. (Annie Haycock)

eradication, and a further twelve arrived the following year. At least eight kits in three different groups were seen in the summer of 2018, and animals are now spreading out from where they were first established.

In contrast, red-legged partridge is a non-native species, brought to the United Kingdom from continental Europe, where it is largely found in France and Spain. Many attempts have been made over the years to introduce them to Pembrokeshire for game shooting, but without continued replacement of birds the species is unlikely to survive here. Other exotic species introduced to the island recently include black swans *Cygnus atratus* and golden pheasants *Chrysolophus pictus*.

Perhaps because of the attractions of the islands to the west, there can be few places in Pembrokeshire so rarely visited by birdwatchers as Caldey. The combination of large numbers of people and lots of cover makes seeing birds quite challenging at times but the rewards, in terms of migrant birds especially, are worthwhile. Rarities recorded from the island over the years include alpine swift *Tachymarptis melba*, woodchat shrike *Lanius senator*, red-backed shrike *L. collurio*, yellow-browed warblers *Phylloscopus inornatus*, hoopoes and two dalmatian pelicans *Pelecanus crispus* (Green & Roberts 2004)! Perhaps the world's largest freshwater bird, the pelican has two main populations. The first breeds in eastern Europe and overwinters in the eastern Mediterranean region, while the second breeds in Russia and central Asia and winters in Iran, Iraq and the Indian subcontinent.

As Caldey has more woodland than the other Pembrokeshire islands, it attracts breeding birds that are absent from those, such as sparrowhawk *Accipiter nisus*. Numbers of sparrowhawks recorded from Caldey have increased in parallel with increases in the mainland breeding population but their secretive nature, especially during the incubation period in May and June, means that they may be under-recorded (Rees *et al.* 2009). The best estimate for the county is 235 breeding pairs.

C. S. Hall's *Guide to Tenby* in 1871 mentions that local people went on an annual spree to Caldey to shoot puffins, afterwards returning to Tenby for the Whitsun Ale (Westerman 2010). The Ale, as its name suggests, was a party based on the consumption of beer. Thomas Dix (1869) noted that 'on Caldey Island it breeds in great numbers, and I am sorry to say that on Whit Monday the boys and men of Tenby slaughtered them by wholesale; it is as much an institution with them as May-day with the sweeps.' Thankfully, this annual destruction does not seem to have had a major effect on the bird populations, and over twenty years later Mathew (1894) recorded that 'there are a good many cliff birds upon Caldy in the summer, that chiefly inhabit its channel, or south, side'. Mathew notes though that Caldey was 'almost the only one of the beautiful Pembrokeshire islands we have not ourselves visited', and he relied instead on the 'many valuable and interesting notes' supplied by 'Mr. Charles Jefferys, naturalist, of Tenby'.

Seabird numbers on Caldey are still low because of the rats, but with their recent eradication hopefully the guillemots, shags, razorbills and puffins that Mathew mentions will return in good numbers. Caldey does however support Pembrokeshire's largest nesting colony of herring gulls. Numbers here have recovered from a low of 675 pairs in 1998, due to an outbreak of botulism in the early 1980s, to just over 2,000 pairs in 2008, making it one of the largest colonies in the United Kingdom (Rees *et al.* 2009).

The presence of gulls was noted in 1662 by John Ray and Francis Willughby, who, as mentioned in Chapter 2, visited Caldey and St Margaret's Island during their tour of Britain. Willughby included 'Caldey Island near Tenby in Pembrokeshire' in his list 'Of some remarkable Isles, Cliffs and Rocks around England where Sea-fowl do yearly build and breed in great numbers' – noting that 'in one part whereof we saw Gulls Nests lying so thick, that we could scarce take a step without setting our feet upon one'. Whether they were herring gulls, or another species, is not clear. This list of remarkable places was later incorporated in *The ornithology of Francis Willughby of Middleton in the county of Warwick Esq, fellow of the Royal Society in three books*, which was edited by Ray following Willughby's early death (Willughby 1678). The book is considered to mark the beginning of scientific ornithology in Europe, since it revolutionised bird taxonomy by organising species according to their physical characteristics.

ST MARGARET'S ISLAND

This small, and rarely visited, tidal island to the west of Caldey, at one time known as Little Caldey, is joined to the larger island by a treacherous ridge of rock at very low tide. Access to the island is therefore extremely dangerous and should not be attempted. Boat trips from Tenby though are a good way to get close. Ray and Willughby noted a small chapel dedicated to Saint Margaret, but in the Victorian period this was converted into housing for the quarry workers, who

FIG 92. The small, and rarely visited, St Margaret's Island, to the west of Caldey, supports one of the largest cormorant colonies in Wales, with around 300 pairs present at its peak. (Annie Haycock)

mined limestone on the island until 1851. One quarry, assisted by storms in the winter of 2001/02, has almost split the island in two. It is now uninhabited and an important nature reserve monitored by the Wildlife Trust (Fig. 92).

Ray and Willughby recorded the presence of the 'sea-swallow' on St Margaret's Island, noting that 'These Birds flock together and breed on Islands uninhabited near to the Sea-shores many together in the same quarter. In the Island of Caldey, adjacent to the Southern shore of Wales, they call them Spurres.' In describing the bird in question, Ray goes on to say that 'Its Bill is long, almost straight, black at the tip, else red.' This clearly identifies the sea-swallow as the common tern, since this is distinguished from the arctic tern *Sterna paradisaea* by the black tip to its bill. The common tern's main breeding area in Wales is now on Anglesey but there was a small colony on Skokholm at the end of the nineteenth century, so they certainly bred in Pembrokeshire in the past. Today the common tern is a rare passage migrant, seen along the coast from the end of March until the end of May. Small numbers are also recorded in the autumn.

The seabird colonies on St Margaret's Island are mainly confined to the northern and western cliffs, the island supporting one of the largest cormorant *Phalacrocorax carbo* colonies in Wales, with over 300 breeding pairs present at its peak. It is one of the best-studied colonies in the United Kingdom, with almost continuous census data for over fifty years and over 5,000 chicks ringed since the mid-1960s. In the early 1970s, 330 of the 450–500 pairs in the county bred on St Margaret's Island, but there has been a significant decline since that date. Ringing studies show that considerable numbers are shot on the rivers, or killed in coastal fishing nets, during the winter (Rees *et al.* 2009). In 2016 there were apparently only 172 occupied nests, so the decline is still continuing.

There are large colonies of guillemots, razorbills and kittiwakes. A few pairs of puffins breed in fissures in the rock, and with the eradication of rats it is hoped that the colony will expand. Other breeding birds include shag, herring gull, lesser black-backed gull and great black-backed gull. For a few years one gannet appeared to be resident on the island, making a nest and even laying an egg one summer. No mate was ever observed though, and successful breeding did not take place, so Grassholm currently remains the only gannet colony in Pembrokeshire. If the spread of the colony across Grassholm continues, however, and the birds eventually run out of space, perhaps gannets will return to St Margaret's and take up permanent residence.

Once, choughs and Manx shearwaters bred on the island, Murray Mathew recording that when Charles Jefferys was there in May 1893, he frightened four or five shearwaters out of holes and fissures. 'They appeared to come from cracks about half-way down the cliffs, and may, or may not, have been nesting there; it certainly looks as if they were.' Some years previously Mr E. W. H. Blagg, who was visiting Tenby in the summer of 1887, told Mathew that he 'saw a large flock of Manx shearwaters flying off Caldy on several evenings'. In relation to choughs Jeffreys said that 'I believe the Chough still breeds at the back of Caldy, i.e., on the Channel side; they did so some four or five years ago, and this spring (1893)'.

The vegetation on St Margaret's Island is dominated by red fescue, with cock's-foot *Dactylis glomerata*, false oat-grass *Arrhenatherum elatius*, nettle *Urtica dioica* and hogweed *Heracleum sphondylium*. On the edges of the cliffs, species include sea beet, common scurvygrass *Cochlearia officinalis* and common mallow. Tree mallow is also present, John Ray noting that it was found 'in great plenty' on the island.

OTHER TIDAL ISLANDS

While St Margaret's is rich in birdlife, the tidal islands on the mainland are unfortunately less interesting. St Catherine's Island, known locally as St Catherine's Rock, is the location of St Catherine's Fort which was built in 1868–70 to defend the coast and the approaches to Milford Haven from the perceived threat from

FIG 93. St Catherine's Island, at the eastern end of the South Beach, Tenby, is dominated by a 'Palmerston fort' dating from 1868–70. (Jonathan Mullard)

FIG 94. Gateholm is situated at the northern end of Marloes Sands. Like other tidal islands, it was connected to the mainland until the sea eroded the cliffs. The island in the background is Skokholm. (Jonathan Mullard)

the French navy (Fig. 93). For many centuries a tiny church was the only building on the island, but its remains were demolished when the fort was constructed. Formed from an outcrop of limestone, the island is honeycombed by tidal caves. In 1856 Philip Henry Gosse noted that a few sheep inhabited the island, describing them as 'half wild sure-footed creatures that run, turn and look, run again and leap from crag to crag almost with the agility of the Alpine Chamois'.

In 1480, Gateholm was recorded as 'Goteholme', which is said to derive from the Old Norse for 'goat island' (Fig. 94). There are, however, no goats there today. In the past this rocky, and now almost inaccessible, sandstone promontory was connected to the mainland. The Channel 4 *Time Team* programme which excavated the site in 2011 found evidence of rectangular huts arranged round three sides of a rectangular courtyard, which is believed to have been a sixth-century Christian monastic settlement. Perhaps they kept goats. Today this uninhabited island is only occupied by nesting seabirds, including fulmars, gulls, cormorants and shags.

FIG 95. The old parish church of St Brynach, Pembrokeshire, prior to its destruction. The engraving is based on a painting by Henry Gastineau (1791–1876). Dinas Island looms in the background. (Wikimedia Commons)

DINAS ISLAND

There certainly used to be wild goats on the slopes of Dinas Island, a small promontory, covering around 160 ha, on the mainland coast between Fishguard and Newport. The 'island' is separated from the mainland by Cwm Dewi, a peat-filled glacial meltwater channel with steep wooded sides. This unusual feature was recognised by George Owen, who described Dinas Island, and another similar feature, Ynys Barri, a headland between Porthgain and Abereiddi, as follows:

> *The two peninsulas or half-islands which I propose to speak of are called islands for in effect they are islands, saving that each of them hath a small valley or bogge between it and the land which the sea possesseth not, but are such as with industry might be cut being bogges, and with small charge the sea drawn about them.*

At the northeastern end of the valley lies the remains of the ancient church of St Brynach (Fig. 95). The building suffered storm damage in 1850 and 1851, when the chancel was destroyed by the sea. When the great storm of October 1859 removed the church roof and damaged the walls, the building was immediately abandoned.

When Skokholm was required by the military during the Second World War Lockley came to Dinas Island, in December 1941, to farm, and he stayed until October 1949 (Robertson 2007). It is easy to see why such a place would appeal to a devoted island man. There is a long history of agriculture on the site, George Owen describing it as 'very good corn land, especially for wheat … also well nourishing sheep'. Lockley, as usual, documented his activities in a number of books: *Inland Farm* (1943), *The Island Farmers* (1946) and *The Golden Year* (1948). He did much to promote the conservation value of Dinas Head, writing in 1942 that:

> *You may see a peninsula rising by a gentle south-facing slope until it is 460 ft above sea level. Its bold cliffs stand out to view from every hill in North Pembrokeshire. Its wild coast is haunted by seals, sea birds and wild goats …*
> *Once or twice I walked its rough fields and thought that one day it might be preserved as a sanctuary for wild creatures.*

Lockley had explored Dinas Island as a boy and remembered it 'for the amount of furze and bracken and stonechats and yellow-hammers I had seen there'. When he took it over the land was divided into 20 small fields, but these have unfortunately not survived. After the war the farm was bought by a local farmer from St Davids, and entered a period of 'agricultural improvement', with fields being enlarged. The area today is still farmed by the same family, although the land is now owned by the National Trust. Many of the habitats present in Lockley's time have been lost over the last eighty years, but there remains a fringe of coastal heath and maritime grassland that still supports choughs, peregrines and a variety of farmland birds. The National Trust is currently working closely with the family to bring nature back into the centre of Dinas Island, reinstating some of the richness that Lockley found when he first arrived.

CHAPTER 8

Grassholm, the Smalls and the Hats and Barrels

... the gannets were the victims of a raid, the particulars of which were made public, and excited at the time no little indignation. Since then, we believe, they have enjoyed peace.

Murray Mathew (1894)

WHILE THE INNER ISLANDS ARE comparatively easy to visit, with thousands of people a year taking the opportunity to travel to them, if only for a day, the outer islands are more difficult to reach, and access is very dependent on sea conditions. The Smalls are the most westerly group, lying 5 degrees 40 minutes west of Greenwich, which is 32 km from the mainland. They are actually further west than the most easterly coast of Northern Ireland, Burr Point on the Ards Peninsula in County Down, and they represent the last visible part of Wales in this direction. Given their remote location, the Smalls are obviously not an easy place to visit. It can take two and a half hours, or more, to get to there and back in the small boats used for wildlife watching, and trips can be cancelled at short notice due to rapidly changing weather conditions. It is well worth making the effort to reach the area though, as the rewards are spectacular. There is a sense of wildness and proximity to nature that is difficult to find elsewhere, even in Pembrokeshire. Diving around Grassholm and the Smalls is popular but requires the more substantial dive boats from Milford Haven and St Brides Bay. The smaller boats, based in St Justinian's, usually do not go out far beyond Grassholm, unless the sea is very calm.

GRASSHOLM

Grassholm, as Gwales, is featured in *The Mabinogion*. It was apparently the site of a fabulous hall where the severed head of Bran the Blessed (the god of ravens and prophecy) was kept miraculously alive for eighty years, while his companions feasted in blissful forgetfulness. Today, however, it is known for its enormous colony of gannets (Fig. 96). Just under 40,000 pairs breed on the island each year, and from January to October Grassholm provides 'arguably the most spectacular birdwatching experience in Wales', the sight and smell of the colony providing a memorable encounter (Lovegrove *et al.* 1994). Grassholm is the third-largest Atlantic gannet colony in the United Kingdom (only Bass Rock and St Kilda are larger), supporting nearly 17 per cent of the UK population – which is around 12.5 per cent of the North Atlantic population, and 7 per cent of the entire world population (Fig. 97). Although Grassholm means 'green island' in Norse, in the summer from a distance the whole northwestern half of the island looks white, stained as it is with guano and covered with the gannets sitting on their nests.

FIG 96. Grassholm is the third-largest gannet colony in the United Kingdom and is a spectacular sight during the summer months. (Janet Baxter)

The name 'gannet' has a long lineage, ultimately deriving from the same root as 'goose' and 'gander' (Greenoak 1979). Indeed, in Scotland the bird used to be known as the 'solan goose'. The word goes back to the Anglo-Saxon period. The *Seafarer*, an Old English poem about a man alone on the sea, includes the lines *Hwilum ylfete song, dyde ic me to gomene, ganotes hleoþor ond huilpan sweg fore hleahtor wera, mæw singende fore medodrince* ('At times the swan's song I took to myself as pleasure, the gannet's noise and the voice of the curlew instead of the laughter of men, the singing gull instead of the drinking of mead'). The seafarer must have been near an active gannetry, since the bird is silent except during the breeding season.

To catch fish, gannets often perform dramatic plunge-dives, achieving speeds of 100 km per hour as they strike the water. They can do this because they have a number of adaptations, including nostrils located inside the mouth, air sacs under the skin on the face and breast, which reduce the impact, and binocular vision, which allows them to judge distances accurately. Gannets can therefore catch fish at greater depths than most seabirds. Despite these adaptions though, they also feed at the surface on small fish, such as sand eels, and on discards from fishing boats, their large size helping them out-compete most of the other scavenging species.

History of the gannetry

Although it is very large today, the gannetry is relatively recent in origin. At the beginning of the nineteenth century Grassholm was noted not for its gannets but for its immense puffin colony, which is thought to have later been abandoned following the destruction of a thick layer of humus that the birds had undermined by their burrowing. Ethelbert Lort Phillips, who camped on Grassholm in 1883, estimated that there were half a million puffins on the island by 1890. Intriguingly, Lort Phillips (1857–1943) was a big-game hunter, Vice-President of the Zoological Society of London and had two mammals, three reptiles and three birds named after him. These include Phillips's, or Somali, wheatear *Oenanthe phillipsi* (Beolens *et al.* 2014). A few years later, in 1893, Robert Drane visited the island with members of the Biological Section of the Cardiff Naturalists' Society. In his subsequent report (1893) he writes:

> *The first wonder was the number of birds. I know of no term or numeral that can adequately convey a conception of it – 'millions' will not do … Let us say miriads, for that implies indefinitely great number, number without ascertainable limitation. The vast majority were puffins …*

Taking the area of the island and allowing two birds to each square yard, Drane estimated there were 689,638 puffins on Grassholm, though he goes on to say that:

FIG 97. Seven per cent of the entire world population of gannets, just under 40,000 pairs, nest on Grassholm each year. (Janet Baxter)

> *Of course, there is no approach to accuracy in this estimate, but I consider it a very moderate one, for the water was as thickly covered with them as the land, and, at the same time, Mr Storrie's photographs will shew that there were as many as 20 in a square yard.*

Lockley recorded only 20 birds in 1956, and there are none present on the island today. The date that the gannets arrived on the island is uncertain, but in Drane's paper he notes:

> *There are upon this island two colonies of gannets occupying the flat summits of precipitous cliffs, which they so whitewash with their excreta that they are its most striking feature. In these two colonies there were about 240 nests, each with one egg, and each egg is a thing of beauty ... Two hundred and forty nests imply 480 birds, and to these, with one young one to each pair, may be added at least 240 birds of preceding years, that is 960, for they do not breed until they are three years old ...*

Murray Mathew (1894) thought that the gannets had been driven to Grassholm by persecution of the colony on Lundy:

The Gannets, of course, form the most attractive feature of Grasholm, which holds a conspicuous place in our account of the county Ornis, as it is the only spot off the coast of Pembrokeshire that furnishes a nesting station to these fine birds that are said to have been originally a settlement from Lundy, where they sustained such persecution from the hands of the channel pilots and other egg stealers, that many of them were driven to forsake that island in search of a more inaccessible, solitary, and peaceful residence.

Certainly, the timings were coincident, even though there is no other firm evidence to support the theory (Lovegrove *et al.* 1994). Comparing dates, Salmon and Lockley considered that Grassholm was a strongly established colony and had been in existence for many years before the persecution which eventually wiped out the Lundy colony was recorded (Salmon & Lockley 1933). It is difficult to determine exactly when gannets began to use Grassholm, but the colony was certainly established by 1860 and there are anecdotal references which suggest it may have been in existence as early as 1820 – in which case Salmon and Lockley may have been correct. Mrs Haydon-Bacon, one of the daughters of Captain Davies, who farmed Skomer, noted in a letter to Lockley that:

My father took over Skomer and Grassholm from my grandfather in 1860, and there were certainly gannets on the latter then, as if the gannets had arrived after 1860, he would have told everyone interested in the fact ... There were always gannets there but few nests. Specimens of their eggs were greatly prized and were always brought back from Grassholm whenever my father took visitors there. (Lockley 1957)

Until the colony was deserted at the end of the nineteenth century, Lundy was the only site where gannets nested in England, and the most southerly of its breeding stations in the British Isles (Ternstrom 1946). Indeed, the colony there was so well established and well known that it featured in *Poly-Olbion*, a topographical poem describing England and Wales written by Michael Drayton (1563–1631) and published in 1612.

This Lundy is a nymph to idle toys inclin'd
And all on pleasure set, doth wholly give her mind
To see upon her shores her fowl and conies fed,
And want only to hatch the birds of Ganymede.

FIG 98. The figurehead of HMS *Gannet*, built in Pembroke Dock in 1857, is a realistically painted, but goose-like, gannet with its wings folded and standing on a scroll head. It is preserved in the National Maritime Museum, Greenwich. (Jonathan Mullard)

It may have been the Grassholm gannet colony though that prompted the naming of HMS *Gannet*, a 151-ft (46 m) wooden screw sloop which carried eleven '32 pounder' guns. Built in Pembroke Dockyard in 1857, she was mainly employed in hydrographic duties, and between 1865 and 1867 carried out survey work around the Bay of Fundy and Labrador. The sloop was sold out of the navy in 1876. Those ships that were given bird names tended to be the lighter and more nimble units of the fleet, rather than the heavyweights. Other Pembroke ships of this type included the *Buzzard, Falcon* and *Vulture* (Pulvertaft 2011). Carved in pine by Frederick Dickerson of Devonport, the figurehead for HMS *Gannet* represents the bird with its wings folded (Fig. 98), though it looks more like a goose – perhaps a 'solan goose'! In contrast, the painting is simple but realistic. It must be remembered, however, that even if the selected birds are shown in a reasonably accurate form, the carvers would not have had access to the same reference books and illustrations that are available to us now. Today the gannet forms part of the collection of figureheads in the National Maritime Museum in Greenwich.

In 1938 the owner of Lundy, Martin Coles Harman, sought the help of Lockley
in obtaining gannet eggs from Grassholm, in order to reintroduce the bird to his
island. In a letter of 11 January 1939 to James Fisher, then Honorary Treasurer
of the British Trust for Ornithology, Harmon describes how 'Lockley, good
fellow, sent me twenty Gannet eggs by arrangement, and will do the same next
year.' These were placed into the nests of cormorants. Although the eggs arrived
rather late, Harman reports in the same letter that he later saw what he believed
to be two young gannets, and although others confirmed that view, he remained
uncertain. He concluded his letter by saying 'in 1939 we will try to be a little
more scientific. I am quite set upon re-introducing gannets.' The following year
there was indeed a second attempt, but Harman reported that the scheme was a
complete failure, the cormorants 'having for the most part thrown out the eggs or
deserted them' (Fisher & Vevers 1943). The outbreak of the Second World War put
paid to any further plans to continue these efforts.

The *Transactions of the Cardiff Naturalists' Society* for 1890–91 include a dramatic
description of a visit to Grassholm in 1890 by Thomas Henry Thomas when he
and three companions stayed on the island and studied the gannets, which then
consisted of around 200 pairs. The visit was, however, cut short by an infamous
and unpleasant incident when a naval party landed, shot large numbers of birds,
and destroyed all the gannet eggs (Fig. 99). Thomas describes the event in the
Society's *Transactions* as follows:

> *I was going to renew my acquaintance with the gannets: among them I was, if not*
> *welcomed, at least permitted, and I began some sketching, I heard the fell crack of*
> *a rifle break in upon our Millennium ... An attack was made upon the settlement*
> *of gannets by a company on board H.M.S. 'Sir Richard Fletcher' followed by a*
> *landing and general battue upon shore, terminating in the slaughter of many birds,*
> *several gannets and, the destruction of the whole of the eggs of the latter.*

This unnecessary destruction led to questions in the House of Commons and
a successful prosecution by the RSPCA. Following this Thomas was elected as
chairman of a committee to consider the best means of protecting wild birds'
eggs at the British Association for the Advancement of Science in Cardiff in 1891.
His work led to the Wild Birds Protection Acts of 1894 and 1896.

Human disturbance on Grassholm has been significantly reduced since
landings by the public were banned in 1997. Tourist boats still visit the island,
and there is a code of conduct agreed with the boat operators to minimise
disturbance from the sea. Disturbance by military aircraft also used to occur
occasionally, but there has been an agreement in place since 1998 to avoid the
area and, except in emergencies, this has been respected. During the Second

The party from the "Sir Richard Fletcher" despoil the Gannet settlement. Chick-smasher at work.

The party from the "Sir Richard Fletcher" leaving Grassholme

Gannet colony with guillemots & Razorbills & Puffins

"How the people in the upper terrace can leave their eggs so long, I cannot think"

M? Gannet - log

Gannets' Nest & Egg two small mackerel ejected by the birds before flying.

Puffin and Gannets - "a friendly visit"

A VISIT TO A SEA-BIRDS' ROCK: GANNETS AT HOME.

FIG 99. The *Daily Graphic* newspaper of 3 June 1890 provided an illustration, derived from the sketches of Thomas Henry Thomas of the Cardiff Naturalists' Society, showing Grassholm and the gannets and the attack on them by a naval party. The bird in the bottom right-hand corner paying 'a friendly visit' seems to be a guillemot, not a puffin as stated. (Cardiff Naturalists' Society Archive)

World War though the island was used by the United States Air Force as a target for bombing practice, leaving bomb craters and fragments which are still obvious today. What effect this had on the gannet colony is not recorded, but since its establishment it has grown gradually, if unevenly. Between 1914 and 1922, for instance, the number of birds was estimated to have grown by around

FIG 100. An aerial view of Grassholm from the northwest, taken in 1980, showing the extent of the gannet colony at that time. It has since expanded further across the island. The West Tump (bottom right) is one of the principal 'club' areas where non-breeding birds congregate during the summer. (Mike Alexander)

16 per cent, while between 1922 and 1924 it grew by 42 per cent, falling to 10 per cent per annum over the next decade. It has been suggested that the only explanation for these irregular but substantial increases is immigration from overflowing colonies elsewhere, such as that on Little Skellig in County Kerry, and those in Scotland.

The first photographic census of the gannet colony on Grassholm was undertaken by Lockley and Salmon in 1933. On one occasion it took them ten hours to row from Skokholm to Grassholm but, despite choppy seas, Salmon completed the census. Lockley independently took another set of photographs and they averaged the results. The pictures were marked off in squares and the birds within each square counted, resulting in an estimate of 4,750 breeding pairs. The results were subsequently published in *British Birds* (Salmon & Lockley 1933).

The original site of the colony was on the western side of the island in the vicinity of the West Tump islet (Fig. 100). As the colony expanded it spread across the remainder of the western side of the island, up to the central ridge and then onto the eastern slopes. In 1984 it was estimated that 28,500 pairs were present. This expansion is continuing today, with just under 40,000 pairs recorded, and it seems likely that gannets will eventually occupy the whole island.

Private life of the gannets

In 1934 Lockley and Julian Huxley camped on Grassholm with the Hollywood cinematographer Osmond Borradaile, who was known for his skills in outdoor filming. The result was a widely acclaimed film, *The Private Life of the Gannets*. It was the first wildlife film to receive an Academy Award, winning an Oscar for Best Short Subject in 1937. Today the 15-minute film is classed by many as the world's first natural history documentary, its academic approach a stark contrast to the 'expedition' format of its predecessors. The filming was supported by the Royal Navy and for one scene Borradaile used a Stranraer flying boat, 'power-diving' onto the colony to simulate the effect of a gannet returning to its nest. These aerial shots of Grassholm provided context, while close-ups of the birds and time-lapse and slow-motion photography manipulated time to reveal the gannets' activities. The description of the film on the British Film Institute (BFI) website gives an idea of its approach and content:

> *The lonely seclusion of an Irish Sea island is intruded upon for this pioneering account of the lives of northern gannets. The hardy seabirds are observed at different life stages: as hatching eggs, daring juveniles and nesting adults. The threat of predatory gulls terrorizes the eyrie, but the gannets are hunters too, plunging at speed from great heights into the sea for fish.*

The film can be viewed for free on the BFI website and is still intriguing viewing, even though Huxley's commentary now sounds very dated. Lockley appears in it twice, the first time holding a gannet in front of him with outstretched wings, supported at the tips by other hands, and later walking, unsteadily, across an area of abandoned puffin burrows before stopping to look out to sea through binoculars. Apparently, the applause at its premiere at a cinema in London lasted longer than that for the accompanying *Scarlet Pimpernel* film. The *Private Life of the Gannets* was subsequently shown at the Eighth International Ornithological Congress, which, as mentioned in Chapter 5, met in Oxford that year.

Gannet-cam

In 2013 several new films on the gannets were made, this time by the gannets themselves (Morgan 2013). Dr Steve Votier, at the University of Exeter, has been carrying out research on the behaviour of the gannets away from Grassholm since

2005. In the early days the gannets carried GPS units alone, but Dr Mark Bolton from the RSPB subsequently proposed trialling a combined video and GPS tag. In addition to the GPS 'fix' this allowed the research team to view the bird in 'real time'. The new approach added even more value to the GPS data by showing whether the bird was in flight, resting on the sea, or feeding. In particular, the bird-borne cameras provided a unique view of the gannet's interactions with fisheries. A total of 20,643 digital images from ten cameras revealed that all the birds photographed fishing boats. Virtually all of these were trawlers, and the gannets were almost always accompanied by other scavenging birds, such as fulmars and a variety of large gulls (Votier *et al.* 2013).

One of the other findings was that male gannets tended to feed more around trawlers than females, and this difference may have conservation implications, now that there is supposed to be a ban on discards, which are unwanted catches. Seabirds spend most of their time at sea away from their nesting sites, making them difficult to study, so the cameras have helped to shed light on the behaviour of gannets when they leave the colony. The approach will hopefully be developed further over the coming years into a useful seabird monitoring tool.

Breeding

Gannets are long-lived seabirds, and individuals up to 35 years old have been recorded. In February each year the air around Grassholm is filled with males returning with nesting materials keen to establish territory in order to attract a female. The nest consists of a mound of seaweed, feathers and other items, including plastic debris. Additional matter is sometimes incorporated into the nest during incubation of the single egg, which is laid in late April or May. Incubation takes between 43 and 45 days and is carried out by both parents, who place their feet on the egg. Chicks begin to hatch in early June and are then fed by the adults for about 90 days until they are fully grown.

The young gannets leave the island in August and September and wander, apparently at random, for a few weeks before heading south to winter off the coasts of Morocco and Senegal. Here they spend most of the next two to three years before heading north again to join the colony. In contrast, both breeding and non-breeding adults disperse into the Irish Sea, the Western Approaches and the North Atlantic, feeding mainly over the continental shelf. They are not absent for long though and the first returnees can be seen inspecting nest sites in January. Numerous birds on passage from the northern colonies are also present off the Welsh coast during autumn, with large numbers regularly recorded off Strumble Head.

Plastic debris

Many of the gannets flying around Grassholm can be seen carrying fragments of discarded fishing net and other flotsam, to incorporate into their nest mounds. The birds collect whatever is floating on the sea, and before the widespread use of plastics that would have been seaweed. Now they increasingly use non-natural material. Over 80 per cent of the nests on the island are estimated to contain plastic waste, the vast majority of which comes from the fishing industry. Synthetic rope is the most frequent item (Fig. 101), followed by monofilament netting and line. As a result, the nest soon becomes a pile of tangled plastic and some of the young birds become ensnared in it as they grow, with no chance of escape. The RSPB estimates that there is now over 18 tonnes of plastic on Grassholm, but removing it would severely disrupt the colony – and even if it was possible the birds would only start rebuilding their nests with more plastic. The only long-term option is to stop the plastic ending up in the sea in the first place, an almost impossible task. There are reports that a recently discovered bacterium, *Ideonella sakaiensis*, has evolved the capacity to feed on PET (polyethylene terephthalate) plastics, but this must not detract from finding a solution to this worldwide problem (Coghlan 2016).

The impact of plastic on Grassholm has become such a concern that a team from the RSPB has started visiting every October, after most birds have left the colony, to save trapped individuals. As the chicks grow and begin moving around on the nest, some are unlucky enough to become entangled in a strand of fibre which slowly winds around the leg the more the young bird moves and tightens like a tourniquet as it grows (Fig. 102). By the time the bird reaches fledging age, around three months after hatching, it will be firmly tethered to the nest. The adult birds keep feeding it but will eventually abandon the site as the instinct to migrate becomes overwhelming. Since 2005 over 600 birds have been freed from an otherwise slow and painful death. The number of birds caught up in plastic varies greatly from year to year. For example, 2017 was considered a 'good year' in the sense that only 26 fledglings and 2 adult birds needed freeing, but in some years over 100 birds have been rescued. The following year only 18 juvenile birds needed cutting free but 20 juveniles and 15 adults were already dead, tethered to their nests by monofilament fishing line and rope.

It is impossible for the team to visit earlier in the year, as their presence would cause too much disturbance and do more harm than good. Healthy young birds would be scattered from their nests, becoming separated from their parents, and many more would die. By October the only chicks that are left (with the exception of a few late individuals) are those that are hopelessly tangled.

FIG 101. An adult gannet on Grassholm with plastic string around its head. The bird would find this almost impossible to remove. (Greg Morgan)

FIG 102. A small gannet chick in a nest constructed almost entirely from plastic debris. Before it fledges it may become entangled, adding to the depressing sight of trapped and dying young birds caught within a sea of plastic. (Greg Morgan)

Without the RSPB's intervention, they would simply starve to death on the nest. The number of birds affected is relatively small, given the size of the colony, but the impact on individuals is obviously significant at a time when seabirds in general are under pressure. It is one more problem they could do without. Their deaths have little impact on the colony as a whole, but the exercise is undertaken for animal welfare reasons and to highlight the issue of plastic in the marine environment. In 2014 the visit was filmed by BBC *Autumnwatch*, and this further publicised the issue.

More worrying even than the plastic in the nests is the floating plastic that is often ingested by the gannets. They mistake it for their usual prey of mackerel and herring, and eventually it kills them, or their chick. Whether it be through the accidental snagging of nets that break, or need cutting free, fishing gear destroyed in storms or the deliberate dumping of old netting at sea, the fishing industry needs to take drastic action to reduce the amount of waste that ends up in the oceans.

Other birds and insects

Despite first appearances, other seabirds nest on Grassholm alongside the gannets, including guillemots, razorbills, shags and kittiwakes, as well as herring gulls, lesser black-backed gulls and great black-backed gulls. Small numbers of storm petrels are also thought to breed among the boulders.

In October 2014 RSPB wardens recorded other birds on Grassholm, most of which were passing through. The highlights were firecrest *Regulus ignicapilla* and tree sparrow *Passer montanus*. In addition, there were single specimens of redwing *Turdus iliacus*, song thrush *T. philomelos*, snipe *Gallinago gallinago*, merlin *Falco columbarius*, turnstone *Arenaria interpres*, skylark, wren and dunnock *Prunella modularis*, together with three starlings *Sturnus vulgaris* and five rock pipits *Anthus petrosus*. There was also a continuous passage of red admiral butterflies *Vanessa atalanta* arriving from the west, probably from Ireland, and flying towards the Welsh mainland. They recorded a minimum of 50 individuals but there were probably many more. The red admiral is a migratory butterfly, arriving in central and northern Europe every year from the south. In autumn, the offspring of these spring arrivals migrate southwards, and this is what was occurring here. There was also one painted lady *Vanessa cardui*, another long-distance migrant, and a single common carder bee *Bombus pascuorum*, one of our most common bumblebees.

Vegetation

Grassholm is aptly named, since the most obvious feature of its vegetation, on areas not yet occupied by the gannets, is a deep 'mattress' of red fescue, with robust specimens of Yorkshire-fog in moist, sheltered hollows, and spear-leaved orache. An unusual type of maritime grassland, created by excessive exposure to salt spray, large amounts of guano and the absence of grazing animals, it similarly occurs at Worm's Head on Gower (Mullard 2006). In 1911 the Belfast naturalist Robert Lloyd Praeger similarly recorded 'deep dense masses' of red fescue on small stacks inaccessible to sheep on Clare Island, County Mayo, which contrasted strongly with the closely cropped grass on the adjacent slopes. On Grassholm, Lockley (1947) stated that during the nesting season he observed a steady stream of gannets plucking tufts of red fescue for nesting material, and later in the season birds in the colony could be seen holding grass in their bills. The amount of material taken is, however, negligible compared with even light grazing by domestic animals.

While the red fescue is the dominant plant on the island, eight other species can be found within the sward at low levels of abundance. The commonest is sea plantain, followed by sea mouse-ear, mouse-ear chickweed

Cerastium fontanum, English stonecrop, rock sea-spurrey *Spergularia rupicola* and curled dock *Rumex crispus.* Specimens of Babington's orache *Atriplex glabriuscula* and tree mallow have been recorded on the more sheltered eastern side of the island. All these plants are dependent on the fescue to create soil, and their tall growth is a direct result of the shading and shelter provided by it. Arable weed seeds and cereal grains are also brought to the island by the gulls, surveys during the late 1940s and early 1950s recording species such as scarlet pimpernel *Anagallis arvensis,* pineapple weed *Matricaria matricariodes,* common plantain *Plantago major,* annual meadow-grass *Poa annua* and common knotgrass *Polygonum aviculare* (Gillham 1953).

THE HATS AND BARRELS

The Hats and Barrels are so 'called from their having that appearance at certain times of the tide' (Fenton 1811). Murray Mathew (1894) describes them as follows:

> *The distance from Skomer to Grasholm is six miles; on the same course, three miles farther on is a cluster of half-tide rocks called the Barrels, and from the Barrels four miles still further westward is reached the Smalls Lighthouse. About half-way between the Barrels and the Smalls lie the Hats, a group of sunken rocks, with eight feet of water over them at low tide. Around these islands and rocks, as if to make some of them still more perilous, is deep water, and between them very strong currents set, in many places forming dangerous 'races'.*

Probably the largest of the Hats and Barrels is Barrel Rock itself, a low-lying rocky reef around 12 m long, surrounded by huge underwater pinnacles. At low water it lies about 3 m above the surface and is often frequented by seals. A survey here by Seasearch in 2014 revealed large amounts of broken wreckage, including ship superstructures, chairs and dinner plates, testimony to the numerous wrecks in the area (Lock *et al.* 2014). Due to the wreck sites and the rich marine wildlife, diving around the Hats and Barrels is very popular, as is reef fishing. Part of the site though was previously used to dump disused explosives and is marked as 'Explosives Dumping Ground' on the marine charts. The whole area is very exposed, and the strong tidal streams mean that boats are only able to get there on the calmest days.

The summary of the Seasearch visit notes that the whole of Barrel Rock was thickly encrusted by scud *Jassa falcata,* a reef-building worm, and there was an abundance of skeleton shrimps *Caprella* spp. Beneath the silty surfaces a diverse community of animals was present, including oaten pipe hydroids *Tubularia* spp., along with sponges and bryozoans. Common starfish were also recorded, together with 15 species of sea slugs, including *Flabellina browni* and *Facelina bostoniensis*, both of which feed on hydroids (Lock *et al.* 2014).

THE SMALLS

Thirteen kilometres west of Grassholm lie the Smalls, a remote group of wave-washed rocks. Along with the Hats and Barrels, they mark the end of a long and treacherous reef. Most are partially submerged, with the main rock only 3.5 m above the highest tides. Seas break over them even on the calmest days and they have been a threat to shipping for centuries. In August 1991 a sports diver found an object, protruding from beneath scattered steel plating from more modern wrecks, which proved to be a Viking sword guard. Cast in brass and decorated with a pair of stylised animals in profile, interwoven with thin, snake-like beasts, it was dated to about 1100 CE. During this period there were frequent attacks, by a fleet based in Dublin, on wealthy settlements such as St Davids. For ships sailing south through the Irish Sea, the Smalls and the adjacent Hats and Barrels reef, along with the vicious tidal rips, were a deadly threat and caused many wrecks before the use of modern navigational aids. The find is therefore likely to represent the wreck of a Viking ship intent on plundering the Welsh coast.

The prominent feature of the Smalls today is the most westerly lighthouse in Wales (Fig. 103). Tall and striking, it was built in 1861 to replace an earlier structure constructed in 1776 by Henry Whiteside, a musical instrument maker from Liverpool. Whiteside designed an octagonal timber hut 4.5 m in diameter, perched on eight legs or pillars, five of wood and three of cast iron, spaced around a central timber post. An incident in 1801, when one of the two men manning the lighthouse died, leaving the other with the corpse for three months before rescue, was responsible for a change in Trinity House policy. Although Thomas Griffiths apparently died in a freak accident, the two keepers had been known to quarrel, so, rather than dispose of his corpse in the sea and risk being accused of murder, the second keeper, Thomas Howell, constructed a coffin and tied it to the outside railings. Sadly, the coffin was soon torn open by the wind and the dead man's arm began to bang against the window, as if he was trying to get in. When the relief boat finally arrived, the crew found that Howell had been driven mad. Subsequently, it was ensured that all offshore lighthouse teams included three men, and this approach continued until the 1980s when lights were automated.

FIG 103. The Smalls lighthouse lies 32 km off the Pembrokeshire coast – the most westerly lighthouse in Wales and the most remote light operated by Trinity House. (Richard Crossen)

Birds recorded from the area include kittiwake, sanderling *Calidris alba* and turnstone, but the Smalls are known particularly for the large number of grey seals that occur there. Since they do not often come into contact with people the animals are very inquisitive and come extremely close, frequently following divers around underwater. From 1947 until 1949 Lockley asked the lighthouse keepers to keep records of the number of seals on the rocks, and these show that there were up to 150 individuals present between April and September. Subsequently, on a visit to the area in July 1955 he estimated that there were between 80 and 100 seals lying out on the reefs to the south of the lighthouse and that around 40 per cent of these were adult bulls, 40 per cent mature cows and the rest immature animals between one and three years old (Lockley 1956). Recent estimates of the population here closely match Lockley's figures, implying that it is relatively stable.

CHAPTER 9

The Mainland Coast

Of all the habitats in Britain in which communities of plants and animals live, none is richer or more interesting than those in the coastal belt.

Countryside Commission (1969)

FROM ST DOGMAELS IN THE north to Amroth in the south, the mainland coast of Pembrokeshire stretches for over 300 km. The cliffs and headlands, beaches and dunes, estuaries and flooded glacial valleys provide a wealth of habitats for numerous species, many of them rare. At its highest point, Pen yr Afr on Cemaes Head in the north of the county (Fig. 104), the Pembrokeshire Coast Path reaches a height of 176 m, while at its lowest point, Sandy Haven near Milford Haven, it is just 2 m above low water. Overlooking the Teifi estuary where it joins the sea, Cemaes Head is, in fact, the highest sea cliff in Wales. In 1840 George Nicholson, in *The Cambrian Traveller's Guide and Pocket Companion*, noted that 'The coasts on each side of the mouth of the river are uncommonly grand, particularly on the Pembroke side, where rises Cemmaes Head, a promontory of immense elevation.'

At the other end of the coast path, Amroth, in the south of the county, is one of the lower stretches of the coastline. This area is known for its sandy beach and, on extreme low tides, the drowned forest described in Chapter 3.

ROCKY SHORES

The rocky shore is a biologically rich environment, which contains many different niches, including steep rocky cliffs, platforms, rock pools and boulder fields. Species that live in this area experience regular changes in their

FIG 104. At 176 m, Cemaes Head, the most northerly headland on the Pembrokeshire coast, is the highest sea cliff in Wales. The yellow flowers of western gorse add to the colour of the rocks. (Jonathan Mullard)

environment and must be able to tolerate extremes of temperature, salinity, moisture and wave action in order to survive. Owing to its accessibility the habitat has been studied for hundreds of years and the species that occur here are well known (Archer-Thomson & Cremona 2019). They can vary greatly, however, from shore to shore, mainly due to the degree of exposure to wave action. Shores sheltered from the direct impact of waves are generally covered by large brown algae, while on the more exposed shores these are replaced by small red algae. The animal communities too change in relation to exposure.

A marine biologist, W. J. 'Bill' Ballantine, working at Dale Fort in the late 1950s, made detailed comparative studies of the fauna and flora of 28 rocky shores in Pembrokeshire, covering the whole range of exposure (Ballantine 1961). His results suggested that an 'exposure rating' for a site could be produced by recording the relative abundance of a number of common indicator species along a vertical transect. The 'Ballantine scale' has subsequently been widely used around the British coast, as a teaching aid to demonstrate ecological differences between contrasting shores. While it provides a simple biologically defined scale it only works, however, for the shores of northern Europe (Hayward 2004).

FIG 105. This replica of a seaweed drying hut on the cliff at Little Furznip, overlooking Freshwater West, was built by the National Park Authority as an example of the many such structures that once occupied this site. (Jonathan Mullard)

Examples of moderately exposed rocky shores can be found between Marloes and Dale. The sublittoral fringe along much of this shoreline is dominated by species of kelp, such as oarweed *Laminaria digitata* and dabberlocks *Alaria esculenta*. In contrast, the area between the lower shore and the middle of the shore tends to be dominated by red algae, including pepper dulse *Osmundea pinnatifida*, dulse *Palmaria palmata*, false Irish moss *Mastocarpus stellatus* and thong weed *Himanthalia elongata*. Thong weed is sold in France and Ireland as 'sea spaghetti', being cut into finger-like lengths and air-dried. When required it is soaked in water and added to salads. In Wales it is purple laver *Porphyra umbilicalis* that is dried in the preparation of laverbread. At Freshwater West women once collected the seaweed from the rocks and hung it up in huts constructed from driftwood and thatched with marram grass (Fig. 105). The laver was then sold to markets in Haverfordwest and Swansea. Washed and boiled, it is usually eaten at breakfast, sometimes rolled in oatmeal.

On shores where the bedrock forms a shallow slope, species such as serrated wrack *Fucus serratus*, or bladder wrack *Fucus vesiculosus* occur, with the 'nonvesiculate' form of bladder wrack, *F. vesiculosus linearis*, being found in more sheltered areas. Above this algal zone the shore is dominated by barnacles, especially Poli's stellate barnacle *Chthamalus stellatus* and the acorn barnacle *Semibalanus balanoides*, the most widespread intertidal barnacle in the British Isles. Even higher up the shore black lichen *Lichina pygmaea* occurs, especially where conditions are more exposed. Finally, at the upper edge there is a band of black tar lichen *Verrucaria maura*, above which a zone of yellow and grey lichens can be found.

Rock pools support species such as sea oak *Halidrys siliquosa*, rainbow wrack *Cystoseira tamariscifolia*, the red alga *Dumontia contorta*, the beadlet anemone *Actinia equina* and a variety of bryozoans and sponges. Shaded boulders support rich carpets of red algae such as *Plumaria elegans*, *Lomentaria elegans* and *Membranoptera alata*, together with sponges such as *Hymeniacidon perleve* and breadcrumb sponge *Halichondria panicea*. Large numbers of gooseberry sea squirt *Dendrodoa grossuleria* and the star ascidian *Botryllus schlosseri* can also be found under overhangs.

SEA CLIFFS

Sea-cliff slopes are one of the few areas in Britain where what can be termed 'climax', or 'natural' vegetation occurs. Succession to scrub or woodland, however, is prevented by exposure to extremes of climate, thin soils and, probably most importantly, salt spray (Oates 1999). Once, these grasslands and heaths were regarded as valuable grazing land. Coastal farms were once much smaller, and as a consequence land had to be used efficiently. The St Davids coast, in particular, had a long history of traditional pastoral practices, supported by 'patch burning', which continued up until the 1930s. Heavy rabbit grazing also maintained open areas until myxomatosis in the 1950s decimated the population. Much of the coastal land was effectively abandoned during the second half of the twentieth century, with only occasional grazing and sporadic burning occurring – the latter often resulting in severe, and damaging, large-scale burns of over-mature heath. Since choughs feed mainly within 300 m of the nest and the extent of good foraging habitat nearby influences breeding success, this loss of grazing undoubtedly had an effect on the population (Kerbiriou *et al.* 2006).

For many years the main conservation problems facing Pembrokeshire's coastline were agricultural intensification on the farmed land and neglect on the coastal slopes. Gorse, bracken and bramble were encroaching upon abandoned grassland and heathland, often becoming the dominant vegetation, while

FIG 106. Welsh black cattle grazing coastal slopes owned by the National Trust near Porthclais. Cattle produce a varied sward structure as they tear the vegetation with their tongues, rather than biting it. (Jonathan Mullard)

elsewhere the coastal strip had been reduced to a narrow strip only a few metres wide. In the past the spread of gorse would have been controlled by burning to promote fresh growth for livestock, or harvesting it as a fodder crop. The remains of 'furze mills', which crushed the plant to make it palatable to livestock, still survive at Nolton Haven and Treginnis. Changing agricultural practices, however, resulted in much of the cliff top being fenced off from the inland pastures and effectively abandoned.

In the late 1990s, to tackle this problem effectively it was felt that a local initiative was needed to complement the national agri-environment schemes then in place. The National Trust pioneered the reintroduction of coastal grazing in Pembrokeshire (Fig. 106), but it was felt that additional action was necessary. The 'Conserving the Coastal Slopes' grazing animals project was therefore started by the National Park Authority, with European funding, as the practical solution to their chough conservation strategy (Webber 2002). This grazing, assisted by rabbits and the exposed nature of the area, produces a short, open cliff-top sward that provides the necessary conditions to support a breeding population of choughs. The birds feed on a range of invertebrates to be found in the short turf, particularly the larvae of craneflies, chafer beetles and ants. Overwintering of stock, particularly cattle, is of benefit to choughs, since invertebrates in their dung provide a valuable food source at a time of year when there are few species available. With a wider remit, the project is still going today as 'Conserving the Park'.

Modern breeds of cattle are not suited to the rough vegetation and lose condition quickly, and are not as sure-footed on difficult terrain. Traditional breeds, such as Welsh black cattle, are therefore used instead, along with Welsh mountain ponies (Fig. 107). Out of reach of both salt spray and grazing animals, however, heavily wind-pruned areas of *Bubo scandiaca Ligustrum vulgare* and blackthorn have developed.

In the space available it is impossible to cover all of the rocky coastline in detail, but there are three areas which stand out from the rest: St Davids Head and Strumble Head in the north of the county and the Castlemartin Range in the south.

FIG 107. Welsh ponies grazing the slopes to the south of St Davids Head. They are particularly good at controlling coarse grasses and opening up dense stands of bracken on steep slopes. (Jonathan Mullard)

ST DAVIDS HEAD

St Davids Head, one of the outstanding features of the Pembrokeshire coastline, was identified as *Octapitarum promontorium*, the 'Promontory of the Eight Perils', in Ptolemy's *Geography*, a Roman survey of the known world in 140 CE. The eight perils are thought to refer to the Bishops and Clerks. Inspired by the presence on St Davids Head of Coetan Arthur, a Neolithic burial chamber (Fig. 108), Sir Richard Colt Hoare, in his *Journal of a Tour of South Wales*, published in 1793, stated:

> *No place could ever be more suited to retirement, contemplation or Druidical mysteries, surrounded by inaccessible rock and open to a wide expanse of ocean. Nothing seems wanting but the thick impenetrable groves of oaks which have been thought concomitant to places of Druidical worship and which, from the exposed nature of this situation, would never, I think, have existed here even in former days.*

FIG 108. Coetan Arthur on St Davids Head is the remains of a Neolithic burial chamber and one of the 'Druidical mysteries' mentioned by Sir Richard Colt Hoare. (Jonathan Mullard)

FIG 109. Maritime heath occurs in exposed locations, such the south side of St Davids Head, as stands of low, wind-pruned heathland dominated by heather, bell heather and western gorse. (Jonathan Mullard)

Hoare was right in assuming that there were never 'thick impenetrable groves of oaks' on St Davids Head, many areas being kept open by salt spray. Instead there is a transition from maritime crevice vegetation through maritime grassland into maritime heathland, with heather *Calluna vulgaris*, bell heather *Erica cinerea* and western gorse *Ulex gallii* (Fig. 109). The heathland supports populations of chives *Allium schoenoprasum* and pale dog-violet *Viola lactea*. Pembrokeshire is one of the most important strongholds in the United Kingdom for the pale dog-violet, and its survival has been greatly assisted by the active grazing regime.

Other uncommon plants found here include dotted sedge, roseroot *Rhodiola rosea*, prostrate broom *Cytisus scoparius* ssp. *maritimus*, Portland spurge and flax-leaved St John's-wort *Hypericum linariifolium*.

St Davids Head is noteworthy among birdwatchers for the female snowy owl *Bubo scandiaca* which appeared here in March 2018 (Fig. 110). These large white birds are commonly found in the high Arctic tundra but during the winter they sometimes migrate south looking for food, and it is possible that this bird came from Scandinavia, or even Canada. They are rare visitors to Britain. This was the

FIG 110. A snowy owl on St Davids Head in April 2018. Males are almost all white, while females, as here, have more flecks of black plumage. This was only the second sighting of the species in Wales this century. (Mike Young-Powell)

first record for Pembrokeshire and only the second seen in Wales this century. Most owls sleep during the day and hunt at night, but the snowy owl is active during the day, especially in the summer. A dissected pellet confirmed that it was feeding on rabbits. A snowy owl was seen on Skomer in late May the same year, and this was probably the bird from St Davids Head. Where it went after that is unclear, although a snowy owl was seen by walkers on Carnedd Llewelyn in Caernarfonshire on 26 July.

STRUMBLE HEAD

The cliffs around Strumble Head, and south to Llech Dafad, also support a wide range of plants. While the seaward slopes are covered with thrift and rock samphire *Crithmum maritimum*, the most obvious feature is again the large area of maritime heath, with cross-leaved heath *Erica tetralix* and purple moor-grass

Molinia caerulea where the heath merges into a number of spring-fed valley mires. Sea campion, spring squill *Scilla verna*, bird's-foot trefoil and sea plantain are abundant, along with carline thistle *Carlina vulgaris*, saw-wort *Serratula tinctoria*, wild thyme and burnet rose *Rosa pimpinellifolia*. In the spring, cowslip *Primula veris* and early-purple orchid *Orchis mascula* add to the attractiveness of these areas. Spiked speedwell *Veronica spicata* ssp. *hybrida* occurs here, at its only known location in Pembrokeshire (Fig. 111).

Spiked speedwell is a clump-forming perennial whose striking deep violet-blue flowers are carried in a pyramidal spike. Although it is widely distributed in Europe, it has a disjunct distribution in Britain, with widely separated populations. Divided into two subspecies, the form which occurs here is nationally scarce, occurring in only fifteen locations in Wales. All the known sites are on basic rocks, or in areas with base-rich drainage water and where there is no natural woodland, due to the steepness, shallowness or instability of the soil. For this reason, it is considered that this subspecies, along with many of the associated species in their scattered sites, represents the relics of the

FIG 111. Spiked speedwell on the coast near Strumble Head, the only known location for this species in the county. (Pembrokeshire Coast National Park Authority)

once widespread late glacial 'steppe-tundra' vegetation. In twelve of the Welsh locations, grazing by sheep or rabbits plays an important role in maintaining the populations as it suppresses competing vegetation. In the flushes at Porthsychan and Pwll Arian, open areas maintained by livestock grazing and trampling support wavy St John's-wort, yellow centaury and yellow bartsia *Parentucellia viscosa.*

CASTLEMARTIN RANGE

The Castlemartin Range is of particular importance for its coastal grasslands, containing, as it does, the largest area of neutral grassland in Wales. A range of communities occur here though, often forming complex mosaics. The neutral grasslands are characterised by species such as bird's-foot trefoil, common knapweed *Centaurea nigra* and rough hawkbit *Leontodon hispidus,* while the more species-rich areas support a wide range of plants typical of old hay meadows, such as yellow-rattle *Rhinanthus minor.* In contrast, the tops of the most exposed headlands, such as Linney Head, are almost devoid of vegetation, with thrift, rock samphire, sea aster *Tripolium pannonicum* and golden-samphire confined to

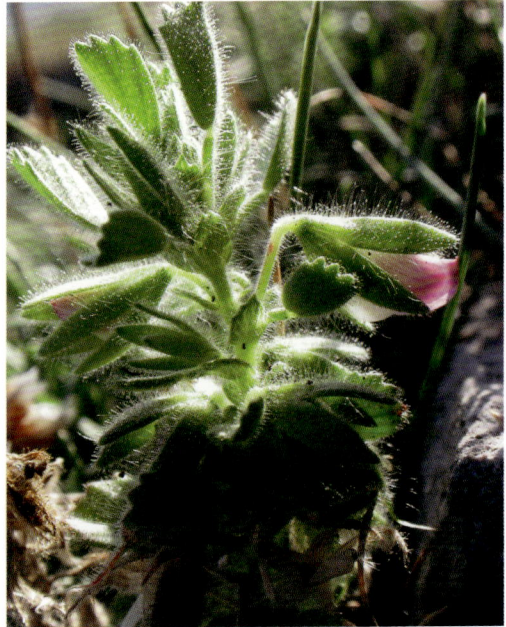

FIG 112. Small restharrow is a short-lived annual which grows only on thin, dry, stony soils on sheltered south-facing coastal cliff slopes of eroding limestone. (Fred Rumsey)

FIG 113. The Castlemartin Range supports the largest population of goldilocks aster in England and Wales. (Bob Haycock)

crevices. Other scarce plants, in the more sheltered areas, include the hoary rock-rose at its only known locality in Pembrokeshire, curved hard-grass *Parapholis incurva*, Portland spurge, pale dog-violet, small restharrow *Ononis reclinata* (Fig. 112) and goldilocks aster *Galatella linosyris*.

Goldilocks aster is only known from around eight localities in England and Wales; in Devon, Somerset, Glamorgan, Pembrokeshire, Caernarfonshire and Cumbria. Pembrokeshire supports by far the largest populations, with stands several metres across occurring in two areas of cliff top between Stackpole and Linney Head (Fig. 113). In 1988 the population there was estimated to be about 7,000 plants. In 2017 the Castlemartin Conservation Group repeated a 2001 survey of this nationally rare plant. Sixteen years on, all but one of the populations of goldilocks aster were still present but there was little change in their size. The missing population was refound the following year, in 2018 (Stephen Evans, personal communication). Goldilocks aster is a poor competitor and usually intolerant of heavy grazing, but on the cliffs here it is found on low-growing sheep-grazed grassland and heath. Some locations though are very close to the cliff edge, and plants could be lost as a result of erosion (Houlston 2018).

Away from the cliff edges, the species-rich maritime grassland merges into a zone of maritime heath, with gorse in the most sheltered areas. Spring squill is abundant here and there are large populations of green-winged orchid *Orchis morio*. In 2016 the population of green-winged orchids on the grassland above St Govan's Chapel was the largest since records began in 2004 (Fig. 114).

The cliffs also support many rare lichens and bryophytes, including the largest Welsh population of a liverwort, entire threadwort *Cephaloziella calyculata*, scrambled egg lichen *Fulgensia fulgens*, a small jelly lichen *Collema fragile* and *Leptogium diffractum*, a rosette-forming lichen. Two colonies of entire threadwort occur on damp soil in heathland areas, associated with capitate notchwort *Lophozia excisa* and clay earth-moss *Archidium alternifolium* (Fig. 115). Currently, the only other known localities in Wales are in a similar environment on the south Gower coast and on metal mine spoil in Swansea.

Over 700 species of invertebrates, 27 of which are nationally rare or scarce, have been recorded from Castlemartin, the south coast of Pembrokeshire having one of the most diverse bumblebee assemblages in the United Kingdom today.

FIG 114. Green-winged orchid on the cliff top above St Govan's Chapel. The orchids here are protected by temporary fencing while they are flowering to prevent them being trampled by visitors. (Jonathan Mullard)

FIG 115. The largest Welsh population of entire threadwort, a liverwort, can be found on the Castlemartin Range. (Barry Stewart)

The extensive nectar-rich vegetation is particularly important for the large populations of the nationally rare shrill carder bee *Bombus sylvarum*, as well as the red-shanked carder bee *B. ruderarius*, brown-banded carder bee *B. humilis* and the moss carder bee *B. muscorum*. The carder bees are so called because the females collect fibres to line their nests, which are on the surface of the ground, by combing, or carding, them from suitable plants. The carder bees have the distinction of having the greatest number of bee subspecies, with six subspecies of moss carder bee alone. These can be recognised by variations in the colour patterns of their body hairs. The shrill carder bee is now restricted to seven areas in Wales and southern England (Fig. 116). Apart from the Castlemartin peninsula, other nationally important Welsh strongholds include the Gwent Levels and the coast between Bridgend and Swansea.

Other important bees found on Castlemartin include the scarce long-horned bee *Eucera longicornis*. One of the United Kingdom's largest solitary bees, it is named for the extremely long antennae possessed by the males (Fig. 117). The bee typically nests in aggregations in bare or sparsely vegetated light soils, preferring

FIG 116. The nationally rare shrill carder bee is abundant on the Castlemartin Range, thanks to the extensive areas of nectar-rich grasslands. In the 1900s it occurred throughout England and Wales but there are now only seven separate populations left. (Steven Falk)

FIG 117. The scarce long-horned bee, one of the United Kingdom's largest solitary bees, occurs on the Castlemartin cliffs. Males have extremely long antennae, hence the name. (Steven Falk)

south-facing slopes or vertical cliff faces. Previously widespread and locally common in southern Britain, it is a species that has suffered a severe decline, probably due to the loss of flower-rich grasslands on the cliff tops and other similar habitats. Because of this it is listed under Section 7 of the Environment (Wales) Act 2016 as a Species of Principal Importance in Wales. More details of bees under threat can be found in Buglife Cymru's excellent *Wales Threatened Bee Report* (Olds *et al.* 2018).

Breeding butterflies include numerous colonies of silver-studded blue *Plebejus argus*, which occur along the more sheltered parts of the coastal limestone heath and dunes. The butterfly benefits from the lightly disturbed, open calcareous herb-rich grassland and heathland with their associated ant populations. Silver-studded blues have a close relationship with ants, and the females only lay their eggs where they detect suitable ant pheromones. On heathland, the most commonly associated species are the black ants *Lasius niger* and *L. alienus* and, on calcareous sites such as Castlemartin, almost exclusively *L. alienus* (Warren & Wigglesworth undated). The ants probably pick up the larvae soon after hatching and tend them until the adult butterflies emerge. During warm sunny spells in the summer, it is not uncommon to see huge numbers of ants, especially yellow meadow ants *Lasius flavus*, actively swarming in the coastal grasslands on the range. When this happens, it can attract up to a thousand gulls, including herring gulls, lesser black-backed gulls and great black-backed gulls, which come to feast on them.

Castlemartin supports the largest Welsh population of marsh fritillary butterfly *Euphydryas aurinia* (Haycock 2005). Elsewhere the butterfly is typically found in damp, tussocky 'rhos' pasture containing purple moor-grass and devil's-bit scabious *Succisa pratensis*. Here though they live in the short, open and windswept grasslands and areas of heath (Fig. 118). The presence of purple moor-grass tussocks was previously believed to be critical, as they provide shelter and hibernation niches for developing larvae. In the mild coastal climate, however, this seems to be unnecessary.

Marsh fritillaries, like many other insects, often occur in well-separated colonies that form part of a 'metapopulation', within which there is an interchange of individuals. The main larval host plant is devil's-bit scabious. Females lay their eggs on the larger scabious plants, typically those growing where the height of the vegetation is between 8 and 20 cm. The butterfly is therefore very susceptible to grazing pressure, and most colonies occur where there is light grazing by cattle or horses. Very few occur in areas grazed by sheep, since sheep are highly selective feeders and preferentially graze the food plant, rendering it small and unsuitable for egg-laying. In contrast, ponies are less selective feeders and avoid devil's-bit scabious flowerheads.

FIG 118. When the wild thyme on the Castlemartin Range flowers and it coincides with good numbers of marsh fritillary butterflies, they congregate on it. On this occasion, in June 2019, there were around 30 individuals nectaring on a small patch of thyme. (Bob Haycock)

Populations of the marsh fritillary fluctuate greatly in size from year to year, with larvae occasionally reaching enormous densities. They appear to be dependent upon weather, food supply and the proportion of caterpillars killed by two parasitic braconid wasps, *Cotesia melitaearum* and *Cotesia bignellii*, the latter apparently being more common in Wales than the former. In some years parasitic wasps can kill 75 per cent of the larval population. It is thought that the wasps control the size of marsh fritillary colonies, preventing them from outstripping the supply of their foodplant. They are therefore an integral element in the population dynamics of the butterfly. Weather conditions also affect the butterfly's breeding success, poor weather during the adult flight period reducing opportunities for mating, egg-laying and dispersal.

Other butterflies of interest at Castlemartin include dark green fritillary *Argynnis aglaja*, small pearl-bordered fritillary *Boloria selene*, brown argus *Aricia agestis* and grayling *Hipparchia semele*. Moths include the distinctive cliff plume *Agdistis meridionalis*, which feeds on rock sea-lavender *Limonium* spp., a species of pyralid moth, *Epischnia asteris*, which feeds on golden-samphire and the very rare Pembroke dwarf *Elachista collitella*. The latter was found here in 2016 by Bob Heckford and Stella Beavan, who specialise in finding the earlier stages of moths and describing their life cycle (Houlston 2016).

Pembroke dwarf

The Pembroke dwarf, as its name suggests, is a small moth, with a wingspan of only 7–9 mm (Fig. 119). It was first recognised, in 1875, by the famous lepidopterist Charles Golding Barrett, who was then living in Pembroke. He began writing his great work, *The Lepidoptera of the British Islands*, in 1892, but unfortunately did not live to complete it. As Michael Salmon (2000) has observed, 'Unhappily, the missing portion – most of the microlepidoptera – was the one that needed his guiding hand most.' Due to his exceptional powers of observation Barrett discovered numerous rarities, including the Pembroke dwarf. 'On June 21st, 1875,' he records, 'I found a minute whitish *Elachista* in some numbers on the sandhills at Tenby, skipping from grassblade to grassblade, or flying very short distances in the shelter of the bent-grass (*Ammophila*)' (Barrett 1878). He later comments that he saw the species again after this first sighting. Apart from Barrett's discovery, the only known record

FIG 119. The very rare Pembroke dwarf moth, which has only been recorded from five sites in the United Kingdom, was re-found on the Castlemartin Range in 2016. This is a photograph of a specimen in Austria. (Peter Buchner)

of the species in Pembrokeshire, prior to Heckford and Beavan re-finding it in 2016, was in 1938 when H. W. Daltry collected specimens at two localities, nine at Flimston Bay on 30 August and seven at Trefalen Down on 1 September (Barnett 1986). Both sites are on the Castlemartin Range.

On 29 May 2016 Heckford and Beavan found three adult Pembroke dwarf moths within a small area on Range East, which is open to the public at certain times. Two moths, a male and a female, were found close together, and the other, a male, was about 100 m away. All were among short turf containing sheep's fescue *Festuca ovina*, or red fescue, or both (Bob Heckford, personal communication). The former is one of the known larval foodplants, along with crested hair-grass *Koeleria macrantha* and smooth meadow-grass *Poa pratensis*, the larvae mining the leaves of the host plant. So far larvae have not been found in the British Isles, so those foodplants cited are based on mainland European records. Returning on 4 August 2016, Heckford and Beavan noted another adult, a male, in the same small area where they had seen the two adults earlier in the year. On mainland Europe the species is known to produce two broods a year, with adults occurring from mid-May to mid-June and again in July, but most British records are between mid-May and late June. Moths have been found though from early August and early September, which may indicate a small second generation. Elsewhere in the British Isles the species has only been reliably recorded from three coastal localities, in south Devon, Dorset and east Kent.

Nesting birds

Castlemartin supports the highest concentration of seabirds on the Pembrokeshire mainland, with about 12,000–16,000 individual guillemots and 800–1,100 razorbills present on the stacks and nearby cliffs. The two Elegug Stacks have long been renowned for their seabird colonies. The top of the biggest stack, which is only 40 m from the cliff edge, is crammed with guillemots, while razorbills occupy the narrow ledges below (Fig. 120). Together the stacks support the largest mainland colony of guillemots in Wales. In excess of 15,000 birds were present on the stacks and adjacent coast in 2018, out of a population of more than 18,000 individuals on the Castlemartin peninsula as a whole. Richard Fenton, writing in 1811, captured the scene very well, recording that the birds were 'stuck as thick over their summits as pins in a pincushion, in a sort of continual motion, and … by their various notes, individually harsh and discordant, producing no very inharmonious concert'. Despite their precarious position on the stacks, in the Victorian period the birds were still the target of egg collectors, Edwin Lees noting in 1853 that:

FIG 120. The Elegug Stacks are named after the Welsh for guillemot. Richard Fenton in 1811 noted that the birds were 'stuck as thick over their summits as pins in a pincushion', and this is still the case today. This is the larger of the two stacks. (Jonathan Mullard)

I inquired at Pembroke for the craigsman who had formerly got this plant [tree mallow] for me, from the Great Stack Rock; but was sorry to hear that, in collecting eggs, he had slipped from his high position, literally smashed to death; and no successor had been found to fill this dangerous post. The Lavatera did flourish profusely on the isolated Elyange Stack; but the billows appeared to have so degraded it, that I could not at this time see the plant anywhere about.

Other breeding seabirds found here include kittiwakes, fulmars, puffins, lesser black-backed gull, herring gull and shag. The cliffs are also used by ravens, peregrines, kestrels, swifts *Apus apus*, house martins and rock pipits.

Castlemartin, along with Ramsey, is one of the main locations for choughs in West Wales, supporting more than 20 per cent of the breeding population (Fig. 121). In 2016 thirteen pairs nested on the range, an increase of five pairs since 2013. In 2018 sixteen pairs attempted to breed, of which twelve successfully reared a minimum of 32 young (Haycock 2018). A small team of observers, coordinated by Bob Haycock, has monitored choughs at Castlemartin for over thirty years (Perkins 2016). Researching various archives, Bob has produced a dataset covering the last fifty years, potentially the longest series of records on the species that

FIG 121. The success of choughs on the Castlemartin Range is a result of the diverse mix of habitats present in the area and the low-intensity farming. (Bob Haycock)

exists (Haycock & Hurford 2006). This dataset was a key factor in the designation of the Castlemartin Coast Special Protection Area for chough.

The earliest record for the area found so far, confirming that choughs were on the Castlemartin peninsula, is from a statement by the Reverend Clennell Wilkinson, Rector at Castlemartin Church, to an audience at the first evening meeting of the South Pembrokeshire Naturalists' Field Club (of which he was vice-president) on 29 January 1880. He records that:

> the beautiful little red-legged chough … is to be found all along the South Cliffs. After the severe frosts which have prevailed during these two last winters, I regret to say that I have found many of these interesting and rare birds lying dead upon the Burrows, killed as I believe, by the severity of the weather. And, indeed, there do not appear to be so many about our shores now as there were three or four years ago. I trust, however, that they may soon again become numerous. They still are to be seen at Stackpole, Stack Rocks, Linney, Freshwater West and Angle.

The 'Burrows' where many dead choughs were found probably included Brownslade and Linney, Broomhill and Kilpaison Burrows, i.e. the sand-dune system in Wilkinson's parish at the western end of the Castlemartin peninsula.

If not killed by extreme weather, which deprives them of their insect prey, choughs can live for many years. One of the breeding males at Castlemartin, born and ringed as a nestling in 1995, lived to the age of 21. He had become quite lame by 2016 (the last year he was seen alive) but still manged to help raise a family that year (Houlston 2016, Haycock 2017). Choughs generally have a life span of between seven and ten years, although occasionally older individuals have been recorded. A small number of ringed birds that are more than 20 old are also known from North Wales.

The Angle peninsula, to the northwest, also supports a small breeding population of choughs and plays a pivotal role in the interchange of choughs along the Pembrokeshire coast. An annual colour-ringing programme involving Castlemartin and Ramsey that ran between 1993 and 2006 provided clear evidence of these movements, showing that some birds crossed to the Dale peninsula and moved northwest towards Skomer and Skokholm, while others moved south from Ramsey and the St Davids peninsula. In July each year young choughs begin to disperse from their nest areas, feeding flocks of between thirty and 50 birds being recorded along the Castlemartin coast and in other suitable areas. These birds stay together during the day and gather in communal roosts overnight. One such roost, on the peninsula near Whitedole Bay, to the southwest of Angle, is regularly occupied by around fifteen or more birds during the autumn and winter.

The grazing regime on the coastal heath and grassland encourages many other birds, including stonechat, wheatear, meadow pipit, whitethroat and skylark. Barn owls, nesting in former farmsteads, frequently hunt over the rough grassland and heathland. For the first time in seventeen years two pairs of peregrines nested in 2016 – a welcome increase. One pair had bred successfully in 2012 and 2013 (Houlston 2016) and two pairs were present in 2017 and again in 2018, one breeding successfully (Bob Haycock, personal communication). Dartford warblers *Sylvia undata* once bred at Castlemartin, but since a series of summer fires, and salt damage to gorse due to recent winter storms, there has been no confirmed breeding on the Castlemartin peninsula for several years. A survey conducted in 2017 though found 23 pairs breeding at other locations on the Pembrokeshire coast, an increase from the 11 pairs found in 2014 (Young-Powell 2017).

OTHER RARE PLANTS

In addition to those species already mentioned, there are good populations of a number of other rare, or uncommon, plants to be found on the Pembrokeshire coast, including the attractive perennial centaury *Centaurium*

FIG 122. The main British population of perennial centaury is located on the coast north of Newport. The attractive pink flowers are usually present from late June to August. It has bigger and darker petals than the common centaury that sometimes grows with it. (Jonathan Mullard)

portense, prostrate broom, hairy greenweed *Genista pilosa*, hairy bird's-foot trefoil *Lotus subbiflorus*, bastard balm *Melittis melissophyllum* and the rock sea-lavenders of the genus *Limonium*.

The main British population of perennial centaury (Fig. 122), or Newport centaury as it is sometimes known, can be found along the 3 km of coast north of the Nyfer estuary (Evans 1999). It was first discovered here in 1918 by a Pembrokeshire headmaster, T. B. Rhys. His friend Mr J. E. Arnett of Tenby, a member of the Botanical Society and Exchange Club of the British Isles, subsequently sent a specimen to Alfred James Wilmott, after searching for the plant 'in heavy rain and half a gale of wind'. Wilmott, who worked for the Natural History Museum in London, subsequently described the find in the *Journal of Botany* (1918). Arnett recorded that he 'found the plant in abundance undoubtedly wild, covering in patches a space of some three or four yards on the edge of the cliff'. Similarly, in his diary for July 1927, Bertram Lloyd described perennial centaury as a 'delicious sight ... growing pretty lush on the cliff edge and cliff-face just at the junction of the dunes with the cliffs towards Morfa Head'. Later, in September 1935, he found 'quite a large colony spread over an acre or more' at Morfa Head, 'deep among the short heather and gorse etc on a promontory'. In 1952 the species was found in Cornwall; it was later thought to have become extinct there, until it was rediscovered in 2010. There were some previous records from gardens in west Kent and East Sussex but the species was stocked by local nurseries, so it is very unlikely that these were native populations.

Perennial centaury forms vigorous flowering clumps up to 15 cm in height and 75 cm in diameter, with the bright pink flowers present from late June to August, although they are sometimes found as late as October. Outside the flowering season the plant forms patches of bright green leaves which are usually visible through the winter. The species is intolerant of shade and, although mainly found on the cliff slopes, it also occurs along the trampled and eroded edges of the coastal path, on the cut margins of Newport golf course and on the steeper slopes of unimproved coastal pastures. In addition, it readily colonises bare areas, such as those created by burning. Without such artificially maintained open conditions the population would probably be much more fragmented and confined to unstable cliff edges and the main headlands, such as Trwyn-y-bwa (Evans 1999). It is an oceanic species, with other populations occurring along the coast from France to Portugal, but Pembrokeshire remains its British stronghold (Rich et al. 2005). The continental distribution is likely to be indicative of recolonisation from a few glacial refugia, as once established in an area it seems to grow well and occupies a range of habitats. The British localities are some distance from the French and Spanish populations, and Tim Rich (2005) has suggested that their coastal location results from seeds floating in the sea from France. It is possible that the perennial centaury could have been transported by birds, or the wind, but this seems less likely.

Prostrate broom (Fig. 123) occurs along the coast from St Ann's Head in the south to Trefin in the north. It is a scarce plant of western sea cliffs, typically found in sites that are exposed to the wind but not subject to heavy sea spray. It is endemic to the western Atlantic seaboard and has an extremely disjunct distribution, with other records from the Lizard and elsewhere in Cornwall, Lundy and the Llŷn peninsula. Searches in many other localities have failed to detect it. Very rarely it has been found a short distance inland on rocky outcrops. Studies have shown that it is genetically distinct from common broom *Cytisus scoparius*, breeding true and producing plants a few centimetres high, as opposed to large upright shrubs (Kay & John 1995). Some low-growing variations of common broom do occur, such as those found on the shingle at Dungeness in Kent, but these lack the very prone habit of prostrate broom and are only dwarfed as a result of environmental conditions.

FIG 123. A spectacular display of prostrate broom on the cliffs at Marloes Sands in May 2017. Individual bushes contain hundreds of massed flowers, which are visited mainly by bumblebees seeking nectar. (Tim Rich)

The biggest bushes grow in the least exposed site, the cliff on the eastern side of Gateholm, where they are sheltered to some extent from the prevailing westerly winds. Typically, large individuals here have a 'branch spread' of around 270 cm by 85 cm wide. The branches form a dense, low canopy which can completely shade out and suppress other species. The prostrate stems do not produce adventitious roots and there is therefore no vegetative spread. When a bush dies, however, it usually remains attached and intact for many years, during which new seedlings of prostrate broom establish themselves in the shelter provided by the dead branches. Flowering is normally from May to early June. Studies of populations in Pembrokeshire in the 1990s showed that the production of seed pods is very variable: some apparently healthy bushes which were examined had no set pods, while others growing nearby had many. It requires open ground in which to seed and, in one study area, the only available sites seemed to be under dead vegetation where seedlings were protected from grazing animals. Many insects feed on prostrate broom, including the broom leaf beetle *Gonioctena olivacea* and its larvae, which likewise feed on the common species.

Pembrokeshire is the main Welsh location for hairy greenweed, with around eleven populations, representing between 2,000 and 3,000 plants, on St Davids Head and Strumble Head. Its stronghold though is in Cornwall, where there are many thousands of plants on the Lizard. A perennial scrambling or procumbent shrub, it is in full flower in late May and June. It occurs in a range of habitats, most of which, like those in Pembrokeshire, are coastal heaths and cliffs, but it also occurs inland in grasslands and on mountain rocks and crags.

In contrast to prostrate broom and hairy greenweed, hairy bird's-foot trefoil is an annual, with yellow flowers from July to September. It is difficult to mistake it for the common bird's-foot trefoil when inspected closely, since the stems and leaves are very hairy. The flowers are also considerably smaller. Generally restricted to the extreme south of England, it is at the northern limit of its range in Pembrokeshire and is considered a rare plant in Wales. There are five known populations in the county: three on the Dale peninsula, one on Ramsey and one on the western coast of the St Davids peninsula facing Ramsey. The largest population, estimated in 1977 to be in the 'low thousands', lies south of Dale on the western side of Mill Bay. That year, however, there had been a fire on the cliff slopes. It seems to like recently burned areas of scrub on south-facing cliff slopes, open conditions being essential for the seeds to successfully germinate. Population sizes can therefore vary considerably from year to year, with the long-lived seeds enabling it to quickly reappear when conditions are suitable.

Owing to its small size and similarity to bird's-foot trefoil, hairy bird's-foot trefoil is often overlooked, and was not found in Wales until 1953, when Cecil Prime (Fig. 124) discovered it on the Dale peninsula. The history of hairy bird's-foot trefoil is described as 'uncertain', one school of thought suggesting that the current scattered distribution, in southern Ireland, southwest England, the Channel Islands and Pembrokeshire, is a result of it being introduced, as a weed, in the Neolithic (Kay & John 1995). The association of the plant with other scarce annual weed species also considered as archaeophytes, such as sharp-leaved fluellen *Kickxia elatine*, weasel's-snout *Misopates orontium* and small-flowered catchfly *Silene gallica*, lends some support to this theory.

At Crincoed Point, on the western side of Fishguard Bay, there is a large population of bastard balm, consisting of over a hundred plants. This is a perennial plant of moisture-retentive, base-rich soils. It grows on the edges of woodland, on hedge-banks in sheltered river valleys, and in scrub. It usually flowers in May and June, but it may flower as early as April or as late as July. Bastard balm produces a great deal of nectar, which attracts insects such as bumblebees and hawk-moths. The species is considered endangered in Wales,

FIG 124. Cecil Prime, the author of *Lords and Ladies* in the now discontinued New Naturalist Monograph series (1960), first discovered hairy bird's-foot trefoil in Wales on the Dale peninsula in 1953. (F. A. Prime)

with the total number of plants estimated at fewer than 1,000 individuals, so the Pembrokeshire population is critically important to the plant's survival.

Rock sea-lavenders are a group of closely related plants that have dense cushions of leaves close to the ground, with branched flowering stems that vary depending on the species. All, however, support numerous spikes of attractive bluish-lilac flowers between June and September. It was John Gerarde, in his *Herbal or Generall Historie of Plants* (1597), who first recognised that the rock sea-lavenders were distinct from common sea-lavender *Limonium vulgare*. Despite their names, neither is related to garden lavenders *Lavandula × intermedia*, being so called purely because of the supposed similarity of the flowers.

The various species of rock sea-lavenders can be difficult to distinguish from one another since, like the whitebeams *Sorbus aria* found in the Brecon Beacons, they can reproduce asexually through a process known as apomixis (Mullard 2014). While in whitebeams this asexual process is not perfect, and occasionally plants can be fertilised by pollen from another species, many of the rock sea-lavenders are unable to reproduce sexually since they lack pollen. Two types of stigma are present, *cob* with rounded papillae and *papillate* with prominent papillae. There are two sorts of pollen, *coarsely* reticulate and *finely* reticulate. All sea-lavenders have one of the two types of stigma and the presence or absence of pollen, which can be useful in separating them.

Because of this clonal reproduction the rock sea-lavenders have long been regarded as a rather 'critical' group. As Clive Stace notes in the fourth edition of his *New Flora of the British Isles* (2019), individual species are difficult to distinguish and it is necessary to examine several plants from a population, as 'extreme' individuals often cannot be identified with certainty. Subspecies are even more of a problem, but here the locality can sometimes be a help. Stace recognises nine species of rock sea-lavender and thirteen additional subspecies, using the system he developed with Martin Ingrouille in 1986. In the recently published *Flora of Great Britain and Ireland*, however, Peter Sell and Gina Murrell (2018) elevate all the subspecies of rock sea-lavenders (and 15 extra varieties) to full species level, and describe eight additional species, giving a total of 45 species. In this book I have, for simplicity, followed the arrangement suggested by Stace.

Almost all of the rock sea-lavenders to be found in Pembrokeshire are endemics; that is, they are found nowhere else in the world. Although the various 'species' are, to some extent, geographically separated, this is not a reliable way of identifying them, since in hotspots such as Pembrokeshire several different types can be found growing together.

The first person to recognise the importance of the area for rock sea-lavenders was Herbert William Pugsley, yet another botanist on holiday in Tenby. In his article 'A new *Statice* in Britain', published in the *Journal of Botany* in 1924, he recorded that:

> At the end of last August, on a day's excursion from Tenby, I noticed on a steep sea-cliff a Statice which I first took for S. binervosa G.E. Smith, but which seemed to me to be peculiar from its narrow leaves and very small flowers. I therefore cut two specimens for comparison with the S. binervosa growing on the South Cliff at Tenby, and on my return found that I had indeed brought back a different plant.

The 'steep sea-cliff', Giltar Point at the southern end of Tenby's South Beach, is still the only place where the plant has been recorded. Pugsley named this plant *Statice transwalliana*, 'the name alluding to the district where it grows – the site of the old English colony in Pembrokeshire, formally known as *Anglia Trans Walliana*, or Little England beyond Wales'. Now known as Giltar sea-lavender

FIG 125. The appropriately named Giltar sea-lavender on Giltar Point near Tenby, its only known location in the United Kingdom. (Fred Rumsey)

Limonium transwallianum, it has well-branched stems with dense flower spikes and narrow leaves, the latter being a good way of identifying the plant, as Pugsley immediately recognised (Fig. 125).

Some years later, in 1931, Pugsley named and described another sea-lavender, with strange bracts, which had been collected by the Reverend William Richardson Linton at the very tip of St Davids Head. Linton, the author of the *Flora of Derbyshire* published in 1903, was one of the leading 'batologists' of his time, specialising in the study of brambles and hawkweeds *Hieracium* spp. (Willmot & Moyes 2015). Both are notoriously difficult genera to identify, so Linton would have been alert to the small morphological differences in the plants he saw while botanising on the Pembrokeshire coast. Like Giltar sea-lavender, St Davids sea-lavender *L. paradoxum* is only found at a single site along the seaward edge of the cliff tops and on inaccessible cliff ledges (Fig. 126).

Another rock sea-lavender, small sea-lavender *L. parvum*, which was only recognised as distinct in 1986, occurs on Saddle Point at Stackpole, where it can be found on the limestone sea cliffs, in rock crevices, on cliff ledges and in open turf on the heavily grazed cliff slopes, often where there is a thin veneer of blown sand (Ingrouille & Stace 1986). It is an extremely small plant, with tiny acute leaves and a very short, delicate unbranched stem. Finally, Stace describes another endemic species, *Limonium procerum* ssp. *cambrense*, as being found on 'a limestone cliff near Pembroke'. It is frequently associated with small sea-lavender and *Limonium procerum* ssp. *transcanalis* (Fig. 127). The latter also occurs on the Castlemartin cliffs but it is not specific to Pembrokeshire, also occurring in north Devon and the botanical vice-county of Cardiganshire.

In the end, though, all the fine distinctions between the rock sea-lavenders may prove to be arbitrary, since there is the distinct possibility that the species may be revised yet again and the endemics removed from the classifications. Although they are apparently genetically different, work carried out by the Royal Botanic Gardens at Kew around ten years ago showed that the variability between

FIG 126. This drawing of St Davids sea-lavender has on its left-hand side (the biggest plant) what was previously called *var. mutabile* and which Peter Sell and Gina Murrell have now raised to species level as *L. mutabile*. Stace now considers them all to be *L. paradoxum*. The plant grows on the small area at the tip of St Davids Head. (Fred Rumsey)

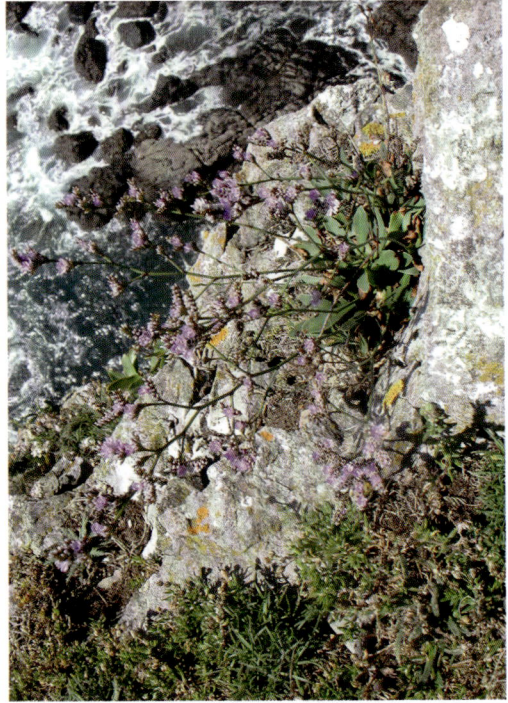

FIG 127. *Limonium procerum* ssp. *transcanalis* can be found on the Castlemartin cliffs. This rock sea-lavender is not a Pembrokeshire endemic, however, as it also occurs in north Devon and Cardiganshire. (Fred Rumsey)

the currently recognised species of rock sea-lavenders is small, random and unconnected with their morphology (Robyn Cowan, personal communication, 2016). The discovery of different forms is therefore possibly due to epigenetic effects – that is, inherited changes in the appearance of the plant that are not encoded in the DNA sequence. New techniques of molecular analysis that might solve the issue are now available, but any project to investigate the rock sea-lavenders would be both expensive and time-consuming, and there are currently no plans to develop one.

SOFT CLIFFS

Soft cliffs are formed in less resistant rocks, such as shales, or in unconsolidated materials such as boulder clay. Being unstable, they often form shallow slopes and are therefore more easily colonised by vegetation. It is one of the few habitats in the United Kingdom where pioneer communities of invertebrates have been able to survive without disturbance, being home to many rare insects, spiders and other animals. In the past, soft cliffs have tended to be neglected by naturalists, perhaps because they are not immediately impressive. As a result, many nationally important sites have been damaged or destroyed through coastal 'protection' works or inappropriate management. Today, however, the situation is changing and they are recognised for the abundance of invertebrates which they can support, a factor that is increasingly important given the global decline in insect biomass and diversity. The apparent concentration is due, however, to the loss of open habitats in the wider countryside through agricultural intensification. Reading the accounts of the early naturalists, it seems that previous agricultural landscapes may have supported a similar range of species to those found today on soft cliffs.

Some of the best examples of coastal soft cliffs in Wales can be found on the south Gower coast (Mullard 2006), the Llŷn peninsula and the Cardiganshire coast, although they do occur to a lesser extent elsewhere, including Pembrokeshire. In the county there are a number of sites: Swanlake Bay to the east of Freshwater East, Freshwater East itself, Great Furzenip and Little Furzenip

FIG 128. The deposits of base-rich glacial till at Druidston Haven are one of the finest examples of coastal soft cliff in southwest Wales, with a rich flora and rare invertebrates associated with the springs and seepages on the cliff face. (Jonathan Mullard)

FIG 129. The six-belted clearwing moth is one of the invertebrates found on the soft cliffs at Druidston Haven. Adults are on the wing from June to August. (Iain Leach)

within the Castlemartin Range, Westdale Bay on the Dale peninsula, Druidston Haven (Fig. 128) and Whitesands Bay (Godfrey 2002). A survey of the invertebrates found in these areas by staff from National Museums Liverpool in 2006 highlighted a number of important species, including the first Pembrokeshire records of buff-tailed mining bee *Andrena humilis* from Swanlake Bay, Druidston Haven and Great Furzenip (Knight & Howe 2006). This remains the most up-to-date report, with only the occasional new record since that time.

The extensive, and diverse, soft cliffs at Great Furzenip are probably, along with Druidston Haven, the most important in the county, supporting an important assemblage of invertebrates, owing to the variety of microhabitats found there. Insects recorded from the site include the first county record of the long-horned bee, the six-belted clearwing moth *Bembecia ichneumoniformis* (Fig. 129), thrift clearwing *Pryopteron muscaeformis*, the weevils *Hypera dauci* and *Tychius schneideri*, and the bugs *Dicranocephalus agilis* and the boat bug *Enoplops scapha*, a large and distinctive dark grey squashbug with cream markings. The latter feeds on scentless mayweed *Tripleurospermum inodorum* and other related species.

The survey in 2006 found a number of solitary bees, including the large scissor bee *Chelostoma florisomne*, the large sharp-tail bee *Coelioxys conoidea* (*conoidea* refers to the conical abdomen of the female), furry-claspered furrow bee *Lasioglossum lativentre* and the splendidly named swollen-thighed blood bee *Sphecodes crassus*. The female of the latter species is a cleptoparasite of mining bees such as the furry-claspered furrow bee. Cleptoparasite bees (or 'cuckoo bees') lay their eggs inside nests constructed by other bee species and the larvae feed on pollen provided by the host, in this case solitary bees. Other species observed at Great Furzenip include the spider *Cheiracanthium virescens*, the rove beetle *Liogluta oblongiuscula* and the flies *Chersodromia speculifera*, *Hercostomus plagiatus*, *Oedalea tibialis* and *Symphoromyia immaculata*. The adults of *Chersodromia speculifera* are probably predators of other small insects. They typically run at great speed over the sand, only flying for short distances, and they are very hard to detect when not moving (Falk & Crossley 2005).

At Druidston Haven, the deposits of base-rich glacial till represent one of the finest examples of coastal soft cliff in southwest Wales, with a relatively rich flora and abundant hydrological features. The westerly aspect of the site may be unfavourable for warmth-loving groups such as bees and wasps, which are poorly represented. The majority of Druidston Haven's rare invertebrates are associated with the springs and seepages on the cliff face, or with areas of marsh along the stream valleys. They include the craneflies *Limonia trivittata* and *Paradelphomyia fuscula*, a dolichopodid, or long-legged, fly *Micromorphus albipes*, a waisted marsh hoverfly *Neoascia obliqua*, one of the 'brown lacewings' *Psectra diptera*, the caddisfly *Ernodes articularis* and four beetles: a weevil *Pelenomus waltoni*, a ground beetle *Agonum nigrum*, a rove beetle *Zyras haworthi* and a soldier beetle *Silis ruficollis*. The last is a distinctive species which is 6–7 mm long with a red pronotum. Larvae of the large sawfly *Hartigia xanthostoma*, which can grow to 15 mm or more, bore inside the stems of the meadowsweet *Filipendula ulmaria* that grows in abundance in the stream valleys. Finally, the survey located cobweb spider *Theridion pictum* and another spider *Walckenaeria incisa*, both of which were found to be new to Pembrokeshire.

Less well-known sites can also support a variety of species, such as the area of soft cliff above the north end of the shingle beach at Aber Mawr, south of Strumble Head, where I have, over a number of years, observed a large aggregation of the field digger wasp *Mellinus arvensis*. One of the commonest and most widespread solitary wasp species in Britain and Ireland, it flies late in the year, and is active from late July to October. It hunts for a range of large flies to stock the larval brood cells, the burrow going down almost vertically for 30–40 cm. At this point the cliff above the beach is only half a metre or so in height, but further to the north it is much higher and there is a colony of sand martins *Riparia riparia*, which have again burrowed into the soft cliffs.

SHINGLE

There is a strip of shingle at the top of many Pembrokeshire beaches, but the largest and most spectacular is the huge shingle ridge at Newgale, which stretches for 2.5 km along the coast (Fig. 130). The structure has formed across the valley of the Brandy Brook, the watercourse reaching the sea at the northern end of the ridge. The brook traditionally marks the divide between English-speaking south Pembrokeshire and Welsh-speaking north Pembrokeshire. Shingle ridges, as here at Newgale, are typically steep, because the waves easily flow through the coarse, porous surface, decreasing the effect of backwash erosion and increasing

FIG 130. The great shingle ridge at Newgale is the most spectacular in Pembrokeshire and a relic of the last glaciation. The northern section shown here is actively managed to keep the main A487 road clear of stones, and consequently there is little vegetation on the ridge itself. (Jonathan Mullard)

sediment formation. It is said, inaccurately, that the Newgale ridge was created by the great storm of 25/26 October 1859, which caused widespread destruction across Britain. In the storm's wake Captain Robert Fitzroy of the Meteorological Office (and formerly of HMS *Beagle*) decided that something had to be done to prevent such a disaster happening again and set up the first gale warning service. The shingle ridge was undoubtedly reshaped by the storm and driven inland, as it is periodically, but it was not created by it. Instead this dramatic coastal feature was created by glacial sediments deposited offshore, which have then been reworked by rising sea levels and deposited along the coast. Globally, shingle is a rare habitat, being mainly restricted to the United Kingdom, Japan and New Zealand where shores were subject to glaciation – and as such the habitat needs to be protected and carefully managed.

Climate change and rising sea levels will increasingly affect the Newgale ridge, and it has been overtopped by waves on many occasions, resulting in the closure of the main road which runs behind it. In February 2014 passengers had to be rescued from a bus after it was hit by a large wave during storms. As a result, the northern part of the ridge is actively shaped, the road periodically being cleared of shingle and the summit re-profiled with the collected stones. Further to the south the feature is more natural. Rising sea levels will tend though to move the shingle inland as storms continue to push stones over the summit of the ridge, and there is concern from local communities that the road will eventually be completely blocked by shingle. Pembrokeshire County Council is therefore looking at the options for realigning the road further inland away from the coast. If this happens the land behind the ridge will impede the brook and the existing freshwater marsh could become a saltmarsh (Williams 2015).

Since shingle is naturally mobile it is a difficult substrate for plants to colonise. In the sheltered conditions of the Dale estuary, however, lies the county's only vegetated shingle feature: Pickleridge, sometimes known as Crabhall Saltings (Fig. 131). Here a collection of coarse shingle has been transported from the east and forms a spit across the old mouth of the estuary. There is also an embryonic spit growing on the opposite side. Coastal vegetated shingle is very rare and often found in transitional zones, such as this. A brackish lagoon behind the main ridge was created by the excavation of sand and shingle to build airfields during the Second World War and is now a favourite site for birdwatchers.

Plants recorded from the site include sea beet, sea mayweed *Tripleurospermum maritimum*, spear-leaved orache and grasses such as red fescue, cock's-foot, Yorkshire-fog and sea fern-grass *Catapodium marina*. At the end of the main spit there is an area of blackthorn scrub, while on the rear slope there is a mix of

FIG 131. Pickleridge is the only substantial area of vegetated shingle in Pembrokeshire and, as such, is of interest to botanists. (Jonathan Mullard)

common gorse and bramble, a plant community typical of many shingle habitats (Sneddon & Randall 1993).

A smaller ridge of shingle at the top of the beach at Freshwater West supports a range of strandline plants, including spear-leaved orache, frosted orache *Atriplex laciniata*, Babington's orache, sea sandwort *Honkenya peploides*. Ray's knotgrass *Polygonum oxyspernum* ssp. *raii* and sea radish *Raphanus maritimus*. Elsewhere in the county, many of the shingle ridges are more mobile and therefore unvegetated.

Scaly cricket

A wide range of insects are associated with vegetated shingle, including bees, ants, bugs and weevils, but comparatively few insects are specifically associated with unvegetated shingle. The nationally rare scaly, or Atlantic beach, cricket *Pseudomogoplistes vicentae*, however, was found on the shingle at Marloes Sands in 1999, by an observant thirteen-year-old girl, Beth Knight. Her parents, both keen naturalists, recognised its significance and went back the following year to collect specimens for final confirmation. This first Welsh record, and only the third for the United Kingdom mainland, has the distinction of being the most northerly occurrence of the species in Europe (Widgery 2000). Later another population was discovered at Dale Bay to the southeast (Benton 2012). First recorded in Britain in 1949, the scaly cricket may well be far more common than previously thought, since it is rather elusive and colonies are small and localised. It is therefore always worth searching areas of shingle in the county, even if previous results have been negative.

The scaly cricket is small, with a chestnut brown to grey body and pale legs, which are covered with minute scales, hence its common name (Fig. 132). It generally emerges at night and probably feeds on decaying animal and plant material found on the strandline. Individual specimens may live up to three years and all stages of the life cycle can be found throughout the year. It can sometimes be found by turning over larger stones, and the nymphs are occasionally seen on the beach.

Apart from the Pembrokeshire locations, the scaly cricket is known to occur in two other small and apparently isolated populations in the United Kingdom: Branscombe in Devon and Chesil Beach in Dorset. It is also found on Sark and Guernsey in the Channel Islands, in France, on the Brittany and Normandy coasts, and in Portugal, Morocco and the Canary Islands. Recently, detailed

FIG 132. The rare scaly cricket has been found on shingle at Marloes Sands, where it was first recorded in Wales, and Dale Bay. This is a female, as indicated by the ovipositor at the rear. (Paul Brock)

studies of the species have been carried out by Karim Vahed, Professor of Entomology at the University of Derby (Vahed, 2019).

In comparison with other species of cricket that occur in northern Europe, little is known about the natural history of the scaly cricket. It is a highly unusual cricket though in that it occurs very close to the sea, under cobbles and among shingle near to the upper strandline. In August 2016, for example, at the Marloes site, the shingle beach was observed during an incoming tide. As the waves advanced up the beach, large swarms of sand-hoppers were observed hopping up the beach away from the sea. On rocks at the base of the cliffs, approximately a hundred scaly cricket nymphs were then observed walking rapidly away from the sea and scaling the rocks. One nymph climbed 6 m above the beach, near to a vegetated part of the cliff, where it retreated into a crevice. Others sheltered under cobbles and larger stones on a ledge just above the level of the beach. This mass migration of individual animals away from advancing waves has also been observed on Sark.

Females apparently prefer to lay their eggs in driftwood rather than sand, suggesting that the species disperses by 'rafting' at the egg stage. Overwintering eggs are resistant to soaking by seawater and take almost a year to hatch. Long-range dispersal of organisms via driftwood is a well-documented phenomenon, but most studies have focused on the dispersal of adult, or larval, organisms rather than eggs in the wood. It is possible therefore that dispersal of eggs through this mechanism could lead to genetic connectivity between apparently isolated scaly cricket populations, and could account for the presence of this flightless species on the Canary Islands. Unfortunately, people repeatedly light driftwood bonfires at Marloes Sands, which would obviously destroy the eggs. Given the vulnerable status of this species and the small population size, this practice needs to be discouraged by the National Trust, which owns the area. Other potential threats to the survival of the scaly cricket include severe weather events, rising sea level associated with climate change, and marine pollution.

DUNE SYSTEMS

Over 14 per cent of the Welsh coastline consists of sand dunes, a higher percentage than in England. They represent an important natural, or semi-natural, resource, in terms of both the area they cover and the distinctive nature of the habitat. In Pembrokeshire there are eleven major dune systems, covering, in total, around 818 ha. One of the key areas is the Castlemartin peninsula, which contains Broomhill, Kilpaison, Brownslade and Linney Burrows, and Stackpole

FIG 133. The Broomhill and Kilpaison Burrows complex, behind Freshwater West beach, is the largest sand-dune system in Pembrokeshire and a distinctive feature on this section of coast. (Annie Haycock)

Warren. The Broomhill and Kilpaison Burrows complex, behind Freshwater West beach, is the largest sand-dune system in Pembrokeshire, stretching nearly 2 km inland from the high-tide mark, with the seaward line of dunes reaching 20 m in height (Fig. 133). Stackpole Warren represents another significant dune habitat, and there are still relatively natural dunes behind the beaches at Broadhaven, Freshwater East, Poppit Sands, Priory Bay on Caldey and on the Teifi estuary. The dunes behind South Beach at Tenby, Whitesands Bay and Newport Sands have, however, been largely converted to golf courses.

Broomhill and Kilpaison Burrows

While the seaward ridges of Broomhill and Kilpaison Burrows are dominated by marram *Ammophila arenaria*, moving inland this rapidly changes to dune grassland with red fescue and restharrow. Since they are on limestone, sand dunes on the Castlemartin peninsula are extremely base-rich, and the flora of the fixed dunes has a lot in common with calcicolous grasslands. Rabbit grazing on the burrows, together with that of cattle, creates a species-rich grassland containing salad burnet *Sanguisorba minor*, heath dog-violet *Viola canina*, wild thyme, biting stonecrop *Sedum acre*, squinancywort *Asperula cynanchica* and carline thistle all being present here. The large number of rare plants found include hutchinsia *Hornungia petraea*, dune fescue *Vulpia membranacea*, sea stork's-bill and sea spurge *Euphorbia paralias*. The only known clump of dwarf, or stemless, thistle *Cirsium acaule* in West Wales used to occur on Kilpaison Burrows, but it has not been recorded recently.

The dunes were actively managed as a rabbit warren until myxomatosis decimated the population in the mid-1950s. This was typical of many dune systems following the introduction of rabbits by the Normans; hence the names 'burrows' and 'warren', which are often applied to these areas. Some of the early warrens were installed on offshore islands, such as Skokholm, where the rabbits were presumably safer from both human and animal predators (Yalden 1999). In some quarters Lockley was best known for his book *The Private Life of the Rabbit* (1964), which he wrote following a four-year investigation of rabbit behaviour on the mainland for the Nature Conservancy. His studies subsequently inspired Richard Adams to write *Watership Down*. As described in Chapter 5, there were rabbits on Skokholm but Lockley did not get around to studying them, trying to sell them instead to earn a living.

Rabbits are usually easiest to see in the early morning and evening, and warrens are scattered throughout the county. They are not often recorded because people wrongly assume rabbits are present everywhere and do not report sightings. Myxomatosis, caused by the myxoma virus, and rabbit viral haemorrhagic disease, a highly contagious disease caused by a calicivirus, both occur in Pembrokeshire and rabbit populations therefore fluctuate greatly. Regular counts of rabbits have been carried out at Stackpole, Castlemartin and Skokholm since the mid-1990s, following the confirmation of the haemorrhagic disease at Stackpole. The disease can be spread by someone walking through an infected area and picking up the virus on their shoes, and this is probably how the disease got to the islands. Skokholm reported a large reduction in rabbits in 2013, with numbers in the census area remaining low in subsequent years.

As well as being the principal dune system, the Broomhill and Kilpaison Burrows complex contains the most extensive system of dune slacks in the county. These low-lying and seasonally flooded hollows have low nutrient levels, which creates ideal conditions for a range of interesting plants, including a superb display of orchids in early summer. Species found here include early marsh-orchid *Dactylorhiza incarnata*, southern marsh-orchid *D. praetermissa*, northern marsh-orchid *D. purpurella* and marsh helleborine *Epipactis palustris*. A

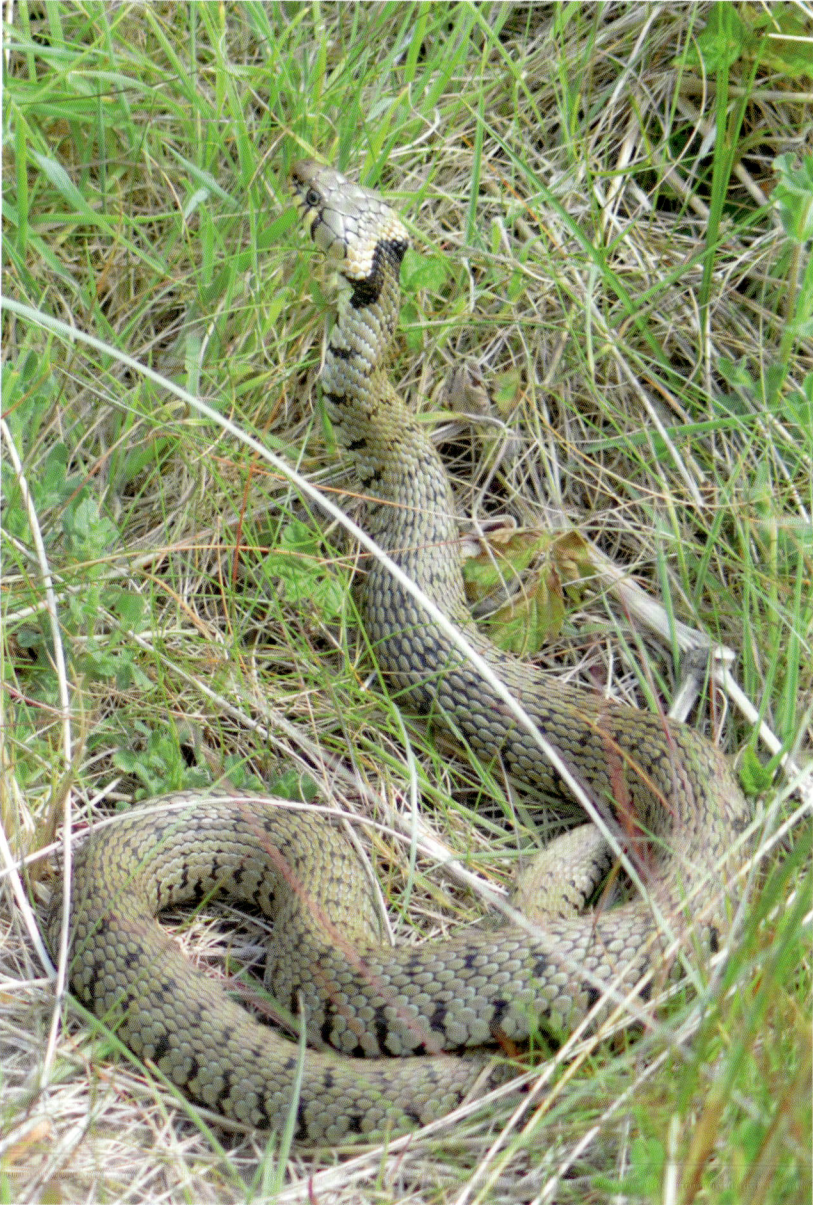

FIG 134. Grass snakes are frequently recorded from sand-dune systems, such as Broomhill and Kilpaison Burrows, along with a variety of other reptiles. (Peter Hill)

long list of other plants has been recorded from this habitat, including small-fruited yellow-sedge *Carex serotina*, knotted pearlwort *Sagina nodosa*, lesser water-plantain *Baldellia ranunculoides*, adder's-tongue *Ophioglossum vulgatum*, black bog-rush *Schoenus nigricans*, variegated horsetail *Equisetum variegatum* and slender spike-rush *Eleocharis uniglumis*.

Another notable species found in the slacks is one of our most distinctive liverworts, petalwort *Petalophyllum ralfsii*. Often compared to a miniature lettuce, it forms green rosettes, or mats, with individual thalli up to 1.5 cm long. It requires moist conditions in winter, but may disappear from May to September, when it is dormant underground, and is often only visible from late October until mid-April. It is sparsely distributed in the United Kingdom, with records from as few as 17 sites (Holyoak 2006). It also occurs on Brownslade Burrows, but there are no records from other likely sites in the county.

Once again there is a rich insect fauna, with the imperial rove beetle *Staphylinus caesareus* and the tortoise beetles *Cassida vibex* and *C. murraea* present, along with two species of bugs, the coastal damselbug *Nabis pseudoferus* and *Pinosomus varius*. Rare moths include scarlet tiger *Callimorpha dominula* and white colon *Sideridis albicolon*. The caterpillars of the latter burrow into the sand in the daytime to hide and so are difficult to find. In the wetter areas, there are two scarce flies *Schoenophilus versutus* and *Aphrosylus ferox*.

Reptiles recorded from the dunes include adder *Vipera berus*, grass snake *Natrix helvetica* (Fig. 134), slow worm and common lizard *Zootoca vivipara*. Palmate newts *Lissotriton helveticus* and toads *Bufo bufo* also breed here. Recently, there were reports in the media that there were two species of grass snake in the United Kingdom. This was, however, incorrect. The suggested change of classification for grass snakes was merely a promotion from subspecies, *Natrix natrix helvetica*, to full species rank (Kindler *et al.* 2017).

Brownslade and Linney Burrows

Immediately south of Broomhill and Kilpaison Burrows, at the western end of the Castlemartin Range, lies the large calcareous dune system of Linney and Brownslade Burrows. Owing to their location in a military danger area, they are amongst the least disturbed dunes in Pembrokeshire, and this is reflected in the richness and variety of species present (Fig. 135). Behind the high dunes fringing the shore the fixed dune grasslands contain uncommon plants such as autumn lady's-tresses *Spiranthes spiralis*. Moonwort *Botrychium lunaria* also occurs here, at its only known location in Pembrokeshire, but the species is easily overlooked and may therefore be under-recorded.

FIG 135. Frainslake Sands, with Brownslade Burrows behind. Together with Linney Burrows to the south, this is one of the least disturbed dune systems in Pembrokeshire. (Jonathan Mullard)

Like Broomhill and Kilpaison Burrows, the dune slacks hold large populations of petalwort, and the Ministry of Defence is currently creating additional areas of open sand to encourage its spread. There is a large population of marsh helleborine, along with several species of marsh-orchids, adder's-tongue fern and variegated horsetail. In places the dune slacks grade into a rich fen vegetation containing blunt-flowered rush *Juncus subnodulosus*, lesser pond-sedge *Carex acutiformis* and great reedmace *Typha latifolia* among other plants.

The dune systems on the Castlemartin peninsula support rich assemblages of bugs and are the United Kingdom stronghold for the shieldbug *Odontoscelis fuliginosa* and the seedbug *Pionosomus varius*. Both species are associated with bare and partially vegetated sand, which supports their preferred foodplants. The ground-dwelling shieldbug forms small, discrete colonies, burrowing in the sand close to the roots of stork's-bill *Erodium* spp. on the open semi-fixed dunes. Once found more widely, it is now restricted to Castlemartin and Sandwich Bay in Kent. In contrast, the seedbug is widely distributed in Europe, where it is thought to be associated with little mouse-ear *Cerastium semidecandrum* (Judd 2006).

FIG 136. Just behind Frainslake beach is Frains Lake Mill, an old mill house that was flooded when the stream was dammed in the twentieth century. The emergent vegetation here is an ideal habitat for hairy dragonfly. (Jonathan Mullard)

The scarce blue-tailed damselfly *Ischnura pumilio* and the hairy dragonfly *Brachytron pratense* breed in the vicinity. As its name suggests, both the males and females of the hairy dragonfly have a downy thorax which, in the male, includes a pair of green stripes on the dorsal surface. It prefers sites with still water and large amounts of emergent vegetation, such as the nearby Frains Lake (Fig. 136).

Strandline beetle

Frainslake beach once supported Pembrokeshire's only known colony of the scarce strandline beetle, or beachcomber, *Eurynebria complanata*, but there have been no sightings since 2013 (Fig. 137). Even then only three individuals were recorded, a tiny number compared with 2000 when 341 beetles were present. While regular monitoring at Frainslake indicates that the number of adults can fluctuate from year to year, the lack of recent sightings is worrying.

In Britain the beetle is restricted to coastal sites around the Bristol Channel, although it is widely distributed along the Atlantic coast of Europe and the Mediterranean. As at Frainslake, it is primarily associated with strandlines on sandy beaches that are backed by coastal dunes (Howe 2016). In Wales it has been recorded from fourteen dune systems, but numbers are declining rapidly. The only other location that still seems to support a reasonable population is Whiteford Burrows on Gower (Mullard 2006). Most of our knowledge of the beetle results from studies at another Gower site, Nicolaston Burrows, between 1964 and 1968, but even here the last record was from 1986.

The adults of the strandline beetle are nocturnal, feeding primarily upon sand-hoppers, mainly *Talitrus saltator*, although they also eat beetle and fly larvae. During the day, they shelter under large debris, such as driftwood. The adult beetles emerge towards the end of May, numbers peaking from June to September. Most overwinter as larvae, moving inland into the dunes. It is thought therefore that the population crash is a consequence of severe winter storms, which have removed their overwintering sites in the foredunes. This theory is supported by the fact that adult beetles have not been found at Frainslake since 3–5 m of the foredunes were lost during the winter of 2013/14. Only three individuals had been seen the previous summer, so this happened at a low point in the population and, on this basis, it is unlikely to recover. Ministry

FIG 137. The strandline beetle, or beachcomber, was once numerous at Frainslake beach but has not been recorded here since 2013. This photograph is of a beetle on Gower in 2006. (Harold Grenfell)

of Defence staff are continuing to monitor the situation, however, and there is a slight chance that some beetles have survived unseen. If severe winter storms are a major factor in the population losses, it may be impossible to prevent the extinction of the strandline beetle in Wales. Climate change is making these events more likely, so even if the population were to recover it could well suffer again. Given its rarity, records of the beetle are very welcome, so if you find it on a beach please note the location and the number of individuals present. Information should be sent to the West Wales Biodiversity Information Centre.

Stackpole Warren

At the eastern end of the Castlemartin peninsula there are two sheltered sandy bays, Broadhaven and Barafundle, together with both open and wooded dune systems, as well as Stackpole Warren, an older dune formation on top of the coastal plateau. The name, of course, is another reference to the presence of rabbits. Edward Donovan (1805) noted that there was a warren here between Stackpole Court and the sea 'consisting of a vast tract of burrows formed by mountains of sand consolidated by that valuable plant morhesg, sea-weed grass; abounding with rabbits, a valuable appendage to a great man's residence'.

FIG 138. The rare cutpurse wasp, a highly specialised spider-hunting wasp that preys on the purse-web spider. It received its common name through a competition in the *Guardian* newspaper in 2012. (Jeremy Early)

The vegetation of the Warren is a mosaic of bracken and species-rich turf. Notably, there are large populations of pyramidal orchid *Anacamptis pyramidalis*, Portland spurge and dune fescue *Vulpia fasiculata*. Disturbed areas support extensive stands of viper's-bugloss *Echium vulgare*, slender thistle *Carduus tenuiflorus* and dwarf spurge *Euphorbia exigua*, alongside yellow-wort *Blackstonia perfoilata* and autumn lady's-tresses.

The sheltered dunes and open habitats of the Warren and Barafundle provide ideal conditions for warmth-loving insects, including the cutpurse wasp *Aporus unicolor* (Fig. 138). This is a highly specialised spider-hunting wasp that preys on the purse-web spider *Atypus affinis*, which it locates and paralyses within the spider's silken burrow. Until relatively recently the wasp did not have a common name, but 'cutpurse' was the overall winner of a competition run by the *Guardian* newspaper and Natural England in 2012. The aim of the competition was to raise awareness of rare species by asking people to invent English names for them. The body shape of the wasp appears to be adapted for gaining entry to the host's nest, with the head and thorax rather elongated, the head flattened and the forelegs enlarged. Once it hatches, the wasp larva eats the paralysed spider and pupates among its remains.

The rare square-jawed sharp-tail bee (or cuckooflower bee) *Coelioxys mandibularis* has been recorded from Stackpole, together with a small and attractively marked ground bug, *Pionosomus varius*, which is covered in bristly dark hairs. The dune systems in Pembrokeshire, and on Gower, are a United Kingdom stronghold for species such as this. Stackpole Warren also supports the only currently known British population of the Stackpole seed-eater *Harpalus melancholicus*, an endangered ground beetle (Fig. 139). While the beetle does not yet have an approved vernacular name in English, or Welsh, the name 'seed-eater' for members of the genus *Harpalus* is reasonably well established. 'Melancholy seed-eater' would be an alternative name, but 'Stackpole seed-eater' accurately describes the current known range of this species.

Prior to 1980, the Stackpole seed-eater was known from a scattering of mostly coastal localities in England and Wales. It was found at Conwy in 1857 and at Tenby in 1894. There were no records from anywhere in Britain between 1964 and July 1992, when the beetle was discovered on Stackpole Warren 'under stones on more or less bare sand with *Thymus* sp., in an area of sand dunes where the limestone bedrock formed an exposed outcrop' (Harrison 1994). A survey in 2017 only recorded adults 'from two small patches of habitat where the toe of a vegetated sand-dune encroaches over outcropping limestone', so it is ecologically highly specialised (Telfer 2018). The survey confirmed that the beetle is a seed-eater, as it appears to feed on lesser hawkbit *Leontodon saxatilis* and probably

FIG 139. Stackpole Warren supports the only currently known British population of the Stackpole seed-eater, an endangered ground beetle. This is a male specimen in the Natural History Museum. The female has a duller body and shorter front legs. (Keita Matsumoto/ Natural History Museum, London)

thyme as well. Much more work remains to be done, however, on feeding preferences. On the whole the ground beetles are about the best-known beetles, after ladybirds, with a good history of published monographs and identification guides, and an active recording scheme – but entomologists are puzzled by the Stackpole seed-eater. There seems to be something very special about the habitat on the Warren that the beetle needs, but it has not yet been possible to work out exactly what this is.

The only other recent British record of Stackpole seed-eater is from Bewl Water, on the border of Kent and East Sussex, where a single adult was found in a light trap on the night of 12/13 August 2003. This may have been a migrant, or a dispersing individual, and efforts to locate a population of the Stackpole seed-eater at this site have gone unrewarded. Stackpole therefore supports the only known British population, so identifying the beetle's status, distribution and ecological requirements here is essential for the conservation of the species. From observations to date, it seems to be a strongly nocturnal species, most sightings being made early in the morning, especially between 05.00 and 06.00. Other ground beetles found at the site during a torchlight search were *Harpalus anxius, Calathus erratus, C. fuscipes, C. cinctus* and *Amara tibialis.*

The beetle fauna also includes glow worm *Lampyris noctiluca* and there are large populations of the great green bush-cricket *Tettigonia viridissima* (Fig. 140). At 40–54 mm long the cricket is one of the largest insects to be found in northern Europe and is known for its extremely loud song, which can be heard for some distance. It can be found along the coast from Marloes to Tenby and its strident song is audible from the early afternoon until well after dark, between mid-July and the first frosts of October (Saunders 1986).

Manorbier

The small dune system at Manorbier is notable for records of the Red Data Book species yarrow broomrape *Phelipanche purpurea*, which was first found here within the fixed dune vegetation, and on the banks alongside the church wall, in 1991. Unfortunately, it has not been seen here for several years. Flowers can reappear

FIG 140. Large populations of the great green bush-cricket occur on Stackpole Warren. It prefers the light, dry soils found here, into which the females lay their eggs using their long ovipositors. (Bob Haycock)

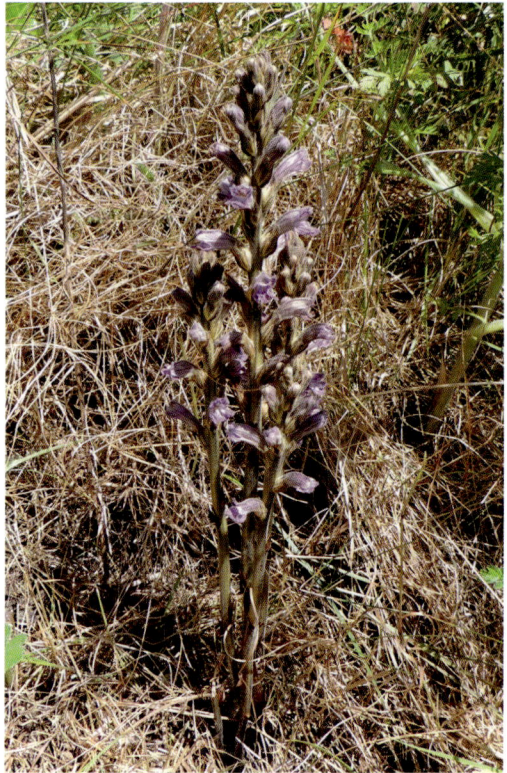

FIG 141. Yarrow broomrape has a very localised distribution and can reappear after decades of absence. Apart from the site at Manorbier it has also been recorded near Tenby, where this plant was photographed. (Stephen Evans)

after decades of absence, however, implying that the seeds are long-lived, or that the species can persist without flowering for many years. Yarrow broomrape is an attractive plant, with dark purple flowers and a pinkish stem (Fig. 141). The flowering spike can reach up to 30 cm in height, but is usually much shorter. There are three bracts surrounding each flower, which separate it from other broomrapes. Despite the abundance of its host plant, yarrow *Achillea millefolium*, yarrow broomrape has a very localised distribution. It almost certainly depends upon root damage, or other stress factors, in the host plant. Rabbit grazing, grass mowing and drought are the main factors here.

The vegetation of the fixed dunes is dominated by red fescue with restharrow and lady's bedstraw *Galium verum*. On the foredunes and strandlines there is the usual mix of marram, sand couch *Elymus junceiformis*, sand sedge and sea rocket *Cakile maritima*.

Tenby Burrows

Since the beginning of the nineteenth century, the dunes at Tenby have spread north, closing off the mouth of Tenby Backwater. Once, they were famous for the abundance of the white-flowered burnet rose, but the creation of the golf course in 1888 reduced it to a few isolated patches, a faint relic of the spectacular sight that once greeted visitors to the area. Tenby was the 'birthplace' of Welsh golf, and landscaping, along with drainage works, has severely affected the amount of suitable habitat for native plants. Instead there are large stands of the introduced sea-buckthorn *Hippophae rhamnoides*, which was planted to stabilise the dunes, and the usual botanically uninteresting mown grassland. Many of the dune slacks are also dry and dominated by creeping willow *Salix repens*, but occasionally bog pimpernel *Lysimachia tenella*, water mint *Mentha aquatica* and glaucous sedge *Carex flacca* occur.

Nearly a hundred years ago the situation was very different. Dune gentian *Gentianella amarella* ssp. *occidentalis* was first found 'in a damp sandy pasture around Tenby' (a dune slack on the Burrows) by Pugsley in September 1923 (Pugsley 1924). The first British record of this small purple-flowered plant was from Braunton Burrows in Devon, some time before 1849, but Tenby Burrows has always been the only recorded locality for the species in Pembrokeshire (Rich & McVeigh 2019). Before 1946 it was reported as abundant in some years, but unfortunately it was last recorded in 1999, scrub having overgrown the slack where it occurred. Dune gentian is a British endemic, confined to dune slacks in North Devon and South Wales, the epithet *occidentalis*, meaning western, referring to its occurrence in the west of Britain. Plants are typically found around the margins of open dune slacks, where seeds are washed by the higher water levels in winter, or on paths and low hummocks. Dune gentian has now been reassessed as an annual subspecies of autumn felwort *Gentianella amarella*. Autumn felwort itself is uncommon in Pembrokeshire and shares the coastal grassland at Stackpole with the only Welsh population of early gentian *Gentianella amarella* ssp. *anglica* numbering, in good years, several hundred plants. The field gentian *Gentianella campestris*, one of the most rapidly declining species in the British flora (Wilson 2017), with its much larger flowers, can also be abundant on the Castlemartin Range.

Sand-dune fungi

Fungi are essential to the creation and survival of dune systems, facilitating colonisation by plants, both by the physical action of their hyphae binding particles of sand and as mycorrhizal symbionts of the major dune-building grasses such as marram and, later in the succession, dune shrubs. Wales is a stronghold for dune fungi (Evans & Roberts 2015), the late Maurice Rotheroe, known for his pioneering work on sand-dune fungi, noting in 1993 that:

> *There can be few other ecosystems in Wales which provide so rich a list of rare or unusual macrofungal species as do coastal sand dunes … Welsh sand-dune habitats are a refuge for rare and unusual species.*

Apart from the rarities, Rotheroe pointed out that dunes in Wales contain fungi typical of old hay meadows, or grazed, unimproved, grassland, large areas of which have been lost to agricultural intensification (Rotheroe 1993). Like the soft cliffs described earlier, they are therefore refuges, as well as ecosystems in their own right. Many of the species to be found in sand dunes are macrofungi, with fruiting bodies easily visible to the naked eye, and they include a number of specialist dune species that are rarely, if ever, found in other habitats. Mycorrhizal microfungi, however, are equally important to dune ecosystems, but since they are not immediately visible they are seldom recorded, even though they dominate the strandline and foredunes.

The first visible macrofungi typically appear with marram grass in the mobile dunes and include a number of agarics that feed on the dead grass. Among these are found species such as dune inkcap *Coprinopsis ammophilae*, dune cavalier *Melanoleuca cinereifolia*, dune brittlestem *Psathyrella ammophilae* and marram oyster *Hohenbuehelia culmicola*. The last of these is an almost sessile (barely stalked) yellow-brown oyster mushroom that grows on the base of the marram grass. It is now regarded as two distinct species, *H. culmicola* and *H. bonii*, which have the same, or very similar, habitat preferences (Ainsworth *et al.* 2016). Both species can occur on the same site, making identification difficult for the non-specialist.

Further inland, the semi-fixed dunes support larger numbers of other fungal species, including *Agaricus devoniensis*, the winter stalkball *Tulostoma brumale* and the scaly stalkball *T. melanocyclum*. In the older texts, the stalkballs are commonly referred to as 'stilt puffballs', and it is now known that puffballs and stalkballs are closely related. The greatest diversity though is found in the dune slacks, where symbiotic macrofungi form associations with plants such as creeping willow. Typical fungi found here include the dune stinkhorn *Phallus hadriani*, which in Britain is found only in sand dunes. It closely resembles the common stinkhorn *P. impudicus*, which also occurs frequently in dunes, but differs in having a pink 'egg' (Spooner & Roberts 2005).

In the fixed dunes, the grassland often supports species such as the dune waxcap *Hygrocybe conicoides*, the limestone waxcap *H. calciphila* and the dung

FIG 142. Until 2005 the only British records of the dung bird's nest fungus were from sand dunes in Wales, and there are still very few records from elsewhere. (David Harries)

bird's nest *Cyathus stercoreus* (Fig. 142). Bird's nest fungi, so called because of their resemblance to a miniature nest containing eggs, usually obtain their nutrients from decaying vegetation, in this case weathered rabbit droppings. Up until 2005 the only British records were from sand dunes in Wales, even though it is common and widespread elsewhere in Europe. The nests, properly known as 'splash cups', use the force of falling drops of water to dislodge and disperse the fungal spores.

Dune scrub may contain macrofungi associated with birch and willow, as well as earthstars *Geastrum* spp. and saddle fungi *Helvella* spp., while dune woodlands on sandy soil, including plantations, are often rich in *Russula* spp. and other agarics. These species, and others, are illustrated in a useful booklet on sand-dune fungi published by the Pembrokeshire Fungus Recording Network (Harries *et al.* 2015).

CHAPTER 10

Waterways and Estuaries

*It is a rich, fertile, and plentiful Country, lying on the Seacoast, where it has
the Benefit of Milford-Haven, one of the greatest and best Ports in Britain. Mr.
Camden says it contains 16 Creeks, 5 great Bays, and 13 good Roads for Shipping,
all distinguished as such by their Names; and some say, a thousand Sail of Ships
may ride in it.*

Daniel Defoe (1742)

GEORGE OWEN CLAIMED THAT MILFORD Haven was the most famous
port in Christendom, while Nelson considered it one of the greatest
natural harbours in the world (McKay 1989). Currently, it is Britain's
third-largest port and the biggest in Wales (Fig. 143). Famous for centuries, it
is the setting for some of the action in Shakespeare's *Cymbeline*. Imogen, on
receiving a letter from her husband Posthumus telling her that he is 'in Cambria,
at Milford-Haven', asks 'how far it is to this same blessed Milford: and by the
way tell me how Wales was made so happy as to inherit such a haven' (Act III,
Scene 2). Geology then was in its infancy, but, as described in Chapter 3, we
now know that Wales was 'made so happy' by the drowning of the valleys of
the Cleddau rivers when sea levels rose during the last glacial period, around
20,000 years ago. Milford Haven's other claim to fame is that Henry Tudor, Earl
of Richmond, launched his invasion of England from here in 1485, landing at
Mill Bay near St Ann's Head. He had been born at Pembroke Castle on 28 January
1457, and used his Welsh ancestry to gather support before the battle with
Richard III at Bosworth Field, which led to him becoming king.

FIG 143. An aerial view of Milford Haven, the largest ria-estuary complex in the United Kingdom, covering an area of 55 km². One of the greatest natural harbours in the world, it has been famous for centuries. (Mark Richards/Aurora Imaging)

The Milford Haven Waterway is the largest ria-estuary complex in the United Kingdom, covering 55 km², 30 per cent of which is intertidal. As such, it dwarfs Solva harbour, the only other large ria in Pembrokeshire. The water depth ranges from 28 m in the main entrance channel, off St Ann's Head, to 6 m at the junction of the Western and Eastern Cleddau (Carey et al. 2015). The deep, sinuous and steep-sided main tidal channel and tributary estuaries are characteristic of a drowned river valley and reflect the underlying geology and the major east–west rock faults and folds. There are also several large embayments, such as Angle Bay, and numerous creeks, or pills, on its shores. Virtually all of the land area of Pembrokeshire lies within the catchment of the waterway, and this has implications for water quality, as discussed later in this section.

The area has been industrialised, in some form, for centuries. Excavations at South Hook, Herbrandston, prior to the construction of a liquid natural gas storage facility, revealed an early medieval settlement, with an associated iron smelting site and crop processing complex (Crane & Murphy 2010). Substantial changes to the movement of sediment in the waterway, and its contamination, only began, however, with the use of steam power in the mining and shipping industries in the nineteenth century. Alterations have included the construction of docks and shipyards and the dredging of channels. The bombing of the Royal Navy's fuel-oil storage depot at Pembroke Dock during the Second World War, in August 1940, started the largest 'single-seat' fire in the United Kingdom's history (Scott 1980). It released an unknown volume of fuel oil into the waterway and the underlying ground, traces of which are still emerging in groundwater (Galperin & Little 2014). Fishing was another major activity, and by 1906 Milford Haven was the sixth-largest fishing port in Britain, with over 500 people working either in the industry itself or in related trades. The fishing fleet continued to thrive throughout the first half of the twentieth century, only declining when fish stocks in the Atlantic began to reduce noticeably in the 1950s. Home to a twice-daily ferry service to Ireland, Pembroke Port accommodates a variety of general cargo and heavy lift vessels (Port of Milford Haven 2016). The waterway is also busy with recreational activities such as sailing, birdwatching, diving and fishing.

Although the significance of Milford Haven as an oil port has declined since the 1970s, it still handles around 29 per cent of the United Kingdom's seaborne trade in oil and gas. The oil tankers once presented the greatest threat to the flora and fauna of the area. Indeed, there was a serious oil spill from the *Esso Portsmouth*, the first oil tanker to unload there in 1960. The *Sea Empress* oil spill, which occurred in February 1996 at the entrance to the Haven, is, however, the largest accident to date. With the loss of 72,000 tons of crude oil, it was, at the time, Britain's third-biggest oil spillage and had a significant effect on the area's wildlife. The damage, however, was less than initially forecast, since at the time of the accident many migratory birds had not returned to breed, and by 2001 marine species had mostly recovered. Thankfully there have been no further incidents on this scale. The *Sea Empress* was a single-hulled tanker, and these are now being phased out. New tankers are built with double hulls, which greatly reduces the likelihood of leakage, although there are additional risks associated with double-hulled vessels, such as corrosion in the space between the hulls. Shipping management has improved greatly though since these major spills and oil pollution has, as a consequence, reduced dramatically over the years. That does not mean, however, that the danger has completely disappeared, since smaller spills still have the capacity to cause significant damage. As recently as January 2019 there was a leak from pipework on a jetty near the Valero oil refinery which resulted in booms being used to protect the saltmarshes around Sandy Haven and the Gann estuary and a clean-up operation being put in place (Malloy 2019).

There is another threat to native wildlife from the shipping, however, as it results in the introduction of non-native species. As listed in Table 3, the majority of these are seaweeds, which almost certainly arrived on the hulls of ships (Jones & Whitmore 2015). A total of 35 non-native species were recorded in Milford Haven between 1953 and 2015. Water quality, particularly hyper-nutrification, is also an issue, as is the presence of heavy metals and various persistent organic pollutants (Blaise Bullimore, personal communication).

The SWEPT project (Surveying the Waterway Environment for Pollution Threats) used citizen-science volunteers to collect information on water quality within the area between November 2018 and February 2019. Extensive slurry spreading within the catchment is increasing nitrogen levels in the waterway, resulting in excessive algal growth and green mudflats. Slurry is viewed as a waste product by livestock farmers, and there is currently no legislation in place to address the issue, only a Code of Practice. Natural Resources Wales routinely collect data on water quality but information on localised pollution sources is lacking. The data produced by volunteers therefore supplement the statutory monitoring process and assist in tackling the issue.

Like the seas around Skomer, Milford Haven is well studied. The waterway has attracted naturalists for many decades, and many aspects of its marine biology have been thoroughly and repeatedly described. The initial researches and surveys were carried out by Swansea University in the 1950s and later by Dale Fort Field Centre, which published the *Dale Fort Marine Fauna* in 1966 (Crothers 1966). The presence of the Field Studies Council's Oil Pollution Research Unit from 1967 to 1999, which carried out several studies as part of the South West Britain Sublittoral Survey in the late 1970s, helped to ensure that the area's profile

TABLE 3. Non-native species recorded between 1953 and 2015 within the Milford Haven Waterway. Based on Jones & Whitmore 2015.

Anotrichium furcellatum	Red seaweed
Amphibalanus improvisus	Barnacle
Antithamnionella spirographidis	Red seaweed
Antithamnionella ternifolia	Red seaweed
Aplidium cf. glabrum	Sea squirt
Asterocarpa humilis	Sea squirt
Austrominius modestus	Barnacle
Botrylloides cf. diegensis	Sea squirt
Botrylloides violaceus	Sea squirt
Bugula neritina	Bryozoan
Bugula simplex	Bryozoan
Bugula stolonifera	Bryozoan
Caprella mutica	Skeleton shrimp
Codium fragile fragile	Green seaweed
Colpomenia peregrina	Brown seaweed
Corella eumyota	Sea squirt
Corophium sextonae	A mud shrimp
Crassostrea gigas	Pacific oyster
Crepidula fornicata	Slipper limpet
Feldmannophycus okamurae	Red seaweed
Ficopomatus enigmaticus	Tube worm
Grateloupia turuturu	Red seaweed
Haliplanella lineata	A striped anemone
Mya arenaria	Soft shelled clam
Mytilicola intestinalis	Parasitic copepod
Mytilus galloprovincialis	Bivalve mussel
Perophora japonica	Sea squirt
Polysiphonia harveyi	Red seaweed
Potamopyrgus antipodarum	Mud snail
Sargassum muticum	Brown seaweed
Solieria chordalis	Red seaweed
Spartina anglica	Common cord-grass
Styela clava	Sea squirt
Tricellaria inoptiata	Bryozoan
Undaria pinnatifida	Brown seaweed

remained high (Baker 1987, Dicks 1987). Initially the majority of the surveys focused on ornithology, rocky shores and sediment macrobenthos. Subtidal areas only accessible by diving, however, have been surveyed since the 1960s when one of the earliest-documented systematic SCUBA surveys was carried out in the Daugleddau (Bailey *et al*. 1967). Until recently, most of the diving surveys were focused in the upper parts of the Haven, or at a small number of locations, the only comprehensive diving survey being over thirty years old (Little & Hiscock 1987). Conditions in the waterway and estuaries are challenging, and careful dive planning is needed. Poor underwater visibility is common, but the richness of the waterway's marine wildlife makes the effort worthwhile. Since Milford Haven is frequently turbid it is generally only used by divers as a fall-back option when the weather is too bad to go elsewhere, so photography is often difficult.

The coexistence of its extremely varied habitats and species with the diversity, and intensity, of human activity results in an ongoing need for the long-term monitoring of potential human impacts and environmental change (Bullimore 2013). The Milford Haven Waterway Environmental Surveillance Group, formed in 1992, plays a coordinating role and provides funding for environmental surveys, such as those carried out by Seasearch – a volunteer project for both recreational and professional divers. Thirty Seasearch survey days were completed in Milford Haven between 2004 and 2015, involving 104 volunteers who completed 287 survey forms for 43 site areas. The report summarising this activity is well worth reading for a detailed account of the species recorded by the project in this important area (Lock & Bullimore 2018).

HABITATS

The habitats in Milford Haven and its tributaries range from subtidal and foreshore rock through mixed sediments of every kind, to soft muds in the embayments and tributary estuaries. High-salinity water and rocky substrates penetrate far upstream, supporting communities characteristic of fully saline conditions in sheltered waters. A wide range of subtidal and intertidal rocky habitats are present, from rocky reefs and boulders to biologically rich under-boulders, crevices, overhangs and pools. The multiplicity of habitats supports a correspondingly wide range of plant and animal communities. Particularly interesting are the seagrass and maerl beds, both of which provide microhabitats for a wide variety of other species,

Milford Haven can be divided into five separate sections: the outer waterway, which lies seaward of a line between South Hook and Thorn Point; the central waterway downstream of Cosheston and Barnlake Points; secondary bays such as West Angle Bay and Sandy Haven; the Daugleddau estuary and the tributary estuaries that drain into it, the Eastern and Western Cleddau and the Carew and Creswell Rivers; and the smaller rivers, notably the Garron, Cosheston and Sandy Haven Pills, the Gann and the Pembroke River. The whole complex forms part of the Pembrokeshire Marine Special Area of Conservation, while the shore is covered by the Milford Haven Waterway Site of Special Scientific Interest.

OUTER AND CENTRAL WATERWAY

The coastline within the waterway is extremely varied, with broad rocky shores, large sandy beaches in the bays near the entrance, and mudflats and saltmarshes in the more sheltered areas. Plants found on the lower saltmarsh include common cord-grass *Spartina anglica*, common saltmarsh-grass *Puccinellia maritima* and red fescue, along with sea-purslane. These grade into upper saltmarsh communities containing sea rush *Juncus maritimus* and saltmarsh rush *J. gerardii*.

FIG 144. The nationally scarce lax-flowered sea-lavender is found on the upper parts of the saltmarsh in the Milford Haven Waterway. (Stephen Evans)

In some areas there is a transition zone from the upper saltmarsh into areas of reedbed dominated by common reed *Phragmites australis*. Species found on the upper saltmarsh include the nationally scarce lax-flowered sea-lavender *Limonium humile* (Fig. 144) and one-flowered glasswort *Salicornia disarticulata* (Prosser & Wallace 2003). This latter species is an annual, reproducing exclusively by seed. Mature branches, containing seeds, becoming detached from the plant and distributed by the tide. The saltmarsh also supports a number of scarce invertebrates, including comb-footed spider *Enoplognatha mordex*, a ground beetle *Bembidion laterale*, short-winged mould beetle *Brachygluta simplex*, two weevils *Polydrusus pulchellus* and *Notaris bimaculatus*, a hoverfly *Platycheirus immarginatus* and a cranefly *Limonia complicata*.

Rocky reefs and shipwrecks

One of the special, and unusual, characteristics of Milford Haven is that it includes large areas of rocky reefs rich in sea squirts and sponges, and these extend far inland into the Daugleddau. There are also numerous smaller reefs in the middle of the outer waterway, such as the Mid-Channel and Chapel Rocks complex and Stack Rocks. As reefs are predominately a feature of exposed coasts, the variety of reefs in the sheltered conditions of the Milford Haven Waterway is exceptional (Countryside Council for Wales 2009). The diversity of the wildlife found on them varies according to the nature and type of the rock present and is strongly influenced by a number of physical characteristics, in particular, the degree of exposure to tidal currents. Strong tidal streams often increase species diversity, although some communities do require still conditions. The reefs are particularly rich in densely packed and massive growths of sponges, especially shredded carrot sponge *Amphilectus fucorum*, mermaid's glove *Haliclona oculata* and breadcrumb sponge.

It is not possible in the space available to describe all the reefs in detail, but there is, for instance, an extensive area of reefs between Rat and Sheep Islands on the eastern side of the entrance to the Haven. Sites immediately south of Rat Island have ridges of bedrock 1–2 m high, with cobbles in the intervening gullies. There is a diverse range of sponges here, with a total of fifteen species recorded. Notable records include the staghorn sponge *Axinella dissimilis*, brain sponge *A. damicornis*, prawncracker sponge *A. infundibiformis* and mashed-potato sponge. Sea squirts include the red sea squirt *Didemnum pseudofulgens*, which has only been recorded from a small number of sites in Wales, the pinhead sea squirts *Pynoclavella aurilucens* and *P. stolonialis* and the non-native leathery sea squirt *Styela clava* (Lock & Bullimore 2018). The latter is thought to have been transported on the hulls of warships following the end of the Korean War in 1953.

A survey of the Martello reef to the west of Pembroke Dockyard by Seasearch in 2007, in an area of high tidal flows, found a steep vertical wall covered in anemones, sponges, hydroids and sea squirts (Lock 2007). The dominant anemones here were the plumose anemone *Metridium senile* and the elegant anemone *Sagartia elegans*. Other species present included the solitary sea squirts *Ascidiella aspersa* and *Ciona intestinalis*, breadcrumb sponge and the shredded carrot sponge. Conversely, on the Dakotian East reef in Dale Roads the rocks were lightly silted and sparsely covered in red algae and encrusting bryozoans, including the potato crisp bryozoan *Pentapora foliacea* and the orange pumice bryozoan *Cellepora pumicosa*. Reefs often support a diverse range of territorial fish, but the most notable feature here was the presence of large numbers of the long-clawed squat lobster *Munida rugosa*, a northern species not often seen in England and Wales.

There are a number of shipwrecks in the outer waterway which act as artificial reefs. They support both schooling fish, such as pollack *Pollachius pollachius* and

FIG 145. A school of pouting on the wreck of the *Dakotian*. This was a merchant ship sunk by a mine while dropping anchor in Milford Haven on 21 November 1940. (Blaise Bullimore)

FIG 146. Oaten pipes hydroid on a wreck near Stack Rock. The backscatter from suspended particles highlights the poor underwater visibility in Milford Haven. (Blaise Bullimore)

pouting *Trisopterus luscus* (Fig. 145), and a variety of territorial fish which use the wreckage as a refuge. While upward-facing steel plates are frequently covered in thick silt, some support dense stands of red algae, or the finger bryozoan *Alcyonidium diaphanum*. In contrast, steep, vertical and overhanging surfaces are often home to the oaten pipes hydroid *Tubularia indivisa* (Fig. 146) and dead man's fingers. Initial visits to the wreck of the *Lochshiel*, which is located near Thorn Island, have found that the silt-covered plates support Devonshire cup-corals and reasonably large numbers of fish including ballan wrasse *Labrus bergylta*, pollack and leopard spotted goby *Thorogobius ephippiatus*.

Other wrecks, such as the inverted remains of a Second World War landing craft between Angle Bay and Thorn Island, are also covered in a rich variety of attached marine life. Here plumose anemones *Metridium dianthus* and feather stars *Antedon bifida* are abundant along with elegant anemones, antenna hydroid *Nemertesia antennina* and numerous painted balloon nudibranchs *Eubranchus tricolor*. Common bryozoans include *Bugula plumosa*, *Scrupocellaria* sp. and white claw sea moss *Crisia* sp. Nine species of sea squirt have been recorded on the wreck, including orange sheath tunicate *Botrylloides violaceus* and, again, the non-native leathery sea squirt.

Oyster and slipper

Until the end of the nineteenth century there was a thriving oyster *Ostrea edulis* fishery in south Pembrokeshire, in the Haven and around Caldey Island, Tenby and Stackpole. George Owen tells us that the oysters from Milford Haven, 'most delicate and of several sorts', were transported by sea to Bristol and the Forest of Dean. The principal oyster beds were those at Lawrenny, Llangwm, Castle Pill and Pennar Mouth, those of Lawrenny and Llangwm being considered the fattest and sweetest. The oysters were dredged 'with a kinde of iron' which was:

> dragged at a boates end by two rowers which rowe up and down the channel, and so the bagge of leather being made apte to scrape up all manor of things, lyeing in the bottome, gathereth up all the oysters that breede there over certaine knowne beds.

As elsewhere around the Welsh coast, overfishing led to a steep decline in numbers. By the time Edward Donovan recorded the situation in 1805 it seems that the Caldey oyster fishery was in a bad way, the beds being nearly exhausted. He suggested that the fishermen should be encouraged to replenish the Caldey beds with spat from Stackpole, the beds there apparently being still well stocked. There are many relict beds of dead shells, as a reminder of the native oyster's heyday. Today, the Haven is the only known location for live oyster beds in Pembrokeshire, with individuals scattered throughout the central waterway (Lock 2011).

There are moves to increase the oyster population and enable the development of 'biogenic reefs' to improve water quality and enhance existing ecosystems. In due course, once a viable population is re-established, a sustainable fishery might be possible. The first diving survey to assess the distribution and abundance of the species was carried out in 2002 and there have been a number of subsequent surveys, the latest in March 2017. As oysters are regarded as a 'sensitive species' by Natural Resources Wales the results of these surveys are confidential (Lock 2017). The oysters do though face competition for food and space from the non-native slipper limpet *Crepidula fornicata* (Fig. 147).

The first known occurrence of slipper limpet in Europe was in 1872 in Liverpool Bay, but the modern British population originates from an introduction to Essex between 1887 and 1890 in association with eastern oysters *Crassostrea virginica* imported from North America (Fretter & Graham 1981). Although slipper limpets can tolerate a wide range of environmental conditions, populations are particularly well developed in wave-protected areas such as bays, estuaries or sheltered sides of wave-exposed islands (Blanchard 1997). First recorded in Milford Haven in the 1960s, it is typically found attached to shells

FIG 147. First recorded in Milford Haven in the 1960s, the slipper limpet is an invasive non-native species which competes with other filter-feeding invertebrates for food and space. It tends to live in low-visibility, turbid and silty environments. (Blaise Bullimore)

and stones on soft substrates. Often occurring in curved chains, with up to twelve animals stacked on top of each other, and forming dense aggregations covering the seabed, it leaves little room for other species.

The slipper limpet's success is probably due to a lack of predators and its unusual method of reproduction, which relies on individuals settling upon each other and being assisted by their close proximity. It competes with other filter-feeding invertebrates, and in waters with high concentrations of suspended material, such as Milford Haven, the dense aggregations encourage the deposition of mud, owing to the accumulation of faeces and pseudofaeces (Barnes *et al.* 1973).

Seagrass

Another defining feature of Milford Haven is the extensive area of seagrass, sometimes known as eelgrass. Seagrasses are the only truly marine flowering plants found in British waters, producing seed while entirely submerged under the water. There are two species, common seagrass *Zostera marina* and dwarf seagrass *Z. noltii*. It used to be thought that there was a third species, narrow-leafed seagrass *Z.*

angustifolia, but this is now considered to be just a different growth form of common seagrass. The larger form typically occurs at, and below, the low-water spring tide level of relatively sheltered sandy bays, whereas the *angustifolia* form is more often found in shallow pools on muddy shores at low tide. Dwarf seagrass occurs at three sites within the Haven: Angle Bay, Sandy Haven and Pembroke River. The extensive beds within the Pembroke River are the largest in Pembrokeshire.

Both common seagrass and dwarf seagrass are an important habitat, supporting a wide range of other species and acting as nursery areas for fish. A total of eleven commercially important fish species have been recorded, as juveniles, from seagrass meadows in Wales, including cod *Gadus morhua*, pollack, plaice *Pleuronectes platessa* and bass (Unsworth *et al.* 2015). Resident fish include greater pipefish *Syngnathus acus*. In turn, these communities are an important source of food for other marine animals and birds, including grazing wigeon *Mareca penelope*. Seagrass meadows also absorb and store large amounts of nutrients, helping to keep coastal waters clean.

Over 80 per cent of the seagrass beds in Britain have been lost due to pollution, coastal development and other factors (Unsworth 2015). Particular losses are thought to have occurred in Milford Haven and along the South Wales coast (Kay 1998). A study of the environmental health of seagrass meadows around the British Isles highlighted specific sites of concern due to high nitrogen values, including Gelliswick Bay in Milford Haven and North Haven on Skomer (Jones & Unsworth 2016), the latter being the only site in the county where boat anchoring and other damaging activities are prohibited. These two seagrass beds, growing in a nutrient-rich environment, may also suffer from a lack of light in the turbid waters. With increasing sea temperatures due to climate change, it is imperative that seagrass beds have sufficient light to enable them to grow properly. The study suggested, however, that many seagrass meadows are potentially already close to their light thresholds, and that the resilience of these systems is being severely reduced. In relation to this, recent data shows that although seagrass beds in Pembrokeshire have not reduced in size in recent years, the density of shoots has decreased in a number of areas. As a result, the so-called common seagrass is classified as a nationally scarce plant in Wales. There is some good news, however, as new beds have established themselves at Picton Point. So, if the pressures were reduced, there is every indication that the seagrass could recover and, indeed, spread.

Fan mussel

In October 2018 it was reported that divers from Natural Resources Wales had found a fan mussel *Atrina fragilis* while monitoring a seagrass bed in Milford Haven (NRW 2018). Targeted searches had previously only uncovered a few broken shells, so this is currently the only known living example in Wales (Fig. 148).

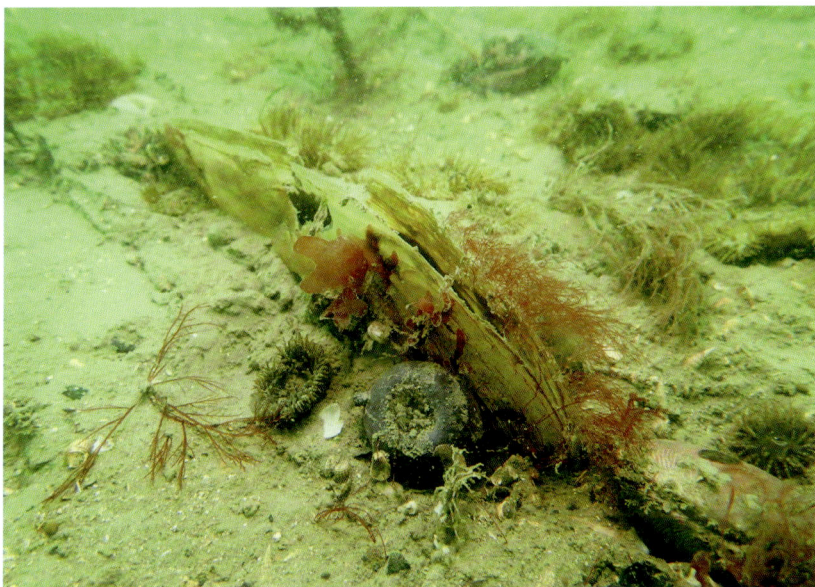

FIG 148. In October 2018 divers from Natural Resources Wales found a fan mussel while monitoring a seagrass bed in Milford Haven. This individual is currently the only known living member of its species in Wales. (Matthew Green/Seasearch)

The species was last found alive in Welsh waters in the 1960s. In the United Kingdom, the fan mussel is found mostly in the southwest of England and in western and northern Scotland. Its range extends, however, as far as southwest Spain and Portugal and the Adriatic Sea. The fan mussel is one of the largest bivalve molluscs in Britain; it can grow to over 45 cm in length, and may be ten or twelve years old. Despite its large size, the triangular shell is very fragile, and it is considered one of our most threatened molluscs.

One end of the fan mussel's shell is pointed, and this is embedded in sediment, or attached to small stones by fine byssal threads, between a third and two-thirds of the shell being buried. The broad end of the shell protrudes and may support growths of sea anemones, barnacles, hydroids and sponges. It is therefore very susceptible to physical damage from mobile fishing gear, anchors and mooring chains, and there is some damage to the upper shell of this specimen. Fan mussels used to occur in beds containing large numbers of animals, but recent records, as in this case, involve only single individuals or small groups. They are recorded from fewer places than was the case in the past, and once their populations are reduced it is hard for them to recover – because without sufficient numbers of other animals nearby, their eggs cannot be fertilised effectively. This declining population has also affected other species, since, as mentioned above, fan shells provide a point of attachment for many marine organisms, creating reef-like microhabitats in the sediments in which they live.

Maerl beds

Maerl is a Breton word used to describe accumulations of unattached living and dead coralline algae (Lemoine 1910). It develops when crust-forming coralline red algae, impregnated with calcium carbonate, become free-living due to fragmentation. Maerl beds have considerable conservation value because of the very high diversity of the associated organisms, some of which are more or less confined to the maerl habitat. Maerl beds can be extremely old, with some accumulations off the coast of Brittany estimated by radiocarbon dating to be over 5,500 years old (Grall & Hall-Spencer 2003). Owing to its very slow growth rate, it is considered to be a non-renewable resource (Barbera *et al.* 2003). The main maerl-forming species are the coral maerls *Lithothamnion corallioides* and *L. glaciale* and coral strands *Phymatolithon calcareum*.

In Wales, small areas of maerl have been recorded around the Pembrokeshire islands and the Llŷn peninsula, but the most extensive beds, formed by coral strands, are found on the north side of Milford Haven (Birkett *et al.* 1998). Even here though the maerl only covers around 1.5 km², of which only 0.5 km² contains live maerl (Bunker & Camplin 2007). It is an ancient bed, a sample of fossil maerl from the site having an estimated date of between 84 BCE and 12 CE by radiocarbon dating (Blake 2005). It is formed on waves of sandy, or shelly, sediment, with the majority of the live maerl occurring on ridges while the troughs act as collecting areas for dead shell and coarse sediment. Although it has existed for at least 2,000 years and is the only living maerl bed known in Wales, the site has undoubtably been affected by the industrialisation of the area, particularly over the last fifty years. It was, for example, damaged as a result of refurbishments to the South Hook Liquid Natural Gas jetty. An 83 per cent decline in live maerl was recorded here between 2005 and 2010 (Bunker 2011).

Many coralline algae produce chemicals which promote the settlement of the larvae of certain herbivorous invertebrates. The herbivores then graze off the epiphytic, and often fast-growing, algae which might otherwise overgrow the coralline algae, competing for light and nutrients. Two rare algae, endemic to maerl, occur in Milford Haven, *Gelidiella calcicola* and *Cruoria cruoriaeformis*, while other uncommon algae found here include *Ptilothamnion sphaericum* and *Spermothamnion strictum*.

WEST ANGLE BAY

Careful searching in the rock pools in West Angle Bay will reveal a very small cushion star, *Asterina phylactica* (Fig. 149). The bay is the 'type locality' for the species, since it was first found here by Roland Emson and Robin Crump, then head of the Orielton Field Centre, in 1978 (Emson and Crump 1979). Until that date there was thought to be a single species, the common cushion star, or gibbous starlet, *A. gibbosa*, on the low shore and in the rock pools of Britain. While it closely resembles the common cushion star, the second species can be separated by its small size, up to 1.5 cm across, and distinctive colour pattern, which consists of a red-brown star on a dark green background. Its behaviour during reproduction is also different, since the young are brooded under the ventral surface of the parent until they are mobile. The common cushion star, in contrast, merely sheds its eggs into the sea where they develop into planktonic

FIG 149. The small cushion star was first found in the rock pools in West Angle Bay by Roland Emson and Robin Crump in 1978. The species is probably under-recorded, rather than particularly rare, as it is small and very cryptic. (Blaise Bullimore)

larvae. At first, Emson and Crump regarded the animals they found as merely a colour variant of the common species, but large numbers of the brooding form were later found, and further studies led to the conclusion that it was actually a separate species (Crump and Emson 1983). Both animals are microphagous feeders, everting their stomach through the mouth over the rock surface and digesting the film of bacteria and diatoms.

A short-lived animal, surviving only two years, the small cushion star is a male in its first year and both male and female in its second year. From early spring onwards, bright orange eggs can be seen inside the animal, and in late May it seeks out a crevice among boulders or the holdfasts of seaweeds to lay them. About two weeks before spawning, individuals form into groups of between five and ten individuals. They brood the eggs by lying on top of them, remaining together as a group for a period of three weeks until the eggs hatch. The juveniles then crawl out from under the parent, which subsequently dies. It nearly became extinct in West Angle due to oil pollution from the *Sea Empress* disaster in 1996. A year afterwards only five individuals could be found at the site, but after ten years the population had increased to over 1,000, due partly to its ability to self-fertilise and protect its young. The species is now also known from sites in Devon, Cornwall, Anglesey, Ireland, Scotland and the northern Mediterranean, so its survival seems assured.

SANDY HAVEN

The artist Graham Sutherland called the small inlet of Sandy Haven 'an estuary of exultant strangeness', being fascinated by the gnarled scrub oaks growing on the cliffs of the western shore (Fig. 150). Sandy Haven Pill, which is the dominant freshwater influence on the area is reduced to a trickle at low tide. There used to be a ferry across the tidal creek, but now it can only be crossed at low tide by a footbridge, known as the 'crab bridge', since it has been used by generations of children as a convenient location for collecting shore crabs *Carcinus maenas*.

For birdwatchers, Sandy Haven is probably the best place in the county to see wintering spotted redshanks *Tringa erythropus* and greenshanks *T. nebularia* (Green & Roberts 2004). Spotted redshanks are a relatively scarce wintering species in the United Kingdom, with more than half the population found at fewer than ten sites, making them an Amber List species. In their summer plumage the adults are almost entirely black, except for the white spots on their wings, which give them their name, a white 'wedge' on the back and a barred tail. In winter though they have a grey back and paler underparts, with a more

FIG 150. The artist Graham Sutherland was attracted to Sandy Haven Pill and the gnarled oaks growing along its edge. As a result, many of his paintings feature the inlet, which is reputed to be the best place in Pembrokeshire to see spotted redshanks and greenshanks in the winter. (Jonathan Mullard)

prominent eye-stripe and a longer and finer bill than common redshanks *Tringa totanus*. Sandy Haven is also a good site for Mediterranean gull and kingfisher *Alcedo atthis*. It is estimated that there are around fifty pairs of kingfishers in Pembrokeshire – but, surprisingly for such a recognisable bird, the true numbers are unknown, as there have been no systematic counts.

DAUGLEDDAU ESTUARY

The Daugleddau forms the uppermost part of the Milford Haven Waterway, immediately southeast of Haverfordwest. It includes the lower, tidal reaches of the Eastern and Western Cleddau, as well as the Millin, Sprinkle, Westfield and Garron Pills. Underwater the upper reaches of the Daugleddau are spectacular, with large areas of the bedrock and shell, or cobble, substrates covered with large numbers of anemones, sponges and sea squirts. Shallow rock communities at Pembroke Ferry, east of the Cleddau Bridge, for example, typically include breadcrumb sponge, shredded carrot sponge, mermaid's glove, the branching sponge *Raspailia hispida*, goosebump sponge *Dysidea fragilis*, coral worm *Salmacina dysteri* and finger bryozoan. At Castle Rocks, near Benton Castle, there is a rugged rocky reef, with steep faces up to 3 m in height, covered in breadcrumb sponge, another species that is also known as mermaid's glove *Isodictya palmata* and oaten pipes hydroid. There are dense areas of hydroids, including *Obelia* species and anemones, including orange anemone *Diadumene cincta*, horseman anemone *Urticina eques*, dahlia anemone *U. felina* and elegant anemone. Areas of cobble and pebble are unfortunately also carpeted with the slipper limpets.

The lower shore on the upper reaches of the Daugleddau and Pills consists mainly of extensive mudflats characterised by polychaete worms, such as the common ragworm *Hediste diversicolor* and catworms *Nephtys* spp. Other species present include peppery furrow shell *Scrobicularia plana*, mud snails *Hydrobia* spp. and cockle *Cerastoderma edule*. In places the mud is mixed with muddy gravels, cobbles and shingle. Species associated with the muddy gravel include sand mason worm *Lanice conchilega*, keel worms *Pomatoceros* spp. and daisy anemone *Cereus pedunculatus*. Areas of consolidated pebbles and shingle provide a suitable substrate for mussel *Mytilus edulis* beds, along with species such as edible periwinkle *Littorina littorea* and barnacles. To the south of the Daugleddau's main channel, areas of bedrock, boulder and mixed sediments, characterised by oarweed and serrated wrack, replace the mudflats on the lower shore. Where these communities are more exposed to tidal water movements it results in greater numbers of animal species, including breadcrumb sponge, gooseberry sea squirt *Dendrodoa grossularia* and sea mats *Bryozoa* spp.

Thin bands of saltmarsh, characterised by common cord-grass, dominate the upper shore in places. Sea-purslane and glasswort exist in very small patches, or in a mosaic with common cord-grass, but they are most often found on the banksides of the numerous winding creeks that occur throughout the site. Stands of sea club-rush *Bolboschoenus maritimus* are a feature of the rivers near their tidal limits and there are also areas of reedbed dominated by common reed.

WILDFOWL AND WADERS

Owing to the sheltered location and the availability of open mudflats, the Milford Haven Waterway and Daugleddau/Cleddau estuary complex provide an important wintering ground for wildfowl and waders. Counts of the birds here have been carried out annually, between September and March, since 1982/83, initially as part of the Birds of Estuaries Enquiry and Wildfowl Counts, and latterly as part of the

FIG 151. Milford Haven Waterway and the Daugleddau/Cleddau estuary complex is also of national importance for its populations of wintering birds, including wigeon. (Annie Haycock)

national Wetland Birds Survey. Since 1993, these counts have been incorporated into the annual programme coordinated by the Milford Haven Waterway Environmental Surveillance Group. Annual summer shelduck *Tadorna tadorna* surveys have been carried out since 1992. The Grey herons *Ardea cinerea* breeding alongside the estuary are also counted in most years (Haycock 2016).

As a result of these counts, the complex is considered to be of international importance since it regularly holds an average of over 20,000 birds, of around sixty species, each winter. Although this appears to be a recent phenomenon, a closer look at the figures shows that gulls, grebes, divers and herons were not included in the counts until the mid-1990s. Adding the 4,000–5,000 individuals of these species to the earlier totals suggests that a large number of birds has been using the area for many years. The area is of national importance for its populations of wintering wigeon (Fig. 151), teal *Anas crecca* and greenshank and for migrating curlew *Numenius arquata*.[*] In addition to the wintering and migratory birds, the area supports a small summer breeding population of shelduck and other wetland birds.

---

[*] A wetland in Britain is considered nationally important if it regularly holds 1 per cent, or more, of the estimated British population of one species, or subspecies, of birds.

FIG 152. A Grey heron preening its feathers. The total breeding population in the county is estimated at anything between forty and seventy pairs. (Annie Haycock)

The number of birds present can vary greatly between years, and within a single winter. Most of the changes in bird populations here are reflected at other sites in Wales, or the United Kingdom as a whole. Large numbers of birds are more likely to visit during periods of extreme weather, while in milder weather they prefer the conditions on the east coast of England where there is generally a better food supply. The Daugleddau/Cleddau complex though continues to be of national importance for wintering and migrating wetland birds, and it is important that the undisturbed feeding areas and high tide roosts are maintained.

Of the nationally important birds, there are nearly 6,000 wigeon present each winter. Their diet consists of common seagrass, algae and grasses gathered on the mudflats and saltings. Once the wigeon have eaten the leaves and seeds of the seagrass they generally move out of the estuary system. They have been observed eating common cord-grass seeds, but there is not the area of saltmarsh in Pembrokeshire to sustain large winter populations. One of their main locations in Britain is on the Somerset Levels and many of the birds appear to go there when their food runs out in Pembrokeshire.

Numbers of teal using the area have fluctuated over the past decade, but with more than 2,300 birds present the species is just over the threshold for national importance. A recent run of mild winters, however, and an increase in protected areas on mainland Europe have enabled larger numbers of birds to remain further north and east.

The greenshank found in Britain are a very small proportion of those using the East Atlantic flyway, as shown by counts of over 10,000 individuals at some Dutch Wadden Sea staging posts. Small numbers have always remained on the Cleddau complex for the winter, and until 2005/06 the estuary system ranked about fifth in Britain for wintering greenshank. As the total numbers wintering in the country have increased, the area has dropped down the list, but it still remains in the top twenty. It is by far the most important Welsh site for the species, even though winter counts only average around 20–30 birds.

Mid-winter peak counts of curlew have similarly decreased, with a five-year average count of 644 birds. Although the species no longer reaches levels of national importance during the mid-winter period, it does exceed the threshold (1,400) during autumn migration and the area remains an important migratory stopover. Up to 500 curlews have been recorded from other sites in the county, notably the Nyfer estuary in the north, and Castlemartin Corse and Freshwater West in the south. The decline in wintering curlew since the early 2000s may be associated with a climate-induced northeasterly shift in distribution which has coincided with stable or increasing wintering numbers in the Wadden Sea. Another factor may be that curlew have been protected from hunting in Denmark since 1994.

The population of Grey herons seems to be increasing in all parts of the United Kingdom, coupled with a general increase in wintering birds, except in Wales for some reason, where there has been a slow decline since 2001/02. Herons currently breed at two main sites in the complex, Eastwood and Upton, with around 10–20 occupied nests each year in the 1990s (Fig. 152). Previously they were also recorded nesting at Slebech, where they have been known since at least 1894, but in the 1990s some birds began to move to a new site a few kilometres away at Eastwood. By 1996, there were no nests at Slebech. Eastwood has no public access, and is not counted regularly, but herons are apparently still in residence there. Three other small heronries are known to be regularly used in the county, together with a few sites used intermittently by one or two pairs. Counts of herons are sporadic though and no conclusions can be drawn about the breeding population. Murray Mathew recorded that 'Although there are no large heronries in the county, there are numerous small breeding stations, and the bird is generally distributed and fairly common.' This still seems be the case today.

Although it might seem strange to us now, as mentioned in Chapter 2, George Owen recorded herons nesting on the cliffs, as did Mathew (1894):

> There are some twenty to thirty pair of Cormorants about Lydstep Head, near Tenby, as Mr. C. Jefferys informs us; a colony nests on trees at Slebech; and some Herons that nested at Poyntz Castle on St. Bride's Bay, were driven from their nests upon the cliffs by Cormorants, who took possession of them for themselves … The ejected Herons are stated to have migrated to Slebech, where they have formed a heronry. Mr. Tracy mentions another at Linney Head, where the Herons nest in company with Cormorants and Guillemots. The nests, from six to twelve in number, are arranged side by side on the ledges of the rocks, and are quite inaccessible.

In his *British Birds* article on the pioneering 1928 survey of active heron nests, Nicholson (1929) notes that 'cliff sites exist at Solva and Porthllisky, and an extinct one at Linney Head'.

Another heron, the little egret *Egretta garzetta*, has become common in recent years, and the species is now a familiar visitor to many estuaries in South Wales during the winter, with a distinct influx in late July and early August. Little egrets were considered to be rare vagrants to Britain until 1989, when the first major movements of birds occurred. Since then there has been a dramatic increase in numbers thought to be linked to breeding success at French colonies. Since the mid-1990s little egrets have also been present in small numbers in the Waterway and the Daugleddau during the spring and summer (Rees *et al.* 2009). The first sign of the species attempting to breed was in March 1995 when a pair was seen displaying in a heronry in Pembroke. The following year three pairs were observed displaying at the same site, two of which appeared to have built nests, but these attempts were apparently unsuccessful (Green 2002). It was not until 2003 that the first breeding record was confirmed on the Daugleddau (Green & Roberts 2004).

Although some shelduck do breed along the estuary, most of the adults leave in March, either to breed elsewhere or to return to their mass moulting grounds – the closest sites to Pembrokeshire being Bridgwater Bay in Somerset and, more recently, off the Dee and Mersey estuaries. A recovery one January of an adult ringed in Germany the previous August suggests that at least some birds travel to the Wadden Sea to moult.

During the last few years the number of Canada geese *Branta canadensis* breeding in the upper reaches of the waterway has increased dramatically. Over the winter period their numbers practically double as geese from other areas come in to take advantage of the conditions. Large numbers of these birds cause erosion, increased nutrient levels and overgrazing of the saltmarsh, bankside vegetation and neighbouring fields, while also disturbing migratory waterfowl. Further research is needed to determine the long-term impact that the expanding population of geese is having on other bird species.

EASTERN AND WESTERN CLEDDAU

The Eastern and Western Cleddau flow south from their sources, at the foot of Mynydd Preseli and near Mathry respectively, through a lowland agricultural landscape. There are significant areas of marshy grassland, wet heath, blanket bog, fen and willow scrub along their lengths which add to the diversity of species found here. Unusually, both rivers cut across the structural orientation of the underlying rocks, which range from the Precambrian to the Silurian, while their nineteen main tributaries follow the orientation, or are controlled by structural features, such as faults and folds. 'Cleddau' is said to be derived from *cleddyf*, meaning sword, which alludes to the way they cut into the landscape (Owen & Morgan 2007).

Both rivers support a rich in-stream flora. The exact mix of species varies along the watercourses and their tributaries, but there are extensive beds of submerged aquatic plants, often dominated by various species of water-crowfoot *Ranunculus* spp. Typical species found here include stream water-crowfoot *R. penicillatus* ssp. *penicillatus*, alternate water-milfoil *Myriophyllum alterniflorum*, common water-starwort *Callitriche stagnalis*, intermediate water-starwort *C. brutia* ssp. *hamulata*, the liverworts endive pellia *Pellia endiviifolia* and great scented liverwort *Conocephalum conicum*, brook-side feather-moss *Amblystegium fluviatile* and greater water-moss *Fontinalis antipyretica*. The slower-flowing headwaters support water-pepper *Persicaria hydropiper* and extensive beds of common reed.

On the Eastern Cleddau (Fig. 153) the nationally rare hybrid pondweed *Potamogeton* × *rivularis*, a cross between a narrow-leaved species, small pondweed *P. berchtoldii*, and a broad-leaved species, bog pondweed *P. polygonifolius*, has been recorded north of Canaston Bridge at its only known site in the United Kingdom. The plant was discovered by the botanists R. V. Lansdown and T. J. Pankhurst in 2002. It grows in shallow, rapidly flowing water over a stony substrate alongside intermediate water-starwort, alternate water-milfoil and water-crowfoots. In places there are numerous scattered patches which are just under 1 m². Otherwise this hybrid is known only from sites in France. Rather than representing an occasional and inconsequential aspect of plant behaviour, hybridisation is a fundamental feature of plant evolution and one of the main

FIG 153. The lower reaches of the Eastern Cleddau viewed from Landshipping. (Jonathan Mullard)

ways by which new species evolve, so it will be interesting to monitor the progress of the plant over the coming years (Stace *et al.* 2015).

Emergent species along the edge of the river channel typically include hemlock water-dropwort *Oenanthe crocata*, branched bur-reed *Sparganium erectum* and unbranched bur-reed *S. emersum*, reed canary-grass *Phalaris arundinacea*, floating sweet-grass *Glyceria fluitans* and water-pepper. Hemlock water-dropwort is perhaps the most poisonous indigenous plant in Britain. The lower stem is usually thick and below this there are clusters of fleshy tubers that give rise to its popular name 'dead man's fingers'. The whole plant is poisonous, and there have been a number of fatalities over the years (Down *et al.* 2002). On the densely wooded stretches of the river, emergent vegetation is scarce and the banks are dominated instead by great wood-rush *Luzula sylvatica* and ferns, often with large patches of bluebells.

Two rare invertebrates can be found in the Western Cleddau, the cased caddis *Ylodes simulans*, and an ageing population of the nationally scarce freshwater pearl mussel *Margaritifera margaritifera*, which is now recorded from only nine Welsh rivers. Caddis larvae, living in a case they have made from small particles of debris, are some of the most familiar freshwater insects. They are thought to derive their name from Elizabethan street hawkers called 'caddice men' who sold braid and had samples stitched to their coats (Wallace 2003). The larvae of *Ylodes simulans* are found among river vegetation, and apparently overwinter at an early instar stage.

PEARL MUSSELS

There is historical evidence of large populations of freshwater pearl mussels in Welsh rivers, especially the Conway and Wye, as well as records from Pembrokeshire, Carmarthenshire, Cardiganshire, Glamorganshire, Denbighshire, Snowdonia, and some tributaries of the River Severn (Joint Nature Conservation Committee 2007). George Owen wrote that the 'chief rivers' in Pembrokeshire for pearl mussels were the Taf and Nyfer, but today they are only recorded from the Western Cleddau and the Afon Brynberian. Interestingly, Owen states that they are rarely found:

> *The river mussels are not for meat, but are chiefly taken for the pearls that are found in them, the fish being great and long, of seven or eight inches, and are so rank they are rejected for the meat, and of the country people termed for their bigness, horse mussels. They are but rarely found, and in most of them are found pearls, some one, some two, some three in a piece, some four and orient but most commonly cornered and dark, which makes them of less account.*

The main destruction of freshwater mussel beds in Wales is, however, generally thought to have occurred between 1926 and 1936. During this period a group of Scottish pearl fishers visited during July and August, when the water was low and fishing was relatively easy. They travelled widely and covered the rivers in rotation, retaining the services of an advocate to challenge any cases against them in the courts to prevent them fishing, but the outcome of the action was often immaterial. By the time judgment was given sufficient time had elapsed to allow them to 'clean up' the river in question. (Jones 1973) As a result few mussels remain today, and it is estimated that there has been a 90 per cent decline in Welsh populations in the last hundred years. Although pearl fishing is now illegal, the level of the populations at most, if not all, of the sites still reflect the impact of this historic activity. There is recent evidence, however, to suggest that pearl fishing is still occurring at a low level in practically every British and Irish river (Young *et al.* 2000). This illegal activity will have a serious effect on the small remaining populations.

The freshwater pearl mussel is one of the longest-lived invertebrates known, and individuals living in colder northern rivers can be up to 120 years old. They live buried, or partly buried, in coarse sand and fine gravel in clean, fast-flowing rivers and streams, drawing in water through their exposed siphons to filter out the minute organic particles on which they feed. In locations where the species was formerly abundant, it is possible that this filtration helped to clean the river water, benefiting juvenile salmon and trout (Ziuganov *et al.* 1994).

Mussels today grow up to 140 mm in length, just under six inches, which is smaller than the seven or eight inches quoted by Owen. The larval, or glochidial, stage is spent attached to the gills of salmonid fishes, thereby allowing the mussels to move upstream. Each female ejects between one and four million glochidia in a sudden, highly synchronised event, usually over 1–2 days. Almost all the glochidia die, but a few are inhaled by juvenile salmon or trout. At this stage they resemble tiny mussels, but their shells are held apart until they encounter a suitable host, when they snap shut onto the host's gill filaments.

Larvae attach themselves during mid to late summer and drop off the fish in the following spring to settle in the riverbed gravel. They then spend between five and seven years living within the gravel, before emerging as filter-feeding mussels. The huge losses involved in this unusual life cycle make the freshwater pearl mussel particularly vulnerable to adverse conditions. Declines have been caused not only by pearl fishing. Acidification, siltation and declining salmonid stocks have also taken their toll, and it is now a rare species throughout its range. A number of organisations are now working together to tackle pollution in the rivers and moving mussels to hatcheries, with the aim of creating viable breeding stocks and replenishing the rivers.

OTTERS AND FISH

The rivers in the Milford Haven complex are one of the best locations in Britain for otters *Lutra lutra*, the dense riverside vegetation providing excellent cover for their holts, or dens. Where there is no suitable vegetation, however, otters have been found using the roots of bankside trees, stone-filled gabions, or even piles of flood debris. They feed on the small fish, particularly salmonids and eels *Anguilla anguilla*, which are to be found in abundance along the rivers, the numerous tributaries with coarse substrates, shallow gradients and riffles providing an excellent habitat for them. Lampreys are widely distributed across the system, with significant populations of brook lamprey *Lampetra planeri*, river lamprey *L. fluviatilis* and, in the lower reaches, sea lamprey *Petromyzon marinus.* Lampreys have similar requirements to trout and salmon for spawning; namely clean gravels with a through flow of water. After hatching, the larvae drift downstream until they reach beds of silt and sand. Here they live as blind, wormlike, filter-feeders for several years before metamorphosing into eel-like juveniles. River and sea lampreys then migrate downstream to estuarine waters and the sea, where they spend several years before returning to the river to spawn, whereas the brook lamprey lives its whole life in freshwater.

Other fish present include bullhead *Cottus gobio*, three-spined stickleback *Gasterosteus aculeatus*, minnow *Phoxinus phoxinus* and stone loach *Noemachilius barbatulus*. Stone loaches are bottom-feeders, sensing the presence of food, such as mayfly larvae and freshwater shrimps, by means of the sensitive barbels around the mouth. The Cleddau are locally important salmon rivers with small populations of both Atlantic salmon *Salmo salar* and sea trout (sewin) *S. trutta*. Non-migratory brown trout *S. trutta* are also common.

Although a marine species, certain stocks of herring spawn in estuarine areas (Ellis *et al.* 2012). The fish that spawned in the upper reaches of the Daugleddau estuary were exploited by people from Llangwm and Hook, using 5 m open rowing boats and compass nets. The names of the nearby settlement of Black Tar, and Blacktar Point, record the substance the boats were painted with to waterproof them. Llangwm was renowned throughout Pembrokeshire for its fisherwomen, who walked many miles, dressed in their traditional costume, selling cockles, oysters and prawns, as well as herring. The fishery continued into the twentieth century, until trawling further downstream brought it to an end.

The compass net fishery at Little Milford on the Eastern Cleddau still survives, but now only six licences are issued each season. On a falling tide small boats are moored on ropes stretched across the river and, using an A-shaped frame suspended in the flow, the netsmen wait for salmon and sewin to swim into their nets (Fig. 154). This traditional method of netting is the last surviving example of its form in England and Wales. It was apparently introduced to the area by two Gloucestershire miners who were working in the coal mines around the estuary. At the fishery's peak in the nineteenth century around a hundred compass nets were present. Today, due to the decline in migratory fish, the activity is closely controlled by Net Limitation Orders and byelaws designed to conserve stocks. Salmon stocks across the Atlantic are affected by a number of factors, including marine survival and barriers to migration. The reduction in the survival rate of salmon in the northeast Atlantic means that the number of smolts leaving rivers now will produce far fewer returning adults than would have been the case in the 1980s. As a result, the abundance of salmon has reduced by around 50 per cent over the last forty to fifty years (Cefas/EA/NRW 2017).

FIG 154. The compass net fishery that operates at Little Milford on the Eastern Cleddau is the last surviving example of this traditional fishing technique in England and Wales. (Godfrey Thornberry)

Apart from the compass net fishery, there is little traditional net fishing left in the Pembrokeshire rivers, apart from a single seine net which is still operational on the Nyfer estuary. This involves a boat laying a net with floats at the top and weights at the bottom, in a crescent shape. The nets used are quite large and require a team of five to pull them in.

MINOR ESTUARIES

Among the smaller rivers entering the Daugleddau and Milford Haven, the estuaries of the Gann, the Pembroke River and the Carew and Cresswell Rivers have been the most studied, particularly the Gann.

The Gann

The shingle spit, tidal lagoon and lower reaches of the Gann estuary have been a favourite location for generations of botanists and ornithologists (Fig. 155). Articles such as 'The Gann Flat, Dale: studies on the ecology of a muddy beach' (Bassindale

& Clark 1960), 'Feeding patterns of wading birds on the Gann Flat and river estuary at Dale' (Edington *et al.* 1973) and 'The Gann Flat, Dale; thirty years on' (Edwards *et al.* 1992) in the journal of the Field Studies Council testify to the continuous use of the area by students visiting Dale Fort Field Centre. Indeed, divided sedge *Carex divisa* was first recorded by the old limekilns here in 1982 by Juliet Brodie, then one of the staff, but now employed by the Natural History Museum in London. A perennial sedge of coastal grassland, it was originally known from only one other location in Pembrokeshire, on Skokholm (S. Evans *et al.* 2012). Its stronghold is in southeast England, especially the Thames and Solent estuaries.

The shingle ridge at the mouth of the estuary, known as Pickleridge, has its origins, like other similar structures across the county, in the huge deposits of sand and gravel which built up at the southern edge of the Irish Sea ice sheet about 17,000 years ago. It supports plants such as spotted medick *Medicago arabica*, musk stork's-bill *Erodium moschatum*, Duke of Argyll's tea plant *Lycium barbarum*, chamomile *Chamaemelum nobile* and rough clover *Trifolium scabrum*. The tea tree was introduced, as its name suggests, by the Duke of Argyll in the 1730s as a hedging shrub, especially in coastal districts. Often naturalised close to settlements, its red berries are particularly attractive to birds, but probably poisonous to humans.

FIG 155. The Gann estuary can only be crossed on the Coast Path, using the boardwalk in the foreground, three and a half hours each side of low tide. (Jonathan Mullard)

The lagoon beside the ridge is one of three saline lagoons in the Haven, all of which are artificial. The others are the old millpond formed by the damming of the Sageston Pill, a tributary of the Carew River, within the grounds of Carew Castle, and a pond with a weir at Westfield Pill near Neyland. Saline lagoons are a rare and unusual habitat, especially in Britain, supporting a number of characteristic species that are rarely found elsewhere. Species found in the three sites include the tentacled lagoon worm *Alkmaria romijni*, the crustacean *Gammarus chevreuxi* and a thin-shelled variety of lagoon cockle *Cerastoderma glaucum* that is only found in brackish lagoon habitats. The adults of the tiny lagoon worm, less than 5 mm long, live in tubes made of mud, the tops of which protrude above the surface of the sediment. Previously thought to be a lagoon specialist, it has recently been identified as a brackish-water species. It has also been recorded from estuaries, lagoons and other transitional water bodies in the German Bight, Denmark, the Baltic and Portugal, where it can reach high densities (Arndt 1989, Gilliland & Sanderson 2000). Due to its small size and the need for specialist identification it is probably under-recorded in the United Kingdom and may be more widespread than the current records suggest. All three species require salinity levels between 10 and 35 per cent, and these conditions enable the lagoon to support other marine invertebrates, such as mud shrimp *Corophium volutator* and the common ragworm.

The beach, known as the Gann Flat, is one of the best examples of sheltered muddy gravel sediment in Wales, but it is used severely affected by digging for ragworms to use as fishing bait. Here it is the king ragworm *Alitta virens* that is sought, as it is considered the best live bait for targeting cod, whiting *Merlangius merlangus*, pollack, wrasse and flatfish. Other activities include the collection of shore crabs by turning boulders and, in some places, by placing artificial habitats on the shore to act as a refuge for the crabs. Because of this pressure Natural Resources Wales has, after a series of consultations, implemented a zoning plan restricting bait digging to certain areas, to protect wildlife and allow the habitat to recover (Bean & Appleby 2014). While the upper muddy/gravel shore had the highest levels of bait digging it was the lower, sandier, shore that supported the highest number of species, including peacock worms *Sabella pavonina* and sand mason worms, so keeping bait diggers away from the lower shore was the best option.

All these invertebrates are a rich food source for birds, so it is no surprise that the Gann is one of Pembrokeshire's best birdwatching sites. During the summer shelduck, oystercatchers, snipe, redshank and greenshank are often seen, and in autumn whimbrel and hundreds of migrant finches and pipits visit the area. Winter bird highlights include the large group of several hundred curlew that roost here. Rarer species such as curlew sandpiper *Calidris ferruginea* also occur from time to time.

Pembroke River

The Pembroke River rises near Manorbier Newton, passing through Lamphey and flowing by Pembroke Castle on the way to its estuary at Pennar Mouth, on the south side of the Haven. In Pembroke what was once a tidal creek is now two ponds, due to the construction of another tidal mill in the thirteenth century and then a dam below the castle in 1975 (Fig. 156). At times the pond around the base of Pembroke Castle supports large numbers of mute swans *Cygnus olor* and black-headed gulls *Chroicocephalus ridibundus*. Small numbers of diving ducks winter here, mainly tufted ducks *Aythya fuligula* and pochard *A. ferina*. There are no records of the tentacled lagoon worm, though it may be present here, as it is at Carew.

The upper pond is more secluded, and is managed as a nature reserve by the Wildlife Trust. The vegetation here represents a transition from estuarine to freshwater conditions. Lesser duckweed *Lemna minor* and horned pondweed *Zannichellia palustris* occur in the open water, with the emergent vegetation dominated by common reed. Estuarine species present include sea club-rush, saltmarsh rush and sea arrowgrass *Triglochin maritima*. At the eastern end of the site is an area of fen and carr vegetation dominated by grey willow and alder,

FIG 156. The lower pond around the base of Pembroke Castle was created by the building of a nearby dam in 1975. It supports large numbers of mute swans and black-headed gulls. (Jonathan Mullard)

with an understorey of yellow iris *Iris pseudacorus*, hemp-agrimony *Eupatorium cannabinum*, purple-loosestrife *Lythrum salicaria* and water forget-me-not *Myosotis scorpioides*. There is a wide variety of aquatic invertebrates present, including water louse *Asellus aquaticus*, common ram's horn snail *Planorbis planorbis* and Jenkins's spire snail *Potamopyrgus antipodarum*. The last of these originates from New Zealand, from where it was introduced into Australia. It was then carried accidentally to the United Kingdom in drinking-water barrels on ships. The snails were probably released while washing or refilling the barrels and, because they can survive in brackish water, they have no problem in estuarine areas such as Milford Haven (Ponder 1988). They also thrive in freshwater, and the spire snail has since become the most common freshwater gastropod in Britain.

Little grebe *Tachybaptus ruficollis*, grey heron, mute swan, mallard *Anas platyrhynchos*, teal, moorhen *Gallinula chloropus*, cormorant, coot *Fulica atra*, kingfisher, tufted duck and pochard occur throughout the year. In autumn up to fifty black-tailed godwits *Limosa limosa* can be seen along with over thirty greenshanks. Common sandpipers *Actitis hypoleucos* and green sandpipers *Tringa ochropus* are regular passage visitors. Bitterns *Botaurus stellaris* have been recorded during the winter.

FIG 157. Carew Tidal Mill is the only restored tidal mill in Wales. The millpond behind it supports species such as the tentacled lagoon worm and lagoon cockle. (Jonathan Mullard)

FIG 158. Looking downstream from Cresswell Quay on the Cresswell River, a tributary of the Eastern Cleddau, at low tide, with Scotland Wood on the northern bank. (Jonathan Mullard)

Carew and Cresswell Rivers

In common with the other rivers in the Milford Haven complex, the Carew and Cresswell Rivers support overwintering birds between September and March, and a breeding population of shelduck during the summer months. Up to 200 wintering dunlin roost at high tide along the shoreline, with other waders such as redshank, greenshank and green and common sandpipers commonly recorded.

The main feature of the upper reaches of Carew River, apart from the large medieval castle, is the adjoining millpond, mentioned earlier, which results from the damming of the tidal reaches of the Sageston Pill. Water trapped at high tide would be released through a sluice to drive the mill wheel (Fig. 157). J. M. W. Turner's painting of Carew Castle around 1832 shows a shallow waterway with cattle standing in the water and a group of women kneeling on the shore, so presumably this was at low tide after the stored water had been used. As in the lagoon behind Pickleridge, species found here in the millpond include the tentacled lagoon worm and lagoon cockle. The current four-storey mill building dates from the nineteenth century and the machinery is intact, though no longer used.

At a promontory on the confluence of the Carew and Cresswell Rivers lies West Williamston, another nature reserve managed by the Wildlife Trust. The saltmarsh here is deeply indented with tidal creeks which were excavated as loading bays for the nearby limestone quarries in the eighteenth century. Many of the creeks are now infilled with sediment and developing as saltmarsh. On the muddy shoreline marsh-mallow *Althaea officinalis* is present at one of its few locations in Pembrokeshire. Large numbers of waders and wildfowl frequent the site, including curlew, little grebe, grey heron, shelduck, mute swan, cormorant and oystercatcher. Cresswell Quay, at the tidal limit of the Cresswell River, was a busy port in the eighteenth and early nineteenth centuries, coal from the many small mines in the area being shipped to Lawrenny further downstream to be loaded onto larger boats. On the steep valley sides there are large areas of deciduous woodland (Fig. 158).

Outside Milford Haven and its extensive tributaries there are a number of other small rivers, which are linked to a variety of important wetlands. These are the subject of the next chapter.

CHAPTER 11

Rivers and Wetlands

In this place, where I purpose to entreat of chief rivers in general, I have determined to speak of none but such rivers and brooks as keep their course and name until they fall into the sea and run at least the two miles, briefly mentioning such rills and brooks as they receive in the way.

George Owen (1603)

T HE MAIN PEMBROKESHIRE RIVERS ARE the Eastern and Western Cleddau, as described in the previous chapter, but there are numerous other watercourses. Some, such as the Gwaun and Nyfer, lie entirely within the county, while the Taf, though rising in Mynydd Preseli, flows through Carmarthenshire and into Carmarthen Bay, and the Teifi originates in the Cambrian mountains. The Nyfer and the Gwaun, like the Cleddau, are typical Welsh rivers: steep, fast-flowing and shallow in their upper reaches and slower, deeper and more meandering towards the tidal limit. In contrast, the coastal river catchments of the Alun, Solva, Ritec, Cresswell, Castlemartin Corse, Westfield Pill and Gann Flats are dominated by surface-water flows, with rapid changes in levels shortly after heavy rain (Natural Resources Wales 2014).

Gerald of Wales may have described the River Alun, which rises near Tretio Common and flows past St Davids Cathedral to enter the sea at Porthclais, as 'a muddy and unproductive rivulet', but watercourses like this once powered woollen mills and other machinery. In 1900 there were 26 woollen mills in

FIG 159. Melin Tregwynt in St Nicholas. The waterwheel here is, unusually, inside the building and now for display only, but once it powered machinery to produce blankets and other woollen items. (Jonathan Mullard)

Pembrokeshire and the county was the centre of woollen production in Wales. There are now only two working woollen mills left: Middle Mill on the River Solva and Melin Tregwynt in St Nicholas (Fig. 159), neither of which still uses the water that flows past.

In addition to running water, there are two large reservoirs in the county, Rosebush and Llys-y-frân, along with numerous lakes, pools, ponds, wells, fens and marshes. All support a rich variety of wildlife.

AFON TEIFI

The mouth of the River Tuerobis (the Teifi) like St Davids Head, is one of the few places on the Welsh coast named in Ptolemy's *Geography*. A short distance downstream from Cenarth Falls in Carmarthenshire the Teifi becomes tidal, and for the rest of its journey to the sea it forms the boundary between Ceredigion and Pembrokeshire. Celebrated for its salmon and sewin (sea trout) fishing, it is considered to be one of the most beautiful rivers in Wales. Around Cenarth there is still a small traditional coracle fishery that exploits the salmon and brown trout, but few coracles are seen these days in Pembrokeshire. The Teifi is a large river, flowing over hard rock, and there are some spectacular gorges in the lower reaches, not least the steep-sided Cilgerran Gorge overlooked by the thirteenth-century castle (Fig. 160). West of Cardigan and St Dogmaels, the river broadens into a wide estuary, with Poppit Sands on its west bank as it enters Cardigan Bay.

Like most of these rivers, in its upper reaches the Afon Teifi is a 'flashy' high-energy watercourse, and as a result the aquatic vegetation is dominated by stream water-crowfoot, the water-starworts *Callitriche hamulata* and *C. obtusangula* and the aquatic moss *Fontinalis squamosa*. Reed canary-grass is a characteristic bankside plant.

As well as salmon and sewin the river supports large populations of brook lamprey and river lamprey. Grayling *Thymallus thymallus* and the rare allis shad *Alosa alosa* have also been reported, and bullheads occur in good numbers. Fish that spawn in these fast-flowing rivers all tend to lay small numbers of eggs and

FIG 160. The incised meanders of Cilgerran Gorge have long been a renowned beauty spot. Samuel Lewis, in 1833, describing the 'sylvan beauties of the scene ... rich groves, alternating with the naked rock, continue to excite the admiration of the traveller.' This is a view from the ramparts of Cilgerran Castle. (Jonathan Mullard)

look after them assiduously. Salmon, for instance, deposit their eggs in excavated nests or 'redds', while bullheads lay eggs under stones, with the male guarding the nest until the fry have hatched. Similarly, lampreys work in pairs, sometimes having the assistance of a second female, to dig out a redd in the stony riverbed, usually about a metre in diameter and about 15 cm deep, where the eggs are deposited. This is in contrast to fish species that occupy lower-energy watercourses, which deposit large numbers of eggs with a reduced degree of care.

As mentioned in Chapter 2, Gerald of Wales noted in the twelfth century that there were beavers on the Afon Teifi:

> The Teify, of all the rivers in Wales and those in England south of the Humber, is the only river where you can find beavers. In Scotland, or so they tell me, there is again only one stream where beavers live, and even there they are exceedingly rare.

Similarly, Daniel Defoe, on his tour through Wales, recorded in 1742 that in Cardigan memories of the beaver were still strong:

> The Country People told us, that they had formerly Beavers here, which bred in the Lakes among the Mountains, and coming down the Stream of Tyvy, destroyed the young Frye of Salmon, and therefore the Country People destroyed them. We thought they only meant the Otter, till I found afterwards, that Mr. Camden mentions also, that there had been Beavers seen here formerly.

Beavers are unaggressive and easy to hunt, and by the fifteenth century it is likely that the species was extinct in Wales. They were sought after for their fur, meat and castoreum (produced by glands at the base of its tail), which was widely used as a medicine in the medieval period as it is rich in salicylic acid – a basic ingredient of aspirin. In the days of Hywel Dda (around 950 CE) a single beaver skin was worth 'six score pence', i.e. 120 old pence (around £1,000 in today's terms). Although beavers have long been absent from Wales, the Welsh Beaver Project is investigating the feasibility of bringing them back. It has been estimated that the Afon Teifi could support around 170 individuals, which would create numerous beaver dams and ponds throughout the catchment (Jones et al. 2012). The effects that beavers have on an ecosystem are the subject of numerous studies, and from these it is clear that that they have a positive impact on wildlife, creating a range of habitats. Otters, in particular, seem to benefit significantly from the presence of beavers on a catchment, using beaver lodges and burrows as holts. Otters, however, are already common along the Teifi and its tributaries, especially where there is cover provided by bankside vegetation (Fig. 161).

FIG 161. The Teifi Otter, a sculpture presented by David Bellamy to Cardigan to mark the golden jubilee of the former Dyfed Wildlife Trust in 1988. The sculptor, Geoff Powell, was a member of the organisation. (Jonathan Mullard)

SMALLER RIVERS

Compared to the Eastern and Western Cleddau and the Teifi, the rest of the county's rivers are relatively short. The Afon Gwaun is typical of these smaller rivers, having a total length of only 14.5 km. It rises on the northern slopes of Foel Eryr in Mynydd Preseli and slowly meanders through Cwm Gwaun to the sea at Lower Fishguard. Generally, the wildlife of these smaller rivers has not been well studied and they would repay further attention. Monitoring of river flies (caddisflies, mayflies and stoneflies) on the River Alun, at Pont y Penyd in St Davids, for instance, revealed a number of mayflies, including the large dark olive mayfly *Baetis rhodani* and the blue-winged olive mayfly *Serratella ignita*, both common and widespread species (Fig. 162).

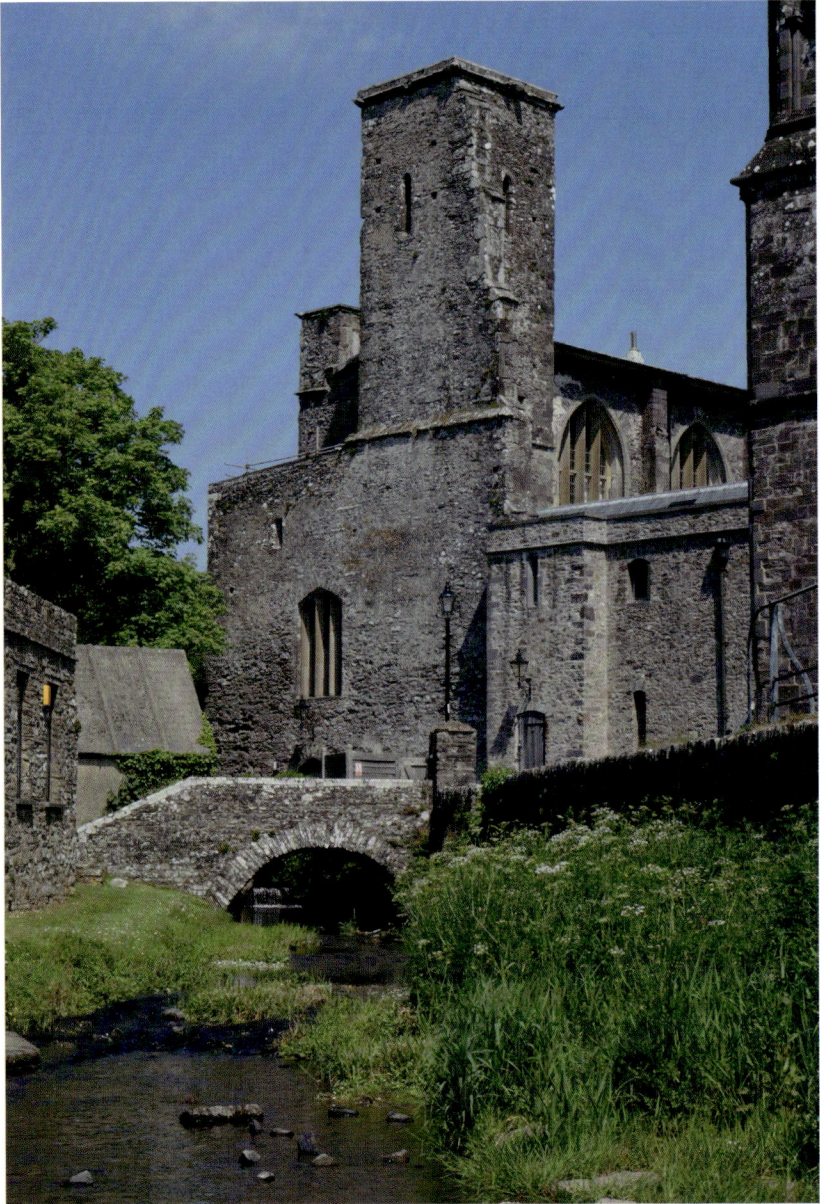

FIG 162. The River Alun, which flows behind St Davids Cathedral, supports a number of mayflies, including the large dark olive mayfly and the blue-winged olive mayfly. (Jonathan Mullard)

FENS, MOORS AND BOGS

In his *General View of the Agriculture of the County of Pembroke*, published in 1794, Charles Hassall wrote that:

> *The fenny parts of Pembrokeshire are not extensive. Castlemartin Corse; some flats near Tenby, some near Mathry and one in Dyffryn Gwein, are the most considerable.*

Despite Hassall's description, at one time extensive fens occupied most of the river valleys in Pembrokeshire, although now there are only the fragments of Castlemartin Corse, Cors Penally, Corsydd Llangloffan and Ritec Fen to remind us of past glories. George Owen provides a tantalising view of some of the birds resident in the wetlands during the fifteenth century when he writes:

> *In the bogges breedeth the crane and byttur, the wild duck, the tele, and divers others of that kinde; on high trees the heronshewes, the shoveller, and the wood quistes.*

The 'heronshewes' are herons, the 'tele' is obviously teal, 'wood quistes' are woodpigeons *Columba palumbus* and 'byttur' is surely the bittern. But Owen's mention of the 'shoveller' nesting in trees refers to the spoonbill *Platalea leucorodia*. This name for spoonbill is the one generally used by early naturalists, and although spoonbills mainly nest today in reedbeds or on the ground, they are also recorded using willows or poplars. In the nineteenth century the species was described as not infrequent in the coastal areas of Pembrokeshire, particularly Milford Haven. As some indication of its past abundance (and the likely reason for its decline) eleven spoonbills were shot in the Haven in 1854–55 (Lovegrove *et al.* 1994.) Thankfully there has been a dramatic increase in records of this species over the past decades, and it seems likely that it will eventually regain its former abundance. The peak arrival time in Pembrokeshire is from mid-May to the end of June and October. A number of the birds turning up in the autumn have overwintered, but there have been no records of breeding to date. Colin Matheson (1932) thought that 'The fact that the Spoonbill returns not infrequently to its old haunts suggests that, with adequate encouragement, this beautiful bird might once more be numbered among our breeding species.'

Castlemartin Corse

Freshwater West may seem a strange name for a beach, but there is indeed a source of freshwater here. Although it is a very popular location for surfing, few visitors realise that at the rear of the dunes lies Castlemartin Corse, the best

FIG 163. Castlemartin Corse contains one of the best calcareous fens in Pembrokeshire, with six different fen and fen-meadow plant communities. The high dunes behind are part of Brownslade and Linney Burrows. (Jonathan Mullard)

example of a calcareous fen in Pembrokeshire (Fig. 163). There is a clue, however, in the name, since it is derived from *cors*, the Welsh word for a marsh. Fed by the numerous springs around St Petrox and St Twynnells to the east, and with the natural drainage impeded by the dunes at the seaward end, this narrow 2.5-kilometre floodplain developed a rich flora and fauna. Unfortunately, much of the area was drained in the eighteenth century, as one of the many agricultural 'improvements' taking place at the time. Reviewing the situation at the end of the century, Charles Hassall explained that the Corse consisted of:

> *a tract of several hundred acres; which within these four years, was a perfect bog, of little or no value. Mr Campbell of Stackpole Court having obtained an act of parliament for draining and inclosing it, such an improvement is already made as must convince every observer, what ground of this sort is capable of, under skilful and attentive management.*

Campbell's agent, John Mirehouse of Brownslade, was apparently awarded the Royal Agricultural Society Gold Medal in 1800 for his work. In the 1970s a further section of the Corse, covering 'several hundred acres' of 'rush, sedge and coarse, unprofitable grass', was enclosed, drained and divided by ditches (Saunders 1986). The natural vegetation that remains occupies the lowest, seaward, end of the valley, which is only about 5 m above sea level. Any rise will therefore inevitably lead to the centre of the valley being inundated by saltwater. The National Trust, as owners of the site, are aiming to manage the transition by extending existing habitats, so that species are able to spread upstream.

Sadly, only 30 ha of Hassall's 'perfect bog' survive today, but even so the Corse contains six different fen and fen-meadow plant communities, the most extensive being a tall fen community. Plants associated with this include meadowsweet, reed canary-grass, hedge bindweed *Calystegia sepium* and the rare tufted-sedge *Carex elata*. The wettest parts support greater tussock-sedge *Carex paniculata*, wild angelica *Angelica sylvestris*, water dock *Rumex hydrolapathum*, yellow loosestrife *Lysimachia vulgaris* and greater pond-sedge *Carex riparia*. The grazed fen-meadow is also species-rich, and plants found here include purple-loosestrife, common fleabane *Pulicaria dysenterica*, hemp-agrimony, marsh horsetail *Equisetum palustre*, and greater bird's-foot trefoil *Lotus pedunculatus*. Fen pondweed *Potamogeton coloratus*, a nationally scarce plant, grows in two of the shallow ditches of the fen-meadows, along with slender sedge *Carex lasiocarpa* and long-stalked yellow-sedge *C. lepidocarpa*. Other notable plants include marsh helleborine, lesser water-parsnip *Berula erecta* and northern yellow-cress *Rorippa islandica*.

The great green bush-cricket is abundant throughout the area, while the short-winged conehead *Conocephalus dorsalis* and the slender ground-hopper *Tetrix subulata* can be found in the fen-meadow.

Cors Penally

Before Cors Penally, the 'flats near Tenby' mentioned by Hassall, was requisitioned by the army in the late nineteenth century, it was traditionally managed for livestock production. Thankfully the grazing was continued, and this prevented the growth of scrub and trees. The inland section of the site is dominated by an area of wetland (Fig. 164). It has long been known for the richness of its flora, but although the botanical records here extend back over 130 years, most date from the mid-nineteenth century. Around 140 plants have been recorded from the site and the majority are still present, with the notable exception of great fen-sedge *Carex mariscus* and greater spearwort *Ranunculus lingua*.

The bog is reputed to contain the greatest concentration of vulnerable vascular plants in Wales (Hawkeswood 2006). These include galingale, which was first recorded here in the 1920s, the nationally scarce marsh fern *Thelypteris palustris* and tufted-sedge. Today the fen is dominated by common reed,

FIG 164. Long known for the richness of its flora, Cors Penally is situated on the opposite side of the tracks to Penally railway station. Plants found here include the nationally scarce marsh fern and galingale, for which it is the only remaining location in Pembrokeshire, as described in Chapter 2. (Jonathan Mullard)

meadowsweet, hemp-agrimony, great willowherb *Epilobium hirsutum*, purple-loosestrife, and yellow iris. Other plants found here, as a result of the calcareous nature of the site, include blunt-flowered rush, yellow loosestrife, water dock, lesser water-parsnip and the locally scarce brown sedge *Carex disticha*. In the wetter areas, there are stands of greater pond-sedge and greater tussock-sedge. Near the centre of the fen, however, there is a more acidic community, indicated by the presence of purple moor-grass, bog-myrtle *Myrica gale*, common gorse, round-leaved sundew *Drosera rotundifolia* and star sedge *Carex echinata*. Creeping willow and royal fern *Osmunda regalis* also occur.

Corsydd Llangloffan

Corsydd Llangloffan, Hassall's fen 'near Mathry', is one of the largest river valley mires surviving in Pembrokeshire. The eastern section is designated as a National Nature Reserve, while the western section is managed by the Wildlife Trust. The wide floodplain here originally contained a river that drained westwards, but glacial deposits blocked and diverted it in an easterly direction around 12,000 years ago. Today's small central watercourse meandering through flat waterlogged ground is the result, and it has created ideal conditions for a wide variety of species.

Reed canary-grass dominates the ground adjacent to the riverbank, while the wetter peaty areas alongside the springs contain abundant clumps of greater tussock-sedge (Fig. 165). The tall fen vegetation includes hemp-agrimony, wild angelica, meadowsweet and purple-loosestrife. The wet woodland, or carr, that has developed east of Llangloffan Bridge is largely dominated by grey willow. Unlike in other wet valley-floor woodlands in Pembrokeshire, downy birch *Betula pubescens* is present along with alder and numerous scarce plants. These include marsh fern and great fen-sedge at its only location in Pembrokeshire. Other notable plants include royal fern, wavy St John's-wort, cranberry *Vaccinium oxycoccus* and marsh stitchwort *Stellaria palustris*. The more mature parts of the carr support impressive communities of epiphytic lichens, with species such as Wirth's beard lichen *Usnea wirthii, Sticta limbata, S. sylvatica* and *S. fuliginosa* present.

FIG 165. The wetter areas of Corsydd Llangloffan contain abundant clumps of greater tussock-sedge, as here on the left-hand side of the boardwalk. The towers the plant develops into as it gets older are said to have inspired John Wyndham's triffids. (Jonathan Mullard)

The birdlife is particularly rich, with corncrake *Crex crex*, quail *Coturnix coturnix* and spotted crake *Porzana porzana* all being recorded in recent years, as well as common winter visitors such as snipe, water rail and a variety of waterfowl. Breeding birds include sedge warbler *Acrocephalus schoenobaenus*, reed warbler *Acrocephalus scirpaceus* and grasshopper warbler *Locustella naevia*. Barn owls and hen harriers *Circus cyaneus* have also been seen hunting across the reserve. Mammals include otters, water voles *Arvicola amphibius*, water shrews *Neomys fodiens*, polecats *Mustela putorius* and badgers *Meles meles*.

Ritec Fen

Ritec Fen once occupied the upper reaches of the Ritec estuary, but by 1810 Charles Hassall was celebrating the fact that drainage and enclosure had converted the marshes near Tenby from 'dreary brown' to 'gay and everlasting green'. Despite these early drainage schemes and the canalised course of the Ritec river there is still a high water table in summer across much of the valley floor, which is dominated by a tall-herb fen. Extensive stands of greater pond-sedge occur throughout the area along with species such as great willowherb, water dock, great reedmace, meadowsweet, reed canary-grass, branched bur-reed, lesser water-parsnip and water horsetail. A small stand of marsh fern persists within the fen, together with brown sedge. Further up the slopes regular grazing by cattle has maintained wet pasture communities with species such as sharp-flowered rush *Juncus acutiflorus*, hard rush *J. inflexus*, soft-rush *J. effusus*, yellow iris and marsh ragwort *Jacobaea aquatica* in the damper hollows. The most species-rich flush has a small population of quaking-grass *Briza media* and marsh arrowgrass *Triglochin palustris*.

There are few insect records from Ritec Fen, but three notable fungus gnats, greyish-brown flies around 3–4 mm long, have been recorded: *Anatella dampfi*, a gnat usually associated with damp woodlands in upland areas, the nationally scarce *Leia longiseta*, a distinctive species only recently discovered in Britain, and *Mycomya britteni*. The biology of all these species is unknown but larvae have been reared from fungi and are web-spinners living on the surface of vegetation.

Teifi Marshes

Other wetland sites, not mentioned by Charles Hassall, include the Teifi Marshes, a Wildlife Trust reserve on the Pembrokeshire side of the river near Cardigan. The marshes occupy the site of what was probably the wide preglacial channel left by the former course of the Teifi. In 1886, when the Whitland and Taf Vale Railway was extended to Cardigan, the construction of the railway embankment divided the area in two, separating Pentood Marsh to the west from the tidal Afon Teifi and the remaining areas of saltmarsh in the east. Following this, the site developed into an extensive area of freshwater marsh, through which the small, but surprisingly deep, River Piliau now meanders. The barrier of the old railway has not proved totally impermeable, as the western end of the site is reverting to saltmarsh as a result of the incursion of seawater at high tides. At the southern end, alongside the Piliau, however, there is one of the largest areas of common reed in Wales, together with a mosaic of fen and herb-rich flood meadows. The meadows contain species such as floating sweet-grass, reed canary-grass, red bartsia *Odontites verna* and yellow iris, as well as drier grassland species such as Yorkshire-fog, common knapweed, meadow buttercup *Ranunculus acris* and devil's-bit scabious.

Birds breeding on the adjacent riverbanks include dipper *Cinclus cinclus*, kingfisher, sand martin and sedge warbler, while the estuarine and freshwater marshes support important populations of breeding birds; including snipe, water rail, reed bunting, reed warbler and the rare Cetti's warbler *Cettia cetti*. The latter

FIG 166. Water buffalo graze the Teifi Marshes all year round, creating and maintaining ideal conditions for both birds and insects. They like wallowing in the wetter areas of the reserve. (Nathan Walton)

is one of the United Kingdom's most recent colonists, first breeding in Kent in 1972, but it is a scarce resident in South Wales and at the northwest limit of its range. In winter, when the marshes are flooded, they attract large numbers of wildfowl, notably teal, wigeon and mallard.

Water buffalo *Bubalus bubalis* graze the site all year round, creating and maintaining habitats favoured by birds, amphibians and the seventeen species of dragonflies and damselflies recorded here (Fig. 166). These include emperor dragonfly *Anax imperator*, broad-bodied chaser *Libellula depressa*, southern hawker *Aeshna cyanea* and scarce blue-tailed damselfly. Ponies had previously been used to manage the area, but they proved to be selective feeders and bracken, rushes and brambles were spreading onto the marsh. In contrast, the water buffalo eat a wide range of vegetation and create open areas in the marsh by wallowing. They are one of the world's oldest livestock breeds, originating in Asia. The species was native to Europe until 10,000 years ago, when hunting reduced its range to the far southeast of the continent. Water buffalo are aptly named, as they spend most of their time in water, their extra-wide hooves preventing them from sinking into the mud at the bottom of ponds, swamps and rivers. Their use in the Teifi Marshes is an example of a growing conservation trend in Europe, that of using large, exotic herbivores to enhance the diversity of native flora and fauna (Schwägerl 2013).

Other mammals on the marsh, alongside the otters and water buffalo, include red deer and sika deer *Cervus nippon* and, unfortunately, mink *Mustela vison*. Mink escaped from fur farms in the 1950s and 1960s and are now found across most of Wales. They are active predators, feeding on almost anything they can catch, including water voles, which are now the fastest-declining mammal in Wales and a priority for conservation efforts. A 2014 survey of sites in Pembrokeshire previously known to have water voles proved negative, suggesting that the species may now be extinct in the county (Stephens 2014).

Goodwick Moor

Goodwick is now known primarily for the ferry terminal to Ireland, known as Fishguard Harbour, but describing the area in 1811 Richard Fenton wrote that:

> *The barrier to these sands [Goodwick Sands] consists of a beach formed with pebbles, the aggregate of ages, backed by a high mound of sand, consolidated by sedge and the dog-rose, over which the horn poppy luxuriates, with its delicate, but perishable, yellow bloom. On the land side, an extensive moory flat occupies the whole vale, covered with low rush and the aromatic Myrica galea.*

FIG 167. Goodwick Moor, the 'extensive moory flat' described by Richard Fenton, still maintains an element of wildness despite being situated close to the ferry terminal at Fishguard Harbour. (Nathan Walton)

Fenton's house, Plas Glyn-y-Mel, built in 1797–99, is located nearby in Lower Town, Fishguard, so he knew this area well. His 'extensive moory flat' is now a nature reserve managed by the Wildlife Trust (Fig. 167). Access is through a large car park behind the petrol station on the seafront, but unfortunately the boardwalk which used to run through the reserve has been removed because of repeated vandalism. The site lies between two streams, and when water levels are at their highest, between October and April, the entire moor can often be flooded – especially when spring tides coincide with heavy rain. The access provided by the boardwalk was therefore essential. The Trust have decided, however, that they can no longer afford the high maintenance costs.

The extensive reedbed covers much of the reserve, with sea rush and sea aster plentiful on the seaward side. Species of interest on the floodplain mire include common cottongrass *Eriophorum angustifolium* and royal fern. In the drier areas there is willow and alder carr woodland, with the bog-myrtle, mentioned by Fenton, growing where the willow has been cut back. The numerous ditches contain species such as grey club-rush *Schoenoplectus tabernaemontani*, bogbean

Menyanthes trifoliata, branched bur-reed, water mint and bog pondweed. Dragonflies recorded from the site include the broad-bodied chaser, emperor dragonfly and southern hawker. The reedbeds support breeding populations of reed warblers and sedge warblers, and stonechats are also present.

Esgyrn Bottom

Esgyrn Bottom, situated to the south of Cwm Gwaun near Llanychaer, is noted as the most southwestern raised bog in Britain. Given its proximity to Cwm Gwaun it was almost certainly the area which Charles Hassall called Dyffryn Gwein, since *dyffryn* refers to a wider valley, a description that fits the site well. Raised bogs are dome-shaped accumulations of peat which occupy former lakes, or shallow depressions, the bog here covering the floor of a subglacial meltwater channel. Its centre lies due north of Gelli, where an indistinct watershed between the Cleddau and a tributary of the Gwaun was once known as 'the valley of dead waters' (Fig. 168). The area was formerly renowned for eels, George Owen writing that the 'great moore, or bogge, being of three miles longe, serveth for the nursery of this slippery fish' (Owen 1603).

FIG 168. Esgyrn Bottom used to be known locally as the 'the valley of dead waters'. The woodland on the steep valley sides is predominantly oak but also includes coppiced sycamore and smaller amounts of beech and ash. (Jonathan Mullard)

The thick layers of peat, which in places are nearly 7 m deep, were first investigated in 1973 and showed that, as the climate improved after the retreat of the ice, the area was quickly filled by reed swamp. Later birch woodland was followed by a more open habitat dominated by cottongrass and purple moor-grass before the raised bog dominated by bog-moss *Sphagnum* spp. developed. This dome of bog-mosses was isolated from and independent of groundwater, being totally reliant on rainfall. Until the late 1960s the peat was cut for fuel, producing an uneven surface of hummocks and wet hollows. Today the hummocks are dominated by heather, cross-leaved heath, cottongrass, purple moor-grass, red bog-moss *Sphagnum capillifolium* ssp. *rubellum*, Magellanic bog-moss *S. magellanicum* and soft bog-moss *S. tenellum*. Species found in the wet hollows include feathery bog-moss *Sphagnum cuspidatum* and white beak-sedge *Rhyncospora alba*, while around their edges there is a diverse range of plants such as round-leaved sundew, cranberry, bog asphodel *Narthecium ossifragum*, papillose bog-moss *Sphagnum papillosum* and recurved sphagnum *S. recurvum*. Also present are hare's-tail cottongrass *Eriophorum vaginatum*, oblong-leaved sundew *Drosera intermedia*, white sedge *Carex curta*, narrow buckler-fern *Dryopteris carthusiana* and royal fern. There are a number of interesting bryophyte records, notably bog pouchwort *Calypogeia sphagnicola*, spurred threadwort *Cephaloziella elachista* and veilwort *Pallavicinia lyellii*, a mat-forming liverwort.

Butterflies found on the site include marsh fritillary, dark green fritillary, small pearl-bordered fritillary and green hairstreak *Callophrys rubi*. In addition, there are records of an extremely rare money-spider, *Glyphesis servulus*, and glow worms. The money-spider has only been found at a few sites in Wales and southern England. Where it occurs in acid mires, such as Esgyrn Bottom, it is often associated with tall, tussocky purple moor-grass (Boyce 2004).

RESERVOIRS, LAKES, PONDS AND POOLS

Rosebush and Llys-y-frân reservoirs

The Afon Syfynwy rises in Mynydd Preseli before feeding the Rosebush and Llys-y-frân reservoirs and joining the Eastern Cleddau. The larger of the two, Llys-y-frân, covers 86 ha, the water being held back by a 30 m dam constructed between 1968 and 1972 (Fig. 169). As mentioned in Chapter 1, the surrounding land is now managed as a Country Park by Dwr Cymru Welsh Water. It is the main area of freshwater in Pembrokeshire, but it is deep, and also a well-known fishery, so only a few birds breed here in the summer, including the occasional pair of great crested grebes *Podiceps cristatus*. In the winter months, however,

FIG 169. Llys-y-frân reservoir is the largest area of freshwater in Pembrokeshire. It is an important roost site for gulls, with up to 12,000 lesser black-backed gulls, the largest roost for the species in Wales. (Annie Haycock)

large numbers of wildfowl congregate on the open water and within the wooded creeks. Teal, goldeneye *Bucephala clangula*, little grebe and goosander *Mergus merganser* are usually present, while red-throated diver *Gavia stellata* and great northern diver *G. immer* make frequent appearances. Smew *Mergellus albellus*, a winter visitor from Scandinavia and Russia, have been recorded in the past, along with black-throated diver *Gavia arctica* and black-necked grebe *Podiceps nigricollis*. The main body of the reservoir is used as a roost by large numbers of gulls, including up to 12,000 lesser black-backed gulls, the largest roost in Wales (Green & Roberts 2004). Mediterranean gull, Iceland gull *Larus glaucoides* and glaucous gull *L. hyperboreus* have also been recorded, as well as the only county sighting of Kumlien's gull *L. glaucoides kumlieni*, a large gull which breeds in the Arctic regions of Canada. It is a regular vagrant to Britain and Ireland, small numbers being seen each year.

Both reservoirs have well-developed 'drawdown', or 'inundation', zones, the area at the edge of a reservoir that is frequently exposed to the air because of changes in water level, although Rosebush tends to be kept at a higher level than

Llys-y-frân. These ephemeral habitats can support a range of specialised plant and animal species that survive and thrive in situations with frequent water-level fluctuations. The zone is especially important for bryophytes, and the species found here have special adaptations to enable them to survive in the changeable environment. In particular, they have long-lived spores or tubers that are produced in abundance in years when water levels fall, but which can survive underwater for several years until a drop in the water level again allows the plant to germinate. When water levels are low bryophytes can grow with extraordinary speed, with some species completing their whole life cycle, from germinating spore to mature plants that release their own spores, within two or three months. They can appear in vast numbers, thousands of plants reproducing as rapidly as possible before the water rises again. The species colonise new water bodies by being transported, as spores, on the feet and plumage of ducks and other waterfowl. Where reservoirs are connected, as is the case here, spores are also likely to be carried by the water flow from locations higher up a valley to those further down.

The fact that Rosebush is full for most of the time is probably the reason for its relatively impoverished bryophyte flora compared to the larger water body. Only five species, bicoloured bryum *Bryum dichotomum*, serrated earth-moss *Ephemerum serratum*, a pocket moss *Fissidens* sp., greater water-moss and delicate earth-moss *Pseudephemerum nitidum* growing on the margins, whereas Llys-y-frân supports a total of eighteen species, including rufous forklet-moss *Dicranella rufescens*, common pottia *Tortula truncata*, cylindric ditrichum *Trichodon cylindricus* and green-tufted stubble-moss *Weissia controversa* (Bosanquet 2010.)

Bosherston Lakes

Before 1780 the three narrow valleys leading down to Broadhaven Beach were a tidal inlet. In that year, however, John Campbell II, the first Baron Cawdor, started to create a designed landscape to provide a setting for Stackpole Court. Over a period of eighty years Campbell and his successors flooded the valleys with freshwater by building a series of dams, weirs and sluices, finishing with a dam at the back of the beach. The Eastern Arm, the largest of the three lakes, is fed by small streams and springs, while the Central and Western Arms are supplied exclusively by springs. As the water drains into the underlying limestone they are normally deep in winter and shallow in summer. Today they are famous for the magnificent displays of white water-lily *Nymphaea alba*, which are at their best in June, hence the name by which they are more widely known, Bosherston Lily Ponds (Fig. 170).

As well as the celebrated lilies, the three lakes contain good populations of a number of rare freshwater plants, their distribution reflecting the amount of

FIG 170. Bosherston Lakes are well known for their magnificent displays of white water-lily but also contain good populations of a number of rare plants. (Jonathan Mullard)

eutrophication in different parts of the system. The spring-fed Western and Central Arms contain mainly clear-water communities dominated by bristly stonewort *Chara hispida*, which forms dense beds up to a metre high, with individual plants up to 3.5 m long. While it is submerged it is green, but, as the water level drops in summer, the tops of the plants dry out, leaving bleached patches of vegetation (Fig. 171). In addition, there are good numbers of fragile stonewort *Chara globularis*, delicate stonewort *C. virgata* and common stonewort *C. vulgaris*. Stoneworts are a unique group of complex algae that typically grow in fresh, or brackish, water that is clear and unpolluted. Their name derives from their encrusted appearance, as most species build an external skeleton of calcium carbonate, as a by-product of photosynthesis, instead of using cellulose for structural support like flowering plants. This skeleton is the origin of the name 'stonewort' (Stewart 2004). Most stoneworts are unable to tolerate significant levels of phosphates and nitrates from nutrient pollution and, as a result, they are excellent indicators of water quality.

In contrast, stoneworts are rare in the Eastern Arm, and the vegetation is dominated instead by dense stands of curled pondweed *Potamogeton crispus*, fennel pondweed *P. pectinatus*, spiked water-milfoil *Myriophyllum spicatum* and Canadian waterweed *Elodea canadensis*. Emergent vegetation fringes parts of the system, mostly common reed, great reedmace, common spike-rush *Eleocharis palustris* and branched bur-reed.

From early summer until autumn, the amount of water flowing into the lakes is exceeded by losses from seepage, evaporation and transpiration, and the lakes cease to flow out to the sea. With the falling water levels, the system divides into three sections, the Upper Eastern Arm, the Lower Eastern Arm, and the Central and Western Arms. These are separated by the weirs at the Eight-Arch Bridge (Fig. 172), the Grassy Bridge and the Central Arm Causeway. On the bare margins of the lakes cavernous crystalwort *Riccia cavernosa* and spreading earth-moss *Aphanorrhegma patens* sometimes grow in great abundance (Bosanquet 2010). Both species are early colonists of bare mud at the margins of reservoirs, pools and rivers. Cavernous crystalwort is especially noticeable as it grows in yellow-green rounded rosettes, up to 3 cm in diameter, composed of oblong or rounded lobes which are widest above the middle. As at Rosebush and Llys-y-frân, the inundation zone is especially important for bryophytes.

FIG 171. The spring-fed Western and Central Arms of Bosherston Lakes are an important location for rare stoneworts, particularly bristly stonewort. This is probably the largest population of this species in Wales. (Annie Haycock)

FIG 172. The Eight-Arch Bridge, which crosses the eastern lake at Bosherston, was built in 1797 to link Stackpole Court to Stackpole Warren. (Jonathan Mullard)

There is a healthy population of otters that feed on the eels, pike *Esox lucius*, perch *Perca fluviatilis*, roach *Rutilus rutilus* and tench *Tinca tinca* that live in the lakes. Greater horseshoe bats and lesser horseshoe bats *Rhinolophus hipposideros* are among at least ten species of bat utilising the surrounding woodland and lakeside margins as 'feeding flyways' connected to important summer, winter and intermediate roost sites at Stackpole and other nearby locations. Breeding birds include heron, kingfisher, little grebe and moorhen, while ducks include tufted duck, gadwall and, in the winter, goosander and goldeneye. Large numbers of mute swans are also resident on the lakes.

A separate valley south of the lakes, known as the Mere Pool Valley, contains semi-natural and artificial pools and marshes. Over twenty species of dragonfly and damselfly have been recorded here, including hairy dragonfly, ruddy darter *Sympetrum sanguineum*, scarce blue-tailed damselfly and beautiful demoiselle *Calopteryx virgo*. Like the main lakes, Mere Pool Valley is a stronghold for otters, whose favourite prey includes eels.

Orielton Decoy

Another artificial water body is Decoy Lake, a six-hectare artificial pond on the western edge of the Orielton Estate, formed by the damming of the Orielton Stream, where a duck decoy was constructed in 1868 to capture wildfowl. Most of the decoy 'pipes' are concealed in the marshy coppice where the stream enters the pond. The word 'decoy' is of Dutch origin, originating from two words *eende*, meaning duck, and *kooi*, meaning cage, and they were originally constructed to capture wild duck for the market. Ducks are lured into a long and narrowing semicircular netted pipe which ends in a cage. Ducks are enticed to the entrance either by baiting, or by using a dog. When threatened by a predator on land, such as a fox, ducks all face it and follow its movements at a safe distance, in a type of mobbing behaviour. A decoy dog is trained instead to run in and out of screens alongside the decoy pipe, luring the ducks further and further into the narrowing structure (Stott & Mitchell 1991). There were never many duck decoys in Wales, and only five are known: in addition to Orielton there are two on the

FIG 173. Orielton Decoy in the late 1930s, with the decoyman Stanley Greenslade and his dog. (*Times* newspaper)

Gwent Levels near Bishton, one in Decoy Wood on Gower and one at Lymore Hall, Montgomery. The decoy at Orielton not only had the longest history of all the Welsh decoys in relation to catching ducks for the table but was also the first in the United Kingdom to be used to catch ducks for ringing.

Records from 1877 to 1918 were kept in the *Orielton Decoy Book* and show that a total of 22,830 wildfowl were caught in this period, the majority being wigeon (11,361) and teal (8,291). Average wintering numbers on the lake were in the region of 10,000–12,000 ducks, of which some 90 per cent were wigeon (Saunders 2008). The decoy fell into disuse during the First World War but was refurbished for ringing purposes in 1934 by two noted ornithologists, C. W. Mackworth-Praed (1891–1974), whose family had lived for a time at Orielton, and Captain H. A. Gilbert (1886–1960). The day-to-day running of the decoy was undertaken by Stanley Greenslade (Fig. 173), who worked on the estate all his life and ensured that no one visited the site without written permission from either Gilbert or Mackworth-Praed. Between 1934 and 1940, 7,246 ducks were ringed, providing some of the earliest information on the movements and migration of wintering wildfowl. As previously, wigeon and teal were the main species caught. Many rings were subsequently recovered from across northern and central Europe, as far as the Volga and Arctic Russia. Many of the ducks had been shot for food. As Gilbert noted, 'it is the cook who generally finds the ring.'

A twelve-minute film of the decoy made by Gilbert in January 1938 has been deposited in the National Library of Wales by David Saunders. The first part is mostly general views of the decoy lake, revealing the large numbers of waterfowl which had gathered there, while the second shows Greenslade with his dog in action and includes a brief view of Gilbert. The decoy continued in use until 1941, part of the haul being sold for food to defray expenses, but work finally ceased in the autumn of 1942 when Greenslade had a serious accident while attempting to stop a runaway horse and cart.

Following the Second World War the pipes were still in good condition, but unfortunately they were wrecked by the weight of the snow during a freak storm in January 1946. A rebuilding programme was subsequently undertaken by the International Committee for Bird Preservation. Greenslade, now recovered from his injuries, was again the decoyman, and by 1950 the total number of ducks ringed since 1934 reached almost 11,000. But the costs of running the decoy were becoming excessive, and management of it was handed over to the West Wales Field Society. The Society also soon found that the costs of maintaining and operating the pipes was considerable, and these were gradually replaced by small funnel traps. At the same time the number of ducks was declining, and by early 1960, when ringing finally ceased, barely 200 birds were present. In that final year just 70 ducks were caught. The last duck to be ringed using Orielton rings was a female teal ringed on 17 December 1958, although rings marked 'British Museum, London' were also used, both before and after this date. The bird was subsequently recaught at the decoy, by David Saunders, on 16 January 1960 (Saunders 2019).

Various suggestions have been put forward for the dramatic reduction in the number of wildfowl using Decoy Lake, including changes in the chemistry of the water due to the intensification of agriculture in the surrounding area, increased human disturbance, and the fact that the lake margins were no longer baited. But perhaps the real reason is that from the seventeenth century onwards decoys offered birds protection during the daytime from shooting, the relatively small number of birds caught in the pipes being quickly replaced by others seeking an area free of disturbance. Today the lake at Orielton is virtually empty of ducks, even in mid-winter, a sad end to the years of activity and planning. We should, however, see this as a positive situation, as it is the result of increased protection for wildfowl.

St Davids ponds

Although Pembrokeshire has long been famous for its early new potatoes, the weather can be very dry in spring and, since potatoes need good supplies of water to grow, many farms have excavated their own irrigation ponds. The large-scale Ordnance Survey map of the St Davids area shows these in detail, and while many are on private land and inaccessible, some, such as the relatively large Penberi Pond on the side of the B4583 north of St Davids, can be easily viewed. During the winter, when they are full, ponds such as this provide a refuge for wildfowl, while in the late summer and autumn the muddy pools attract waders (Green 2004). Penberi Pond usually contains something of interest, whether it is breeding little grebes in summer, waders on passage in autumn or wildfowl and gulls in winter. Waders recorded here include local rarities such as green sandpiper, ruff *Calidris pugnax* and black-tailed godwit. Mediterranean gull, Iceland gull and little gull have all been seen, while the more unusual winter wildfowl have included scaup *Aythya marila* and ring-necked duck *Aythya collaris*, which is native to North America.

Dowrog Pool

In 1811, Richard Fenton wrote about:

> *the famous moor called the Ddyfrog given to St Davids by Rhys ap Tewdwr Mawr, most of which is under water, and appears like a considerable lake for seven or eight months of the year, and is seldom entirely dry, affording in winter abundance of wildfowl and in summer a cooling run for the cattle and horses.*

FIG 174. Dowrog Pool, in the southwestern corner of the common, is a seasonal waterbody which provides an important habitat for a wide variety of species, including overwintering birds. (Annie Haycock)

He also mentioned that it provided an abundant supply of clay 'which mixed with culm, the chief firing of the county, cements and prepares it for use'.

Today Dowrog Pool still covers about 6 ha in the winter, but is much reduced in summer (Fig. 174). At its deepest point it is only 1.5 m deep, and the natural accumulation of organic matter and the spread of great reedmace, water horsetail *Equisetum fluviatile*, bogbean, branched bur-reed and marsh cinquefoil *Potentilla palustris* is reducing the area of open water in summer. The margins of the pool are covered by greater tussock-sedge and common cottongrass. Truly aquatic plants include alternate water-milfoil and broad-leaved pondweed *Potamogeton natans*. These species are also abundant in the many small ponds and disused clay and gravel pits, where they occur alongside lesser water-plantain, floating club-rush *Eleogiton fluitans*, lesser bladderwort *Utricularia minor*, unbranched bur-reed and numerous stoneworts.

In the winter Bewick's swans *Cygnus columbianus bewickii* and whooper swans *C. cygnus*, teal, wigeon, shoveler *Spatula clypeata* and mallard can all be found here, along with snipe, water rail, coot and moorhen. The wetland supports a number of breeding birds, including grasshopper warbler, reed bunting and sedge warbler. As at all the water bodies and wetlands in Pembrokeshire, otters are regular visitors, and water shrews have been recorded. The water shrew is one of Britain's least-known mammals and its habitat requirements are poorly understood. Food availability, however, seems to be the major factor determining the distribution and abundance of populations (Champneys 2012).

Sacred eels

Until quite recently eels and other fish found in holy wells and streams were considered sacred, being used to foretell the future and cure the sick. There are many examples across Wales of this belief, and Pembrokeshire is no exception. The well in the churchyard at St Teilo's Church in Llandeloy, for example, once contained sacred eels (Jones 1954). Today the stone-lined spring appears to be empty (Fig. 175), but previously visitors would have waited, for days if necessary,

FIG 175. The holy well in the churchyard of St Teilo's Church in Llandeloy, a former haunt of sacred eels. The redundant church, rebuilt in 1926–27 from twelfth century ruins, is cared for by the Friends of Friendless Churches. (Jonathan Mullard)

until the eels made their appearance. The patient stood in the well, and it was believed that they would be cured if the eel coiled itself around their bare legs. One young girl standing in a well in Caernarfonshire is said to have died of fright when this happened. The movements of the eels could also be interpreted to predict future events, such as marriage. Today it seems inconceivable that this small spring and the rivulet that emerges from it contained eels, but they were once regarded as so common they were looked upon as food for the poor. To give just one example of their former abundance in small streams, this is what Philip Henry Gosse (1856) saw in a small brook near Giltar Point:

> *In the brook my attention was arrested by multitudes of young eels, wriggling along in the direction of the current. I counted a hundred that passed me in about two minutes; and they crowded on uninterruptedly, occasionally diving into the mud at the bottom, when I made a clutch at one. They were all about four or five inches long.*

A flying snake covered with scales is said to have flown from the tower of St Edrin's church, south of Mathry, to the 'marshy Rholwn', one of the small streams nearby, where it lurked during the day (Jones 1954). At night it visited the holy well in the churchyard and coiled up at the bottom. The church, now converted into a private house, stands in isolation among the fields. It is surrounded by a circular 'llan' or banked enclosure, which is indicative of a pre-Christian site. Today the well is dry, but in earlier times the water from it was reputed to cure rabies. It dried up when 'something harmful' was put into it, but its healing qualities were conveniently transferred to the surrounding grass. The grass was traditionally eaten with bread and butter, while cattle at risk from rabies were grazed in the churchyard. Where the story of the flying snake originated is difficult to tell, but eels can survive out of water for many hours and their movement on land is quite like that of a snake. They usually travel at night when it is cooler and they are at less risk from predators, so perhaps once again it was an eel visiting the well.

Marloes Mere

Marloes Mere is located in a hollow on the Marloes peninsula which is lined with glacial silt and clay. Each winter the wet acidic pasture floods to form the mere. The area was common land until 1811 when 'Meer Moor', as it was called, was enclosed. In the same year, the antiquary Richard Fenton described Marloes Mere as a 'common nearly all overflowed in winter, but having in summer the greater part dry for pasture and in a very dry season nearly the whole, about 60 or 70 acres, affording support during the hot months to the cattle of the villagers'.

FIG 176. Marloes Mere once supported large numbers of medicinal leeches, but the species has not been found here for many years and is probably extinct at this site. Today it is a prime birdwatching site, particularly during the winter. (Jonathan Mullard)

At the time the nearby village of Marloes was mainly inhabited by lobster and crab fishermen, who supplemented their income by the sale of medicinal leeches *Hirudo medicinalis* – which were found in 'great numbers' on the mere (Fig. 176). Once common in Britain, the leech has declined throughout its geographical range owing to the loss of wetland habitats and collection. It is found mainly in warm, shallow, still ponds that have abundant amphibian populations, but it will also feed on fish, birds and mammals, especially cattle. The villagers' animals would have been an ideal source of food for the population at Marloes.

The mere is now owned by the National Trust, and consists of a large area of marsh with smaller areas of open water, including two deep irrigation ponds. The marshy grassland on the eastern part of the mere is dominated by soft-rush, while the three western fields, which are regularly cut, are covered by acid grassland. Rare plants include three-lobed crowfoot, which is abundant in the shallow, muddy winter pools, and tubular water-dropwort *Oenanthe fistulosa*, which grows on the margins of ditches and ponds. Insects recorded from the site include great green bush-cricket and emperor dragonfly.

The mere is also a key site for birdwatchers. Wintering wildfowl include wigeon, shoveler, pintail *Anas acuta*, teal and mallard, while birds of prey recorded from the site include hen harrier, peregrine, merlin and short-eared owl. In the winter and early spring the gulls that feed in the nearby fields, including the occasional Iceland gull, spend quite a lot of their time bathing in the mere. In spring and summer there are warblers and waders, particularly black-tailed godwit, while marsh harrier *Circus aeruginosus* and garganey *Spatula querquedula* occur regularly on migration. Rarer birds are also recorded. In June 2012, for instance, a flock of three glossy ibises *Plegadis falcinellus* arrived. The glossy ibis, a large, heron-like bird with rich bronze feathers and a slender, curved bill, is an increasingly frequent stray from southern Europe. Other sightings include ring-necked duck and black duck *Anas rubripes* from North America, as well regular small groups of whooper swans, white-fronted geese *Anser albifrons* and pink-footed geese *A. brachyrhynchus*, while in May 2016 a golden oriole *Oriolus oriolus* was seen at the west end of the mere.

Although 'Meer Moor' no longer exists, there are still many other lowland commons in Pembrokeshire.

CHAPTER 12

Lowland Commons

The cottager and the labourer in the early days of Parliamentary enclosure exercised no political power ... and the ruling spirit of the times [was] against the interests and claims of the defenceless commoners.

Ivor Bowen (1914)

TODAY REGISTERED COMMON LAND IN Wales covers around 1,750 km², which equals 8.5 per cent of the total land area (Laimann 2018). It has been estimated, however, that commons and 'waste' once covered over 20 per cent of the country, around 4,118 km². Like the rest of Wales, a considerable area of Pembrokeshire remained open and unimproved until the late eighteenth century. Edward Laws (1888) noted that in George Owen's day 'Gorse covered a very great part of the country, and was to be found eight and nine feet high, with the stems twelve inches in circumference; indeed, Owen mentions one that was measured to decide a bet, and was found to be not less than thirty-three inches in compass.' Even private land does not seem to have always been enclosed, Henry Penruddocke Wyndham, a gentleman travelling through Pembrokeshire in the summer of 1774, noting that:

Few inclosures are to be seen in the neighbourhood of St Davids and the property is regulated in a manner different from that of the open fields in England: for here there is no common feed and every proprietor has a private right to the pasture of his own ground only, and to no other. This circumstance is attended with much

FIG 177. Ponies grazing Waun Caerfarchell to the north of St Davids airfield. This is part of a large unenclosed belt of informal 'waste' or 'moor' (*waun*) which was divided between medieval or post-medieval vills. (Jonathan Mullard)

inconvenience both to the owners of the lands and to the traveller. For there being no common shepherd, all the horses, sheep and even poultry, are staked at the end of a line to the ground, in order to prevent mutual trespass. The consequence being that the ropes frequently cross the high road and entangle the horses' feet of the unwary rider.

Open fields like those described by Wyndham were only found in a few places where ecclesiastical and private fields had become intermixed, as at St Davids, where a considerable tract of land was farmed on this system (Bowen 1914). The land was not divided into the long, narrow strips typical of English open-field systems, but was distributed in rectangular 'shares' scattered across a wide area. Each settlement, the medieval or post-medieval 'vill', or parish, had its own system, but most were linked with two small separate areas of land, one known as 'common' and one as *waun* or 'moor', the latter being 'waste' land. Waun Caerfarchell, Waun Treflodan, Waun Llechell, Waun Llandruidon and Waun Vachelich, to the north of St Davids airfield, for example, are part of a large unenclosed belt of informal waste which was divided between at least five vills, each of which had rights to a portion of the land (Fig. 177).

Private ownership of land is actually a relatively recent idea. Over the course of a few hundred years, however, much of Britain has been privatised, that is taken out of collective ownership and management and handed over to various individuals. The most contentious mechanism in previous centuries was enclosure – the subdivision and fencing of common land into individual plots, which were allocated to those people deemed to have previously held rights to the land (Fairlie 2009). An example can be seen in Meer Moor, mentioned at the end of Chapter 11.

In Wales the situation only started to change after English laws were introduced in the sixteenth century, with numerous private Enclosure Acts being promoted. Enclosure developed momentum though with the establishment, in 1793, of a Board of Agriculture. The Board was composed mainly of large landowners and people interested in agricultural projects, and its aim was to promote agricultural improvement. Although it was a voluntary initiative it covered the whole of England and Wales and received financial assistance from Parliament. Wales survived relatively unscathed until the General Enclosure Act of 1845, which drastically reduced the area of common land in the country. While much of the land was divided among the people who already held common rights, some land was sold. Private purchases, at least as recorded in the enclosure awards, seem to have been fairly limited, but public sales of land to cover the legal costs of the process frequently took place. Much of this newly enclosed land appears to have undergone little improvement in the formal sense, but, at least in the view of contemporaries, sheep farming became much more profitable when freed from the problems that were considered to be associated with common usage of the land (Chapman 1991).

The process added 'great tracts of the country as cultivated lands to the rent rolls of the landowners and to the tithe charges of the clergy' (Bowen 1914), and numerous people were deprived of their means of subsistence. In Pembrokeshire, for example, a distinctive area of enclosure is located north of Maenclochog, where enclosure in the 1820s provoked a riot among the dispossessed cottagers and small freeholders, who had lost their right of common and turbary – the rights to graze animals and cut turf, or peat, for fuel. Similarly, when the Rhoshirwaun waste, which covered around 12 km² on the Llŷn peninsula in North Wales, was enclosed in 1802 large numbers of cottagers, who supported themselves through fishing and casual labouring, were evicted. Thrown out of their homes, 'they could only remove to the trading towns, America, or enlist in the army or navy, leaving the old and infirm to exist upon the scanty allowance of the parish' (Bowen 1914).

Common land in Pembrokeshire today covers 56.53 km², around 3.6 per cent of the total land area. The largest area, Mynydd Preseli, extending to 21.32 km², comprises nearly 38 per cent of the total resource in the county. It seems that, in comparison with other parts of Wales, Pembrokeshire lost relatively little open land due to the formal enclosure process, and this is consistent with a return made to the House of Commons in 1904. This lists only five enclosure awards for Pembrokeshire, although there were actually six (compared with 491 awards recorded for the old county of Denbighshire). These awards included land in Narberth Forest, Castlemartin Corse, Marloes and St Davids, among others.

While most surviving Welsh commons are located in the uplands, such as those in the Brecon Beacons, two notable groups of lowland commons exist in South Wales: those on Gower (Mullard 2006) and those on the St Davids peninsula. The shallow, and wet, clay basins to the east of the city support a wide range of habitats including wet and dry heathland, fens, swamps and pools, along with a significant number of rare and endangered species. Indeed, there is more lowland and coastal heathland surviving in Pembrokeshire than in any other part of Wales. There are three important, and connecting, commons in this area, Dowrog Common, Waun Fawr and Tretio Common. Along with the 'waste' north of the airfield, they support numerous rare plants.

The maintenance of traditional grazing by cattle and horses, combined with controlled burning, is essential for the survival of these sites. Up until the beginning of the twentieth century commons were, as described above, an essential part of the rural economy, but grazing ceased on the majority of lowland commons in Pembrokeshire in the 1980s. In the late 1990s, however, the Heritage Lottery Fund's *Tomorrow's Heathland Heritage* initiative provided funds to enable the National Trust to reintroduce grazing with cattle and ponies on its own properties and support the restoration of a number of other inland commons. Although the funding was only for a limited period, since that date there has been an active programme of grazing that has maintained these important areas. Under the title of *Pembrokeshire Heathland Beef/Cig Tir Comin*, for example, a group of farmers is currently working closely with the National Trust to graze these areas with traditional British cattle breeds and produce high-quality beef.

DOWROG COMMON

Dowrog Common, covering around 101 ha, is a large and exceptionally diverse area, with over 300 different species of plants recorded here to date (Fig. 178). Waun Fawr, a small common to the west of Dowrog, and Tretio Common,

FIG 178. Dowrog Common, the largest of the commons in the St Davids area, supports an exceptionally rich flora and fauna, much of it dependent on the conditions created by grazing animals. (Jonathan Mullard)

essentially the northeastern extension of Dowrog, support a smaller, but similar, range of species.

Dowrog means a watery place or marsh (being derived from *dyfr(i)og*, 'watery'). This is particularly apt, since it lies on the upper reaches of the River Alun and there are extensive areas of riverside marsh and fen, as well as Dowrog Pool, described in Chapter 11. On the edge of the fen there is marshy grassland dominated by purple moor-grass and large areas of wet heathland with heather, cross-leaved heath and western gorse, along with scattered specimens of creeping willow.

Rare plants found on the common include three species dependent on open swards and bare ground: pale dog-violet, yellow centaury and wavy St John's-wort. Pale dog-violet cannot generally compete with other plants, and abundance here is probably a result of the frequent winter burning. Grazing with cattle, or ponies, to reduce the sward height and create patches of bare ground is also beneficial. It frequently hybridises with common dog-violet *Viola riviniana* to form large flower-rich clumps which are sterile and reproduce vegetatively. Yellow centaury is another plant for which reduced competition is essential. A short-lived

FIG 179. Lesser butterfly-orchid is one of the rarer plants that can be found on Dowrog Common. It occurs mainly in the southeast corner, near the A487 road. (Stephen Evans)

summer annual, it germinates in spring on bare ground exposed after winter flooding. Like many such annuals, it may be extremely abundant over small areas during favourable seasons, but absent in other years. Similarly, the seedlings of wavy St John's-wort require bare ground for successful establishment: in tussocky grassland, like that on Waun Fawr, the plant tends to occur in open areas, either on the sides of purple moor-grass tussocks or on the ground poached by animals between the tussocks. Periodic soil disturbance is essential, as all of these plants regenerate from a buried seed bank. Localised poaching by cattle is the best way of causing disturbance, and strategically located 'pinch-points' can be used to concentrate them and maintain populations of these plants if overall stocking densities are low.

Bell heather dominates much of the dry heath while more open areas support sea plantain and the uncommon lesser butterfly-orchid *Platanthera bifolia* (Fig. 179). Lesser butterfly-orchid produces copious amounts of seed, which are dispersed by the wind, but they require a specific mycorrhizal fungus to be present in order to germinate. Without this the seeds are not viable, so as a result it can be found on a scatter of sites across Wales, although the majority of locations are in

Cardiganshire. The flowers are pollinated by hawk-moths, which take nectar from the long spur at the back of each flower.

Dr Falconer was first to report lesser butterfly-orchid from Pembrokeshire, in his 1848 *Catalogue of Plants*. This and other early records indicate that the orchid favoured heathy fields and unenclosed moorland, where it was sometimes locally abundant. Sites were lost to ploughing during the Second World War, but in her *List of Pembrokeshire Plants* (1950) Lillian Rees describes lesser butterfly-orchid as 'in profusion, together with the Yellow Star of the Moors (*Narthecium ossifragum*), on Dowrog Moor'. Today it is limited to the managed wet lowland heathlands of St Davids airfield, Waun Fawr Common and Dowrog Common. In 2012, 339 flowering spikes were counted at these three locations, 310 of which were on Dowrog Common and 29 on St Davids airfield. None was found at Waun Fawr Common, where there had been two in 2009 and one in 2010 (Evans 2013).

The first record of the lichen *Cladonia peziziformis* in Wales was made by the lichenologist Alan Orange at Dowrog in 1989 (Fig. 180). Since then it has only been recorded at two other locations, Mynydd Garngwch in Caernarfonshire, in 2002, and South Stack on Anglesey in 2007. It forms cushions of small rounded

FIG 180. When it was first found on Dowrog Common the cup lichen *Cladonia peziziformis* was noted as being frequent over an area of around 1,400 m² on bare peaty patches in the burnt heathland, but it was lost as the vegetation grew back. This is a dried specimen. (Alan Orange)

grey-green squamules – small leaf-like scales. The majority of records of this ground-living (terricolous), lichen are from heathlands close to the coast in the south and west of Britain. The species appears to have a requirement for bare ground exposed during fires, being unable to withstand competition. It is possible that the conditions provided by the nutrients released during burning trigger its appearance from a spore bank in the soil, though no experimental work has been carried out to investigate this. It was only seen at one place on Dowrog Common and was lost as the vegetation recovered. Despite several searches it has not been recorded since. According to the Species Recovery Trust, *Cladonia peziziformis* became extinct in England in 1968, due to 'human disturbance, inappropriate use of burning for land management, the natural succession of heathland vegetation and high grazing levels'.

Because of the rich diversity of habitats, the common supports a wide range of invertebrates, including marsh fritillary butterfly, scarlet tiger moth, small red damselfly *Ceriagrion tenellum*, hairy dragonfly and scarce blue-tailed damselfly. A number of rare flies have been recorded from the area, including the band-winged pygmy marsh, or snail-killing, fly *Colobaea bifasciella*. This is a distinctive fly, having a yellow body with stripes along the top and sides of the thorax, darkened front legs and wings that have two bands. The larvae of this fly, as the name suggests, attack stranded or aestivating water snails, and adults are on the wing from May until August. The related dark-sided pygmy marsh fly *Colobaea distincta* is also found here. Pupae of this species have been found inside an empty shell of the ram's horn snail *Anisus spirorbis*.

There is a rich beetle fauna, notable species being *Bradycellus caucasicus* and the black-margined loosestrife beetle *Galerucella calmariensis*. Both the adults and larvae of the second species feed, as its name suggests, on purple-loosestrife. The beetle is cylindrical, mid-brown and 3–6 mm long. Seen from above it often has two blackish lateral lines down either side. Its larvae are yellow speckled with black and resemble small caterpillars.

Dowrog Common is an important area for birds, with hen harriers roosting here during the winter (Green 2002). In fact, it is the main roost site for the species in Pembrokeshire, with three individuals recorded in the winter of 2000/01 – though they can also be seen on other commons, including Puncheston (Fig. 181). Castlemartin, Marloes Mere and Skomer are also favoured areas. Outside the breeding season the birds move away from exposed upland areas to the lower, and milder, coastal zone, where they can hunt for voles across bogs and marshes (Lovegrove *et al.* 1994). The Reverend Murray Mathew described how hen harriers bred on moorland throughout the county around 1844, but the decline in the species across the United Kingdom, due to persecution, was

FIG 181. A female hen harrier on Puncheston Common, showing all the underside features of this raptor. This superb photo won Jean Dovey, a keen local photographer, a Birdguides Photo of the Week in November 2018. (Jean Dovey)

mirrored in Wales. It was only in the 1950s that the breeding population began to increase again and birds were encountered more frequently in winter. Merlin and short-eared owl are likewise regular winter visitors to Dowrog. Another raptor, Montagu's harrier *Circus pygargus*, which is now a rare summer visitor and passage migrant, formerly bred in small numbers at a number of sites in Wales, including Dowrog Common. In 1954, for example, two nests, each containing three young, were found here. All six chicks fledged successfully, and breeding continued for a number of years.

Willow blister

An extremely rare species of fungus, willow blister *Cryptomyces maximus*, was discovered on Dowrog Common in 1987 by members of the British Mycological Society. It is a conspicuous fungus found growing through the bark on dead, or dying, willow twigs, appearing as a black fruit body, usually with a striking yellow/orange border (Fig. 182). It is visible all year round, but the yellowish edge

FIG 182. Willow blister is a rare ascomycete fungus found on dead and dying willow twigs. Here the fertile fruit body has broken through the bark, producing spores for dispersion by the wind. (Trevor Theobald)

can fade later in the year, making it less easy to spot. Due to its scarcity it has been included in a list of the most threatened species on the planet, published by the IUCN and the Zoological Society of London (IUCN 2011).

Willow blister appears to be transmitted by wind-borne spores, infecting only those trees which have already been damaged, perhaps by hedge-trimming or by animals. The fungus produces a strong scent, though, which some authors have suggested might be attractive to insects that could be involved in its dispersal (Granmo *et al.* 2012). Willow blister may be associated with humid localities, preferring willow trees close to water, but there are never many fruiting bodies and they are only found on some specimens. At Dowrog the only plants affected were some small scrubby bushes in a ditch, adjacent to an area containing several horses. The horses could access the bushes, and there was evidence that they had been damaged by grazing. Earlier observers also noticed an association with injury by horses, but the reason for the link is not clear. Although the affected parts of the willow die, the fungus appears not to be a significant threat to the tree (Harries 2011).

Willow blister has been known for over two hundred years, but it has never been common. James Sowerby (1757–1822), the artist and natural historian, was the first to describe the species, from specimens collected near Cambridge in England during 1801 (Sowerby 1803). Since that date, the fungus has probably been observed fewer than a hundred times, and records have tended to occur in small groups, with long intervals between them. Following Sowerby's description, for example, there were no more records for around seventy years. Before the discovery at Dowrog the Fungal Records Database for Britain and Ireland listed only five records in the twentieth century: 1910 (England), 1925 and 1957 (Scotland), 1964 (Northern Ireland) and 1966 (England). The assumption therefore is that there are small, widely dispersed, or fragmented, populations with minuscule numbers of fruiting bodies produced at infrequent intervals. More recently it has been recorded in northern Europe from Denmark, Sweden, Finland, Ireland and Norway, as well as Pembrokeshire, and there was a single observation from Shropshire in 2015. There are also records from North America but none since 1971. The geographical distribution suggests a preference for cooler climates and more northerly latitudes, from sea level up to 600 m (Minter 2017).

Since 2008, members of the Pembrokeshire Fungus Recording Network have carried out annual surveys at Dowrog to check whether the species is still present (Harries *et al.* 2013). Other suitable sites have similarly been visited, and willow blister has now been recorded from five additional locations in the county: Trefeiddan Common, Porthclais carpark, a footpath near Nine Wells, Corsydd Llangloffan and Goodwick Moor. Several sites have been visited by researchers from Aberystwyth University in an attempt to sample air-borne willow blister spores, but seemingly without success.

ST DAVIDS AIRFIELD HEATHS

Together the five sites to the north of the airfield comprise an extensive area of heathland, with habitats ranging from the wet heath itself to acid marshy grassland, fen and swamp. On all five heaths there are small disused clay pits with bogbean and common cottongrass. Waun Vachelich has extensive areas of wet heath with abundant heather, cross-leaved heath, bell heather, western gorse, purple moor-grass and devil's-bit scabious. Interestingly, despite the area lying some distance from the coast sea plantain can be found on the more open areas. Meadow thistle *Cirsium dissectum* also occurs along with few-flowered spike-rush *Eleocharis quinqueflora* and hooked scorpion-moss *Scorpidium scorpioides*, one of the most dominant species to be found in these mineral-rich flushes. Waun

Llandruidon is similar, but there is more creeping willow, and along with Waun Llechell it is noted for the swamp communities dominated by water horsetail, marsh cinquefoil and grey willow. Greater bladderwort *Utricularia vulgaris* and lesser water-plantain are abundant in the swamp at Waun Llechell.

On Waun Treflodan and Waun Caerfarchell there are also extensive areas of marsh and streamside vegetation, with water horsetail, hemp-agrimony, purple-loosestrife, common fleabane, wild angelica and meadowsweet.

PUNCHESTON COMMON

The numerous springs and small streams on this common have created extensive flushes, which are grazed by cattle and ponies. Much of the area is dominated by a short turf containing common yellow-sedge *Carex demissa*, carnation sedge *C. panicea*, sharp-flowered rush, bulbous rush *Juncus bulbosus* and devil's-bit scabious. Some of the flushes contain marsh arrowgrass, marsh pennywort, marsh lousewort *Pedicularis palustris*, bog pimpernel and marsh violet *Viola palustris*. Others contain star sedge, round-leaved sundew and bog asphodel. Among these wetland plants is a sizeable population of pale butterwort *Pinguicula lusitanica*. It flourishes under the intensive grazing regime, many of the young plants being clustered in the hoofprints. Other wetland plants of note include wood horsetail *Equisetum sylvaticum*, ivy-leaved bellflower *Wahlenbergia hederacea*, white beak-sedge, hare's-tail cottongrass and fen bedstraw *Galium uliginosum*.

The common is another important site for dragonflies and damselflies, species found here including keeled skimmer *Orthetrum coerulescens*, golden-ringed dragonfly *Cordulegaster boltonii*, southern hawker, common darter *Sympetrum striolatum*, southern damselfly *Coenagrion mercuriale*, emerald damselfly *Lestes sponsa*, large red damselfly *Pyrrhosoma nymphula*, blue-tailed damselfly *Ischnura elegans*, common blue damselfly *Enallagma cyathigerum*, azure damselfly *Coenagrion puella* and beautiful demoiselle. There are many other notable insects, including a number of beetles and flies.

TRE-RHÔS COMMON

While Tre-rhôs has a similar range of plants to the other commons, the real interest here is the geomorphology, since the heathland is covered by small basin mires associated with relict 'pingos' of periglacial origin – that is, structures created on the margins of glaciers (Fig. 183).

FIG 183. Tre-rhôs Common, to the west of Wolf's Castle, is covered by small pingo scars associated with 'ground-ice depressions' of periglacial origin. The lush vegetation in the scars contrasts with the surrounding heathland. (Annie Haycock)

The biggest of these features on the common is around 7 m in diameter, with at least 2 m of peat in the centre. It is surrounded by drier ramparts up to 1.5 m high at the edges. The term *pingo* is of Inuit origin and means 'small hill', so the periglacial features here are more accurately pingo 'scars', or 'fossil' pingos (Walmsley 2008). A pingo is formed when water, which has been forced upwards by freezing pressure, itself freezes in a lens-like shape near the surface of the ground, lifting the surface layers and forming an earth-covered ice-cored mound. When this mound thaws, the surface layer melts first and slumps down the ice core, creating a rim of earth. Later, when the core itself melts, it leaves a hollow, surrounded by the raised rim, the pingo scar. If slumping has occurred evenly around the core, the rim will form a complete circle, as is the case with most of the structures at Tre-rhôs.

A number of studies have examined the distribution, morphology, structure and origin of these ramparted depressions at a range of sites in Wales (Ross 2006,

Ross *et al.* 2007). Pingos are perhaps better developed here than anywhere else in the British Isles, apart from Norfolk. The first phase of this work, commissioned by the former Countryside Council for Wales, was undertaken largely in response to the destruction of a number of features by drainage, excavation and levelling for agricultural use. These 'ground-ice depressions' are important habitats in their own right, and sites with these features have been found to support a wide array of plants and animals. As yet, there have been no studies of the species to be found in the scars at Tre-rhôs, but it is hoped that this omission will be rectified in the next round of surveys.

The central and eastern portions of this common consist of mature heathland with the usual mixture of heather, cross-leaved heath and western gorse, together with *Cladonia impexa*, the commonest of the 'reindeer lichens' found on heaths and moors. Gorse and grey willow are invading the common from the margins, and woodland consisting of alder, ash *Fraxinus excelsior*, sessile oak *Quercus petraea* and sycamore has developed in one area.

FIG 184. Trefeiddan Moor is a very wet common and, as a result, was ungrazed for many years. As on the Teifi Marshes, water buffalo are now being used to open up the area. (Jonathan Mullard)

TREFEIDDAN MOOR

Swamp, fen, acidic marshy grassland and lowland heath are all present on this common on the exposed western extremity of the St Davids peninsula. The area of very wet common forms a very distinct landscape which contrasts with the surrounding farms and fields (Fig. 184). Although it is common land, it was not grazed for many years, due to the extremely wet nature of the site, but recently water buffalo have successfully been used. It is likely that at one time the peat in this area was cut to provide fuel, and there are a number of small clay pits. The swamp covers a shallow pool and is dominated by bogbean and water horsetail, with lesser bladderwort, lesser water-plantain and alternate water-milfoil also present. The pool is fringed by marshy grassland, typified by rushes, or tussocks of purple moor-grass. Towards the edges, there are patches of wet heath, with heathers, carpets of bog-moss and sedge and the golden spikes of bog asphodel.

Of the eleven different dragonflies and damselflies recorded, the small red damselfly is the scarcest. Wintering and passage wildfowl include Bewick's and whooper swans as well as small numbers of mallard, teal, wigeon, pintail and shoveler. Twenty species of bird breed on the common, the most interesting being stonechat, linnet, yellowhammer *Emberiza citrinella* and whitethroat in the heath and scrub.

Fascinating as these lowland commons are, the majority of the common land in Pembrokeshire is located in the uplands, such as Mynydd Preseli. These areas are considered in more detail in the following chapter.

CHAPTER 13

Higher Ground

Wall of my boyhood, Moel Drigarn, Carn Gyfrwy, Tal Mynydd, In my mind's independence ever at my back.

Waldo Williams's own translation of the
beginning of his poem 'Preseli'

WHILE MOST OF THE COUNTY consists of a wave-cut coastal plateau, in the north there is higher ground, which mainly follows the distribution of the igneous rocks. The most prominent features are Mynydd Preseli and Carningli, the two being separated by Cwm Gwaun, with an outlier at Treffgarne. There are no real mountains though, Foel Cwmcerwyn, the highest point in Pembrokeshire, reaching only 536 m. It is, however, a complete contrast to the lower-lying land in the south. The landscape, consisting of moorland, heath and grassland, supports a wide range of species, some of which are quite rare. In addition, the area is rich in archaeological and historical sites, and includes part of the Preseli Landscape of Outstanding Historic Interest. The open, unimproved uplands contain a range of relict prehistoric landscapes, with monuments surviving from the Neolithic through to the Iron Age and Romano-British periods. Some of these, such as the triple enclosures with their numerous hut platforms surrounding the three massive Bronze Age cairns of Foel Drygarn, at the eastern end of Mynydd Preseli, are conspicuous and famous monuments (Fig. 185).

FIG 185. Foel Drygarn, at the eastern end of Mynydd Preseli, is one of the most dramatically sited Iron Age hillforts in Wales, with three massive Bronze Age cairns on its summit. (Jonathan Mullard)

MYNYDD PRESELI

Mynydd Preseli consists of a long ridge of open moorland, 300–400 m above sea level, broken up by a large number of impressive rocky crags, or tors. While the mountains themselves are composed of soft Ordovician slates, formed around 470 million years ago, the crags are made of a volcanic rock, known as dolerite, created when molten rock or magma cooled and crystallised beneath the Earth's surface. After crystallisation, the cooling process produced stresses which resulted in fractures known as 'cooling contraction joints', giving the tors their fragmented appearance. The crags are, in addition, often surrounded by extensive 'blockfields' consisting of enormous numbers of frost-shattered boulders. These were formed as a result of debris being carried down the slope by the movement of wet soil above permafrost between 40,000 and 10,000 years ago.

Possibly because of the striking nature of the crags and the relative remoteness of the area there is a pervading sense, especially on the summits and the upper slopes, that the hills are a place apart from the rest of the world.

The Reverend R. Parri Roberts, who successfully led the fight in the late 1940s to prevent Mynydd Preseli becoming a permanent military training area, stated, 'We nurture souls in these areas' (Wyn 2000). Today the area still attracts those who are seeking tranquillity, though visitors are also drawn by the links between Stonehenge and the bluestones, which, for over a century, have been thought to come from the ridge.

The bluestone quarries

There are two main types of stone at Stonehenge, the large sarsens, a type of silcrete found mainly on the Marlborough Downs, and the smaller bluestones. 'Bluestone' is the common name for spotted dolerite, an igneous rock that appears blue when broken or wet, and which contains prominent white spots of feldspar. Spotted dolerite is the most common bluestone at the monument, comprising 30 of the 43 surviving stones, but other bluestones consist of rhyolite, another igneous rock, argillaceous volcanic rocks and sandstone (Fig. 186).

Carn Menyn, a jagged outcrop, or tor, on the south side of Mynydd Preseli, was thought for many years to be the source of the majority of the Stonehenge bluestones (Fig. 187). Although Herbert Thomas, an eminent geologist, is often

FIG 186. The bluestones at Stonehenge are overshadowed by the larger sarsens but have a much more interesting history. (Jonathan Mullard)

FIG 187. Carn Menyn is a natural outcrop of spotted dolerite that naturally fragments into pillars, blocks and screes and was once thought to be the source of the Stonehenge bluestones. It now appears that most of them come from other prehistoric quarries in the area. (Jonathan Mullard)

quoted as the first person to suggest the link, in the 1920s, it seems the theory was originally put forward by Sir Andrew Ramsey in the mid-nineteenth century (Thomas 1923). Ramsey served for forty years, from 1841 to 1881, on the British Geological Survey, and was first stationed at Tenby. As a result, much of his subsequent geological work dealt with Wales.

In 2005 work led by the archaeologists Timothy Darvill and Geoff Wainwright again supported the idea that Carn Menyn was the primary quarry, but, until recently, there was no agreement on the precise source of the bluestones on Salisbury Plain. That changed in 2009 when Richard Bevins from National Museum Wales and Rob Ixer, a consultant petrologist, analysed bluestone fragments that had been collected in 1947 from a field near Stonehenge by a local archaeologist, the appropriately named John Stone. Stored in a shoebox in Salisbury Museum, they had been catalogued and forgotten. It is not widely known but there is a large amount of stone debris buried at Stonehenge and in the surrounding area. Thought to derive from the original shaping of the megaliths, or from destruction of the stones in the post-medieval and modern periods, it is known by archaeologists as the 'Stonehenge layer'. As such it is an extremely important resource for understanding the monument.

Bevins and Ixer identified some of the samples from the museum as rhyolite, but the structure of the rock was not one that Bevins had ever seen in Pembrokeshire, or anywhere else in Britain for that matter. Some months later he remembered that he had visited a Pembrokeshire outcrop twenty years earlier but had never had time to analyse his samples. The next morning, he sent them straight to the laboratory – and amazingly they proved to be a perfect match for some of the pieces in the shoebox in Salisbury. The samples collected by Bevins came from Craig Rhos-y-felin, a striking crag in a small valley north of Mynydd Preseli. When Ixer studied the microstructure of the rock he found that every sample was very slightly different and the structure varied depending on the location on the crag. Only at the north end of the outcrop was there a near-perfect match with the Stonehenge bluestones. Ixer described the rock structure in both samples as 'Jovian' since he felt that the microscopic swirls he could see in the rock resembled the weather patterns on Jupiter (Ixer & Bevins 2011).

FIG 188. Craig Rhos-y-felin, the source of one of the bluestones at Stonehenge. The broken megalith is clearly visible following the archaeological excavations. (Jonathan Mullard)

FIG 189. Carn Goedog, an impressive outcrop on the northern slope of Mynydd Preseli, is now recognised as the source of the majority of the Stonehenge bluestones. (Jonathan Mullard)

The naturally formed pillars and sloping west side of Craig Rhos-y-felin seemed to form an ideal megalithic quarry (Fig. 188), and in September 2011 excavations proved that this was the case (Parker Pearson *et al.* 2016). The first trial trench exposed a 4 m megalith, lying flat, which seemed to have been moved there and abandoned, perhaps because a large block had snapped off the underside. Later the archaeologists found an empty niche from which a rhyolite pillar had been removed. There was also a shallow groove in the rock face where it is thought a hole was made to enlarge a natural fissure behind the stone and lever it away from the outcrop.

Despite apparently ideal conditions at Craig Rhos-y-felin it seems, however, that only one of the Stonehenge bluestones came from the site. It is now agreed that the majority came from Carn Goedog, on the northern slope of Mynydd Preseli (Fig. 189). Five stones are confirmed as originating from this site, while two match outcrops on Cerrigmarchogion, to the southwest. A further four stones match a number of other outcrops, but not Carn Menyn. That does not necessarily exclude Carn Menyn, but there is no geological evidence to support it at the present time. The precise location of the sources of the bluestones is critical for focusing archaeological investigations, as there has been a long-running debate as to whether the stones were transported to Salisbury Plain by the actions of humans or by glaciers, and if the former, by what route (John 2018). The detailed evidence of prehistoric quarrying that is now available would, however, seem to indicate that people were responsible. How the bluestones were moved, over a distance of more than 250 km, remains unknown, but it is probable that they were both carried on boats, or rafts, and hauled over land.

There are a number of theories, some more viable than others, as to why the bluestones were selected. One is that spotted dolerite had some significance to people because the pattern revealed in the cut surface resembled the night sky. Other people have suggested that the bluestones have musical qualities because certain rocks ring loudly when struck. Whether these theories have any substance will never be known. Unfortunately, though, as awareness of its significance continues to spread, more and more pieces of bluestone are being illegally removed. An increasing amount of stone chips and large chunks of rock are disappearing, sometimes for sale.

Radiocarbon dates from both Craig Rhos-y-felin and Carn Goedog indicate that the bluestones were extracted from the quarries three or four hundred years before they were erected at Stonehenge. The question now is, where were the stones in the intervening period? Archaeologists think that they were erected locally, creating a monument very close to the quarries, before being dismantled and moved to Salisbury Plain. In this period, some 5,000 years ago, henges and passage tombs such as Bryn Celli Ddu on Anglesey were being constructed, so it has been suggested that somewhere between the two quarries, perhaps on a tributary of the River Nevern, lie the remains of an unknown passage tomb, or a henge, built using bluestones. Only further surveys and excavations will reveal the truth.

Vegetation

All this activity in prehistory had a significant impact on the natural vegetation of the area. Pollen grains found in peat deposits show that Mynydd Preseli was once covered by woodland, and it has been suggested that this is the reason for the absence of montane plants such as stiff sedge *Carex bigelowii*, crowberry *Empetrum nigrum*, dwarf willow *Salix herbacea* and alpine clubmoss *Diphasium alpinum*. By the late Bronze Age, 1000 to 700 BCE, the trees had disappeared, to be replaced by the landscape we are familiar with today – an open landscape maintained through rough grazing by sheep, cattle and ponies. The flora consists mostly of species-poor acid grassland, with mat-grass *Nardus stricta*, heath rush *Juncus squarrosus*, sheep's fescue and common bent dominating, along with tormentil *Potentilla erecta* and heath bedstraw *Galium saxatile* on the drier upper ridges.

Covering 2,132 ha, Mynydd Preseli is the largest area of registered common land in Pembrokeshire. Over the last decade, however, there has been a dramatic reduction in grazing, following restrictions on cattle movements due to bovine tuberculosis and reductions in the value of ponies. Species which rely on relatively well-grazed vegetation are threatened by these changes, and where grazing pressures have declined dramatically even those species which rely on light grazing are affected. It might be argued though that while this is a cultural landscape of considerable antiquity, a return, via the spread of scrub, to the original forests should be supported. Carn Goedog itself provides an indication of what species such a future forest might contain, at least in its first stages. Among the rocks on the outcrop, protected from the ever-present sheep, are numerous rowan *Sorbus aucuparia* saplings, one or two hazels *Corylus avellana* and a wind-pruned holly *Ilex aquifolium*, which has taken up the shape of the rock behind which it shelters.

The area is particularly noted though for the combination of upland and lowland features found here, and for the extensive and varied flushes. Around and below the numerous springs a diverse range of vegetation has developed. On the wetter slopes purple moor-grass, cross-leaved heath, patches of bog-myrtle

FIG 190. Fertile spikes of marsh clubmoss growing out of the water on Brynberian Moor, which is now one of its main strongholds in Britain. (Stephen Evans)

and heather can be found. Dioecious sedge *Carex dioica*, white sedge, common butterwort *Pinguicula vulgaris*, ivy-leaved bellflower, cranberry and royal fern also occur in these flushed areas. In some of the more base-rich flushes there are often large areas of mosses, such as hooked scorpion-moss, yellow starry feather-moss *Campylium stellatum* and rusty hook-moss *Drepanocladus revolvens*. On the saddles of the upper slopes, above the spring line, these flushes merge into areas of blanket bog, marked by stands of hare's-tail cottongrass.

Rare plants found in the wetter parts include pale butterwort, bog orchid *Hammarbya paludosa* and, on the northern slopes of the mountains, around Brynberian Moor, marsh clubmoss *Lycopodiella inundata* (Fig. 190). Clubmosses are simple plants, related to ferns. Their common name indicates their resemblance to true mosses, while the 'club' refers to the shape of the spore-bearing cones that most produce. Marsh clubmoss is the only British member of its genus. In appearance, it resembles another clubmoss, stag's-horn clubmoss *Lycopodium clavatum*, which is relatively common. Marsh clubmoss forms long leafy prostrate strands that grow on the surface of the soil, and in the autumn vertical spikes develop with pale brown spore cases at their tips. In Wales it has always been relatively uncommon, being found on Mynydd Preseli and around the headwaters of the Afon Conwy in Snowdonia. Formerly widespread on acid soils throughout much of Britain, it has declined substantially during the twentieth century, mainly due to the destruction of its habitat.

Dry dwarf shrub heath is now largely confined to the eastern parts around Carn Menyn and Foel Drygarn, where there are extensive stretches of bilberry *Vaccinium myrtillus* and heather. Clumps of western gorse and common gorse dot the lower, rockier, areas and are nibbled short by sheep and ponies, while bracken is spreading where soils are deeper. Several scarce ferns are associated with the rock outcrops and their blockfields, including beech fern *Phegopteris connectilis*, Wilson's filmy-fern *Hymenophyllum wilsonii* and parsley fern *Cryptogramma crispa*, while the fir clubmoss *Huperzia selago* is frequent among the moss-filled crevices of some of the blockfields.

Mynydd Preseli is known for its rare damselflies, with southern damselfly, scarce blue-tailed damselfly and small red damselfly are abundant around the flushes, shallow streams and fen pools. It has one of the largest populations of southern damselfly in the United Kingdom, with thousands of adults present during the summer and a long history of records from the site (Fig. 191). It is, however, on the extreme northwestern fringe of its European range. The southern damselfly's other main stronghold is in the New Forest, but smaller colonies occur in Devon, Dorset, Anglesey, Gower, Oxfordshire and on the floodplains of the Test and Itchen rivers in Hampshire (Thompson *et al.* 2003). Undoubtedly,

FIG 191. The southern damselfly is primarily a species of base-rich runnels and streams that occur within areas of acid heathland, such as that found on Mynydd Preseli. (Harold Grenfell)

one of the main reasons for the decline of the southern damselfly in Britain over the last forty years has been the change in grazing regimes on some sites (Evans 1989). Grazing by heavier animals, such as cattle and horses, is preferred, as it causes some poaching of watercourse margins and creates the diversity of tussock structure preferred by the southern damselfly. Several populations have become extinct due to failed grazing regimes, including those at St Davids Head and Waun Isaf, near Mynachlog-ddu in the east. The larvae live in flushes and shallow runnels, often less than 10 cm deep, with slow-flowing water. The adults are on the wing from June to August but are weak fliers, tending to stay level with grasses and other vegetation. Females lay eggs onto submerged plants, and the predatory aquatic larvae probably take two years to mature.

The area also supports the largest Pembrokeshire population of the black darter dragonfly *Sympetrum danae*. The males of this small species are the only black dragonflies found in Britain. Females and immature males have a yellow abdomen and brown thorax, with a black triangle on top. Males become black as they mature but some yellow markings remain along the sides.

Birds

In contrast to the plants and insects, the birdlife of Mynydd Preseli is not that rich. Snow buntings *Plectrophenax nivalis* occur, however, on passage, and sometimes overwinter on the highest points, while ring ouzels *Turdus torquatus* occasionally breed in the old quarries. Whinchat *Saxicola rubetra*, one of the most characteristic birds of the Welsh uplands in summer, also occurs here. Typical species in these upland areas though include wheatear, meadow pipit, skylark, raven, buzzard and red kite (Fig. 192).

Formerly widespread throughout Wales, the red kite was almost eradicated between the late eighteenth and early twentieth centuries. As a frequent predator of young domestic fowl and game birds, it was among the least-tolerated raptors and one of the easiest to exterminate. By 1894 Murray Mathew reported that:

> *The Kite is only a rare occasional visitor, and it is long since it has nested in the county. From evidence we have accumulated we think it probable that in the adjoining counties of Cardigan, Carmarthen, and Brecon there may be at the present day at least six or seven pairs of Kites annually nesting, in great danger, we fear, of destruction; we wish such interesting birds could obtain protection.*

FIG 192. Red kite, now a common species again in the upland areas of Wales. They are more common in the north of Pembrokeshire than in the south. (Harold Grenfell)

In 1944 Bertram Lloyd mentioned that the position had still not improved:

*A tiny remnant of these grand and once well-known birds still nest in Wales –
owing to the well-organised efforts of the egg-collecting fraternity or Oologists, a
very tiny remnant!*

As the population of red kites in Wales finally increased again during the
1990s there was hope that they would breed in Pembrokeshire once more. At
long last, in 2002, three pairs bred in the northeast of the county, close to the
Carmarthenshire border. Red kites are still generally scarcer in the south of the
county than in the north. The majority of records come from the area between
Newport and Cardigan, but individuals have, at times, been reported from St
Davids, Ramsey, Haverfordwest and the Tenby area (Green & Roberts 2004). Most
kites breed near their birthplace, however, and they 'infill' rather than pioneer
new ground (Green 2002).

Lloyd recounts an 'amusing old Welsh song' about the red kite which was
sung and written down for him by 'a middle-aged woman in Cardiganshire in
June 1924'. Unfortunately, she only wrote down the words, not the tune. The
lady was a native of Boncath, 'where the lines were still current as a nursery
rhyme in her childhood some forty years ago', i.e. around 1904. The supposed
location of the song, Freni Fawr, is an outlier of Mynydd Preseli close to Boncath.
Bwncath means buzzard in Welsh, but Lloyd supposed that 'the buzzard is rarely
distinguished from the kite by the natives.' The words of *The Boncath Kite* are
translated as follows:

*I placed my hen to sit
On the top of Freni Fawr.
With ten eggs under her
And ten were duly hatched.
On Sunday I went to church
A kite took them away.
Whilst I own a hen with chicks
I will never go to church again.*

Given the tendency of kites to take domestic fowl, literally putting all your eggs
in one basket, on the top of the nearest high hill, would seem to be a sure way to
lose them. Or perhaps the intention is to create an excuse for missing church!

Once golden eagles, as well as red kites, soared above Mynydd Preseli,
but they are now absent – as they are from many areas in Britain, because of

FIG 193. Foel Eryr 'the hill of the eagle', the highest point at the western end of Mynydd Preseli, may have been a former nesting site for golden eagles. (Jonathan Mullard)

persecution in the past. Some wild animals are charismatic enough for their former presence to have been recorded in place names, which can be used by biological historians as an important resource for reconstructing past landscapes (R. J. Evans *et al.* 2012). The golden eagle is one such animal, and Foel Eryr, 'the hill of the eagle', the highest point at the western end of the Preseli ridge, is an obvious example of a place name that might well indicate the existence of a former nesting site (Fig. 193). There are, however, very few locations named after this species in South Wales, and the eagle's stronghold was always Snowdonia, so even in earlier times the population in Pembrokeshire probably amounted to only a few pairs. From the name alone it could have been a site used by *Eryr y mor*, the white-tailed eagle *Haliaeetus albicilla*, rather than *Eryr Euraid*, the golden eagle – but the habitat requirements of the two species are markedly different. White-tailed eagles tend to nest more often in trees, at lower altitudes and closer to water than golden eagles. In contrast golden eagles tend to use sites more than 150 m above sea level, with higher ground close by, but not necessarily close to water or wetlands.

Adult golden eagles occupy a hunting and nesting area known as a home range all year. The eyries, or nests, are traditional and can be used for many years by the same or successive birds, being added to each year they used. As a result, they can be extremely large. Nests on cliffs are often 1–1.5 m across and up to 2 m high, while nests built in trees can be twice this size. Today Foel Eryr is an almost conical hill with relatively smooth slopes on all sides, broken only by a few large rocky outcrops. Although it commands a fantastic view of the countryside around, there are no cliffs here, so golden eagles may have used these outcrops as the base for their nest. The other option, of a nest in a large tree, seems unlikely in this location – but given the ecological history of the area perhaps woodland survived here longer than elsewhere.

CARNINGLI

This range of hills, incorporating Mynydd Dinas, Mynydd Melyn, Mynydd Caregog and Mynydd Carningli itself on the western side of Cwm Gwaun, is dominated by bands of intrusive igneous rock which outcrops to form the rocky tors and associated frost-shattered blockfields (Fig. 194). There are many myths

FIG 194. The summit of Carningli, the 'hill of the angels', surrounded by the most extensive blockfields in Pembrokeshire. (Jonathan Mullard)

and legends associated with Carningli, the prominent mountain above Newport.
In the fifth century Saint Brynach used to climb to the summit to 'commune
with the angels', and as a consequence it is still known as the 'hill of the angels'.
In some old texts and maps the mountain is called Carn Yengly, or Carnengli,
which are probably corruptions of *Carn Engylau*, 'the rocky summit of the angels'
(Charles 1992). There are no records of whether Saint Brynach stayed overnight
but it is said that anyone who spends a night on the mountain either becomes
a poet or goes mad. On the summit, however, is one of the largest hillforts
in Wales, so many people must have slept there in the past. The structure
incorporated the rocky tors and blockfields, substantial stone embankments
being built where the natural defences were less secure.

The blockfields on Carningli are the most extensive in Pembrokeshire and
support a range of interesting plants in the damp, shady crevices between the
boulders. These include lady-fern *Athyrium filix-femina*, hard fern *Blechnum
spicant*, polypody *Polypodium vulgare* and male-fern *Dryopteris filix-mas*, along with
bilberry, foxglove *Digitalis purpurea*, wood sorrel *Oxalis acetosella*, wood sage and
other species, again suggesting that the area used to be covered by woodland.
The lichens found here include crustose species such as *Ophioparma ventosum*
and *Lecanora caesiosora*, which are locally common on base-poor rocks in upland
areas in northern and western Britain. In contrast to Mynydd Preseli, the area
is too acidic to support a wide range of bryophytes, although western frostwort
Gymnomitrion crenulatum and a rock moss *Andreaea megistospora* can be found
here. To the west though, on Garn Fawr, Mynydd Dinas, can be found 22 tufts of
black-tufted moss *Glyphomitrium daviesii*. A notable species, its main populations
are on coastal rocks, mainly along the west coast of Scotland, but it also occurs
in Ireland and at an unusual inland locality in Aberdeenshire. The moss forms
small dark green to black cushions, and spore-bearing capsules are common,
giving colonies the appearance of miniature pincushions.

Away from the blockfields the dominant vegetation consists of heather, bell
heather and western gorse, interspersed with patches of grassland containing
common bent, red fescue, sheep's fescue, sweet vernal-grass *Anthoxanthum
odoratum*, tormentil and heath bedstraw. Soils that are seasonally waterlogged
support areas of wet heath with cross-leaved heath and purple moor-grass.

Lower down the slopes there are numerous springs and wet flushes, which
support plants such as sharp-flowered rush, many-stalked spike-rush *Eleocharis
multicaulis*, common cottongrass, devil's-bit scabious, bog pondweed and
round-leaved sundew. Bog orchid can be found in these areas, along with pale
butterwort, dioecious sedge, white beak-sedge and fir clubmoss. Insects are
numerous, and there is a small breeding population of southern damselfly in

the flushes adjacent to Pont Ceunant, while keeled skimmer and golden-ringed dragonfly have also been recorded from the area.

TREFFGARNE TORS

At the eastern end of Great Treffgarne Mountain, towering above Treffgarne Gorge, are two prominent tors, Poll Carn and Maiden Castle. Visible from the A40 road between Haverfordwest and Fishguard, they are well-known features in this part of Pembrokeshire. Remarkably, the construction of the railway line through the gorge, the building of which defeated Brunel and practically bankrupted the Great Western Railway, was said to have been prophesied in the eighteenth century by a local medium, Sarah Bevan. Experiencing a vision, she described a line of carts moving through the gorge at high speed, with the front cart on fire – suggesting a steam locomotive pulling carriages.

Situated at the eastern end of a ridge of high ground, the tors are all that remains of volcanoes that erupted around 600 million years ago, towards the end of the Precambrian. The ridge is composed of rhyolite lava, which probably

FIG 195. Poll Carn, or Wolf Rock, a prominent tor on Great Treffgarne Mountain, is a key site for both geologists and bryologists. (Jonathan Mullard)

FIG 196. Maiden Castle, also known as Lion Rock or Treffgarne Pinnacles, is a prominent feature in central Pembrokeshire. (Jonathan Mullard)

solidified in the vents of the volcanoes, and tuff, a rock formed from fine ash. The tors therefore represent the remnants of harder, more resistant, rocks. Weaker material, which previously surrounded them, has disintegrated and been removed, possibly by glacial ice. Poll Carn is also known as Wolf Rock, and, together with the nearby settlement of Wolf's Castle, is thought to mark the place where the last wild wolf *Canis lupus* in Wales was slain (Fig. 195). The wolf is also commemorated in Llannerch-y-bleiddiau, 'wolves' glade', in Cwm Gwaun. Other locations for this deed, however, include Coed y Bleiddiau, the 'wood of the wolves', near Maentwrog in Snowdonia. Maiden Castle is also known as Lion Rock, since from certain angles it is said to resemble a crouching lion (Fig. 196).

There are no wolves prowling around the tors now but they are still of great interest to naturalists. Like Mynydd Preseli and Carningli they support a notable bryophyte flora, with over a hundred different mosses and liverworts recorded from the tors and gorge. On the sheltered north side of the tors an oceanic flora has developed, with greater whipwort *Bazzania trilobata*, rock fingerwort

Lepidoza cupressina, straggling pouchwort *Saccogyna viticulosa*, Scot's fork-moss *Dicranum scottianum* and western earwort *Scapania gracilis* present (Bosanquet 2010). Western earwort, one of the most characteristic liverworts found in Wales, is locally abundant around Maiden Castle, found on boulders close to the north side. The thin peaty soil around the tors also supports two uncommon species, matchstick flapwort *Odontoschisma denudatum* and bent-leaved beard-moss *Leptodontium flexifolium*. The latter often persists under bracken on moorland, appearing after the fern has died back in the winter.

Surprisingly, perhaps, there is a population of vernal squill *Scilla verna* near Treffgarne Tors. It is a prominent coastal plant, but its occurrence here, 11 km from the shoreline, has puzzled botanists for years (Evans 2018). There are larger populations on Carningli and Mynydd Dinas, and the species has additionally been recorded from a number of other similar locations, but this is the most inland location in the county. All the sites are on shallow soils close to rock outcrops, or tors. On the St Davids peninsula vernal squill can be found on isolated outcrops such as Carn Warpool, Carn Trefeiddan and Carn Rhosson. Here, however, it is accompanied by a number of other maritime species and, as the outcrops are only 1 or 2 km from the coast, their presence is not surprising. Strong westerly winds could easily carry seeds a short distance inland. It is unlikely though that the wind would be powerful enough to transport seeds as far as Treffgarne.

Explanations for this mystery are varied. Probably the most convincing is that the species was once more widespread before woodland covered Pembrokeshire after the last glaciation and it was then shaded out, except in areas where trees were restricted by shallow soils or exposure. It has been suggested that grazing livestock on coastal farms, which were then moved inland, could have transported seeds in their hooves. Birds have also been suggested as the culprits. Neither of these latter explanations seem very convincing though, so the idea that these isolated populations of vernal squill represent relicts from a former, wider, distribution is the most likely. Without more evidence, however, it is difficult to come to a firm conclusion.

Peregrines once nested on the tors, but they have been absent now for many years, since there is no longer a good mixed bird population in the area for prey (Bob Haycock, personal communication, 2017). The main requirement for a peregrine breeding site is a steep cliff, or similar, with commanding views over open country, so this would have been an excellent location for them. Perhaps one day they will return, as the population in Pembrokeshire is still increasing, with an estimated 315 pairs in 2004 (Green & Roberts 2004). Disturbance may be an issue, however, since the tors are being increasingly used for 'bouldering' by climbers, especially the southwest face of Poll Carn. This activity is damaging the bryophytes, and action needs to be taken to manage the recreational use of the area if its special flora is to survive.

TREFFGARNE GORGE

The value of Treffgarne Gorge for 'lower' plants was recognised as long ago as 1811, when Richard Fenton described the gorge as 'covered with tangled shrubs, lichens and mosses of various sorts and colours'. Down in the gorge, next to the busy A40 road, Tunbridge filmy-fern *Hymenophyllum tunbrigense*, for instance, is present in the small cave where it has been known since 1931, and it also occurs nearby on the many rock surfaces above the cave (Fig. 197). In addition, over fifty species of lichen have been recorded from the area, including the locally notable species of coral lichen, *Sphaerophorus globosus*.

There are many other caves in Pembrokeshire, besides that in Treffgarne Gorge, and these are the subject of the next chapter.

FIG 197. Tunbridge filmy-fern in the small cave near the A40, where it has been known since 1931. (Stephen Evans)

CHAPTER 14

Caves and Cave Life

It is one of the strange connections of the hidden landscape that the clear seas of the early Carboniferous came to house pioneer inhabitants of these islands, living on the edge of a darkness where even troglodytes would not venture.

Richard Fortey (2000)

C AVES ARE ONLY PRESENT WITHIN the limestone belt that extends from Tenby to Pembroke, cavities in the Old Red Sandstone being relatively small and originating mostly in the loss of joint-blocks, or in erosion along fault planes. While sea caves are a common feature of the Pembrokeshire coastline, caves containing archaeological evidence are much rarer. The sites that do contain suitable material though indicate that humans have been in the area from the Palaeolithic onwards. During the last glaciation, the ice sheets did not extend to what are now coastal areas, and the natural caves and rock shelters were used as temporary settlements by hunter-gatherers and occupied by other animals.

There are seven main sites: Hoyle's Mouth, Little Hoyle, Priory Farm Cave and the Wogan on the mainland, and Nanna's Cave, Potter's Cave and Ogof yr Ychen on Caldey Island. In addition, five relatively unknown caves are located high above the waves in the south Pembrokeshire cliffs: Ogof Gofan, Ogof Pen Cyfrwy, Ogof Morfran, Ogof Garreg Hir and Ogof Bran Goesgoch. All of these five are extremely difficult to enter. The entrance to Ogof Garreg Hir, for instance, is 'situated about 21 m down a sheer cliff 46 m high, and access is only possible

FIG 198. Hoyle's Mouth Cave overlooks the Ritec Valley near Tenby. It was occupied by humans on two separate occasions, around 30,000 and 10,000 years ago, but today it is an important bat roost and hibernation site. (Jonathan Mullard)

via a rope descent in an open-sided chimney and a 20 m traverse along a ledge which narrows to 0.5 m' (Davies 1989). Although today these caves are almost unreachable, in postglacial times they would have been easily accessible via scree slopes, formed by a process of alternate freezing and thawing. Subsequently sea-level rise and wave erosion has removed most of the material, and in some cases the outer cave chambers, but the areas of scree which remain in the 'slades' on Gower provide an idea of what these slopes might have looked like (Mullard 2006). In contrast, the well-known sites are by definition accessible, but several support important bat colonies and the deeper recesses should therefore be left undisturbed as much as possible (Fig. 198).

Caves have been described as 'natural museums', because they preserve important archaeological evidence and deposits that provide information about past environments that has been lost elsewhere. The evidence, however, is fragmentary, and excavations in the nineteenth and twentieth centuries were very often superficial and much was missed. Even in 1944 Lionel Cowley, who examined the bones found in a re-excavation of Priory Farm Cave, considered that 'The remains of the other [smaller] animals represented call for no special comment', so we do not have a complete picture of the fauna represented by these finds. Only today are we beginning to piece together the full story.

HOYLE'S MOUTH CAVE

Overlooking the Ritec valley at Penally, near Tenby, Hoyle's Mouth Cave and Little Hoyle Cave, sometimes known as Longbury Bank Cave, penetrate the north-facing slopes of the limestone embankments. Both have been of intermittent interest to antiquarians and archaeologists since the nineteenth century. The remains found in the two caves originate from just before, or just after the maximum of the last glaciation. Collectively they provide some of the most detailed evidence currently available from the Upper Palaeolithic in Wales, some 50,000 to 10,000 years ago.

Hoyle's Mouth Cave has received more attention than any other Pembrokeshire cave, probably because it is easily accessible. It was occupied by humans on two separate occasions, around 30,000 years ago and 10,000 years ago. Along with other caves on Caldey and Gower, it therefore represents one of the first records of human occupation in South Wales. The site was described in a letter by the Reverend H. H. Winwood, of Bath, to the *Geological Magazine* in 1865 as consisting of:

> *a lofty arched entrance, extending about 24 feet into the limestone hill. A tortuous passage, about 79 feet long connects this with a small chamber 8 feet in diameter; another narrow passage, about 32 feet in length, leads into a second chamber, which is dome-shaped, about 11 feet in diameter, and has a funnel shaped roof.*

In 1856 Philip Henry Gosse wrote that:

> *The people talk a good deal of a curious cavern called Hoyle's Mouth, about which they have some strange notions. It opens at the end of a long limestone hill, or range of hills, about a mile inland; and the popular legend is, that it is the termination of a natural subterranean chasm which communicates with the great cave called, the Hogan, under Pembroke Castle, some eight miles distant. It was once traversed, they say, by a dog, which, entering at one end, emerged from the other, with all his hair rubbed off! A gentleman is said to have penetrated to a considerable distance, and found 'fine rooms.' But the vulgar are very averse*

to exploring even its mouth, on the ostensible ground that a Boar, 'a wild pig,'
dwells there; I fear, however, that there are more unsubstantial terrors in the
case. I walked out to look at it; and if I found no dragons, nor giants, nor pigs, I
enjoyed a most delightful rural walk.

While early excavators of the cave did not find any wild pigs, the site is important
for its Pleistocene mammal fauna, which includes mammoth *Mammuthus*
primigenius, woolly rhinoceros *Coelodonta antiquitatis*, horse *Equus caballus*,
reindeer *Rangifer tarandus*, hyena *Crocuta crocuta* and cave bear *Ursus spelaeus*. The
Reverend Winwood noted that:

> *In this last-named chamber, which is at present the farthest part of the cave*
> *accessible, we found, beneath a mass of undisturbed breccia, the right and left*
> *femur, the os innominatum, some vertebra, and other portions of the great cave-*
> *bear: these were extracted in a very perfect state.*

The cave bear was a large animal, weighing 500 kg on average, and largely
herbivorous. It hibernated in the depths of limestone caves, where the remains of
individuals that died during hibernation slowly accumulated over time (Stiller *et*
al. 2010). The species began to decline in Europe around 24,000 years ago, due to
competition with humans for land and shelter, rather than climate change. The
brown bear did not suffer the same fate and has survived until today, as it does
not depend so heavily on caves. Mammals currently inhabiting the cave include
a number of greater horseshoe and lesser horseshoe bats. Natterer's bat *Myotis*
nattereri has also been recorded from the site.

The earliest recorded work on Hoyle's Mouth Cave was carried out by Colonel
Jervis in 1840. He was followed by the Reverend G. N. Smith (1863), the Reverend
H. H. Winwood, mentioned earlier, W. Boyd Dawkins (1874) and E. L. Jones in
1882. The disturbance caused by these explorations and the need to confirm the
contexts of the artefacts unearthed by them prompted rescue excavations by
Hubert Savory, an archaeologist at the National Museum Cardiff, in 1973.

The range of bird remains found in the 'Bear Chamber', located over 30 m
beyond the daylight zone within the cave, demonstrates the rich diversity in the
Ritec valley some 12,000 years ago. The most numerous bones were those of
mallard, but there were other species of dabbling ducks and marginal feeders,
waders and the occasional small passerine and raptor (Table 4). These finds
indicate the presence of quite a complex avifauna that equates closely with other
assemblages of dates around this time from southwest Britain, the Rhineland,
Belgium and Aquitaine. Unfortunately, for a few of the species, the only evidence

TABLE 4. The main bird species found in the Bear Chamber in Hoyle's Mouth Cave. Note that this list is not definitive, owing to the nature of the bone samples. (Modified from Eastham 2016)

Common name	Scientific name
Barnacle goose	*Branta leucopsis*
Lesser white-fronted goose	*Anser erythropus*
Shelduck	*Tadorna tadorna*
Gadwall/wigeon	*Mareca penelope/strepera*
Mallard	*Anas platyrhynchos*
Pintail	*Anas acuta*
Teal	*Anas crecca*
Grey heron	*Ardea cinerea*
Common crane	*Grus grus*
Little ringed plover	*Charadrius dubius*
Woodcock	*Scolopax rusticola*
Redshank	*Tringa totanus*
Black tern	*Chlidonias niger*
Barn owl	*Tyto alba*
Starling	*Sturnus vulgaris*
Linnet/twite	*Linaria cannabina/flavirostris*

is a phalanx bone from the foot. Such remains cannot always be assigned to species, but they are reasonably reliable indicators (Eastham 2016).

There are far fewer published records of cave invertebrates from Pembrokeshire than from the Gower or the Brecon Beacons. The late Geoff Jefferson, in his pioneering summary of Welsh cave biology in the 1980s, only listed seven records for the county, all from Hoyle's Mouth (Davies 1989). They included typical species of the cave-entrance community, such as the orb-web cave spider *Meta menardi* (Fig. 199), the herald moth *Scoliopteryx libatrix*, and the fungus, or cave, gnat *Speleolepta leptogaster*. Other species are likely to be present, however, and the invertebrate fauna of Pembrokeshire's limestone caves probably closely resembles that of the better-known caves elsewhere in South Wales (Mullard, 2006, 2014).

The orb-web cave spider occupies the deeper part of cave entrances, trapping insects on their way in and out. It feeds on virtually any prey and catches both flying and crawling insects. Adaptions to the subterranean environment affect both its life cycle and its life span. Since food is limited, the reproductive effort is reduced in comparison to spiders found outside caves, and the female lays fewer and larger eggs. The eggs are laid in a compact mass and covered with silk to form distinctive sacs, which are hung on stalks to protect them from predators.

The distinctive herald moth often overwinters in caves as an adult and becomes very torpid, resting in the dark zone of the cave covered with beads of moisture (Fig. 200). It appears that a period of suspended development is necessary before the ovaries of the female can produce eggs. The caterpillars can be found from May to July and, in the south of Britain, there is a second brood in August. Herald moths are fairly common and well distributed over much of Britain, though they are less frequent in Scotland.

The cave gnat is particularly widespread and numerous, its thin, translucent larvae, up to 14 mm long, living on the damp walls and feeding on microorganisms and fungal material. Like many animals that live permanently in caves, the gnat can be found in the deep threshold as well as in the dark zone, but it is only rarely recorded from outside caves.

FIG 199. The orb-web cave spider can be found near the entrance of Hoyle's Mouth, where it traps insects on their way in, or out of, the cave. (Jonathan Mullard)

FIG 200. The shadow of an orb-web cave spider looms over a herald moth in Hoyle's Mouth Cave. (Jonathan Mullard)

Large numbers of the common mosquito *Culex pipiens* can found in the cave between September and April. These are females of the autumn generation which have already mated, but which will not lay their eggs until they have left the cave in spring. A large yellowish-brown caddisfly, *Stenophylax permistus*, was also noted as overwintering, along with the short-palped cranefly *Limonia nubeculosa* and *Heleomyza serrata*, one of the fly species most commonly found in British caves.

LITTLE HOYLE CAVE

Little Hoyle Cave lies 350 m south of Hoyle's Mouth, on the edge of a golf course. It is situated in a narrow ridge with entrances on both sides and with a large chimney linking the cave to the surface (Fig. 201). Like Hoyle's Mouth, Little Hoyle has been explored and excavated by antiquarians and archaeologists since the second half of the nineteenth century. Research was begun in 1866

FIG 201. Little Hoyle Cave appears to have been a bear den around the glacial maximum of the last ice age. (Jonathan Mullard)

by Winwood and then G. N. Smith, followed by Edward Laws, Wilmot Power and Professor Rolleston. The majority of the material from their excavations, along with an itemised description of the finds, remains in Tenby Museum. Environmental evidence from excavations indicates that the cave deposits probably span some 50,000 years, although the principal faunal finds, which include bear (probably in this case brown bear), fox *Vulpes vulpes*, reindeer and collared lemming *Dicrostonyx groenlandicus*, are thought to date from the Devensian glacial maximum, around 18,000 years ago. The presence of bear teeth throughout the cave deposits suggests that the cave was a bear den for considerable periods of time.

Among other species, there were numerous remains of barnacle geese and at one stage it was thought that the bears were feeding on a breeding population which was located nearby, the first such record for the species from the Pleistocene in Britain (Green & Walker 1991). Taken together, however, the 87 bones probably represent not more than nine individual birds, so they cannot be considered evidence of a major cull of wildfowl congregating in the valley, and are more likely to represent the utilisation of a temporary resource (Eastham 2016).

Both species of horseshoe bat have been recorded hibernating in the cave, along with Natterer's bat, brown long-eared bat *Plecotus auritus* and Daubenton's bat *Myotis daubentoni*.

BLACKROCK QUARRY FISSURES

Sir William Boyd Dawkins recovered the bones of mammoth, hyena, woolly rhinoceros, lion *Panthera leo*, horse, and, supposedly, hippopotamus *Hippopotamus amphibius* from fissures in the Blackrock quarries, near Tenby, in 1871. The quarry, which is now disused, has been thoroughly examined over the years but no trace remains of any caves. The mention of hippopotamus, an interglacial species, has to be treated with caution, unsupported as it is by similar finds from neighbouring caves. In his explorations Dawkins (1874) noted the occurrence of a strange 'fungoid deposit of calcite':

> When examining the Black-rock quarries in 1871, the workmen pointed out a small opening which they believed to be the entrance of a cave, but which was too small for them to enter. By knocking off, however, a few sharp angles, I got into a small chamber about five feet high, with sides, roof, and bottom covered with massive dripstone. A few loose stones rested on the bottom. The whole surface, even including the stones upon the floor, was so completely covered with these peculiar fungoid bodies, that it was impossible to move without destroying hundreds of them. All were about the same height, 0.2 inches, snow-white, or of a rich reddish brown, and conformed to the unequal surface on which they stood. It is quite impossible to describe the effect of a whole chamber bristling with these peculiar structures.

PRIORY FARM CAVE

Across the mouth of the tidal inlet, to the west of Pembroke Castle, Priory Farm Cave looks out onto a steep slope above the mudflats on the south side of the Pembroke River (Fig. 202). Sometimes known as Catshole Cave, it consists of a relatively long winding passage with a side passage and chamber part of the way along its length. As in the case of Hoyle's Mouth it seems that the cave was occupied by humans from at least the Mesolithic. The cave was excavated in 1908, and finds from inside and from the entrance area include human and animal remains, flints, and a rare Bronze Age hoard which can be seen in the National Museum in Cardiff. The objects, all made of bronze, consist of a saw blade, a

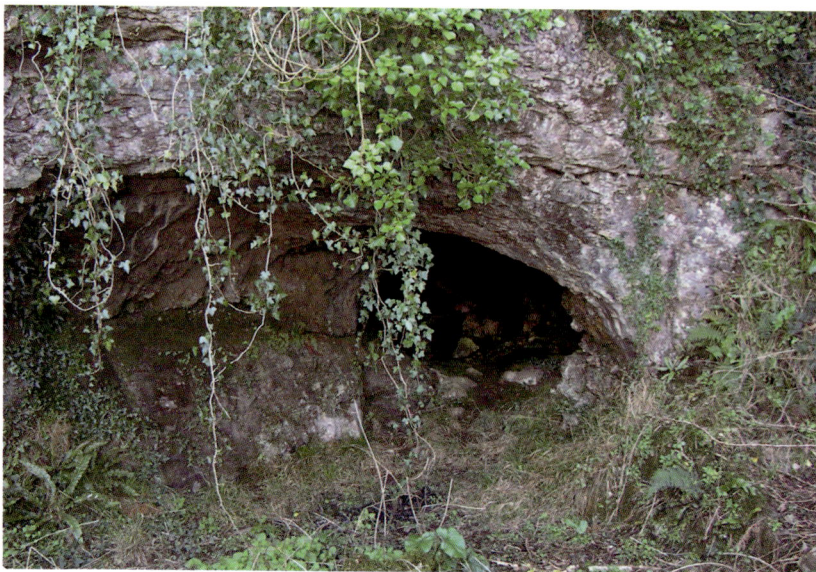

FIG 202. Priory Farm Cave is situated below a small overhanging cliff on the south side of the Pembroke River. A rare Bronze Age hoard was discovered when the site was excavated in 1908, and this is now in the National Museum in Cardiff. (Linda Asman)

tanged chisel, a torsion ring and the blade of a palstave. These probably represent property buried for safekeeping and never retrieved (Laws 1908).

The cave was subsequently re-excavated in the 1940s by William Francis Grimes, a distinguished archaeologist born in Pembroke (Grimes 1944). It seems to have been a hyena den at some point in the past, since there were very few complete bones and some show signs of being gnawed. The animals represented by these bones include reindeer, brown bear, horse and the hyena itself *Crocuta crocuta spelaea*. The human remains consisted of a skull, which was almost complete, and several portions of lower jawbones, limb bones, ribs and vertebrae, along with the upper jaw of a child, estimated to be around seven years old.

THE WOGAN

Under Pembroke Castle itself is a huge cave known as the Wogan, or Wogan's Cavern. A massive single chamber, some 10 m high, with a large natural entrance overlooking the Pembroke River to the north. The mouth of the cave was blocked

when the castle was constructed during the twelfth century and it seems to have
been used as a boathouse. Despite this, many sections remained untouched, and
flint tools since found in the cave suggest that, as with the other caves, it was
occupied by people from at least the Mesolithic and probably earlier. Accessible
via 55 steps from inside the castle, it is heavily visited and today there are few
signs of wildlife within the cave. In 1856 Philip Henry Gosse recorded what he
saw:

> The sides of the cave are ribbed and groined, and painted with grey-green lichens;
> hart's tongue ferns droop in tufts from various points of the roof. Several large
> holes were visible in the sides, but I think none were perforate; there was also
> a great hollow in the roof, into which I peered to try if I could see any sign of
> communication with the world above. I saw none, but everything was so dark, I
> could be sure of nothing. Just within the entrance there is a perpetual dripping, as
> if you stood under a tree in a heavy shower, and the ground thereabout is covered
> with beds of soft mosses of the most brilliant green, and with the branching leaves
> of Marchantia polymorpha [common liverwort].

CALDEY CAVES

For its size Caldey is rich in caves, which are concentrated in two areas on the
limestone: Eel Point in the northwest and High Cliff, to the south of Den Point,
in the northeast. Public access to these caves is discouraged by the monks on the
grounds of safety, although there is a small display of finds in the information
centre adjacent to the monastery. Many were discovered during excavations
organised by Brother James Van Nedervelde between the 1950s and the 1970s.

The most accessible cave, Nanna's Cave, is set high above current sea level
on the eastern coastline and looks across Carmarthen Bay to Gower (Fig. 203). A
relatively shallow rectangular opening, only 5 m deep, it was occupied for over
10,000 years, during a period when the view was over a coastal plain rather than
the sea. There is a long and complex excavation history, starting in 1911, but it is
clear that three adult humans and at least one juvenile were interred within the
cave. Animal remains consisted of pig, sheep or goat, cattle and fox. Some of the
bones were blackened by fire and may have been the remains of a feast. Given
the shallow and open aspect of the cave there are unlikely to be any significant
invertebrate records.

When first discovered by Van Nedervelde in 1950, the two north-facing
entrances of Potter's Cave, some 6 m apart, were blocked by blown sand and
fallen rocks. Excavations which continued until 1970 initially revealed stalagmites

in which tools and animals were embedded. After removal of these finds, three human skeletons were found, two dating to the Palaeolithic and one to the Romano-British period. It was perhaps one of these people who owned the magnificent necklace of 49 blue glass beads, which dated from the first or second century BCE. When the 60-centimetre-thick layer of stalagmite was removed in 1973 additional finds appeared, including horse bones gnawed by hyenas and the bones of woolly rhinoceros.

The best-documented cave, Ogof yr Ychen, had already been truncated by nineteenth century quarrying when Van Nedervelde found it in 1970. The main bone-bearing layer was a periglacial scree mixed with a yellow, silty, clay which contained the remains of wolf, hyena, woolly rhinoceros and, unlike Hoyle's Mouth, wild boar. The remains of three human adults were also found, one of whom, a male, had obviously fallen head first through a shaft above the first chamber with such force that on hitting the cave floor his lower jaw was disarticulated. Other animal remains included ox, or bison, *Bos* sp., red deer and roe deer *Capreolus capreolus*. All of them were complete and ungnawed, suggesting that fatal falls into this cave system were not uncommon.

It may have been Ogof yr Ychen that Dawkins was referring to in his book *Cave Hunting* (1874), when he refers to one of the 'most beautiful stalactite caverns in this country ... discovered some years ago in the limestone cliff and explored by Mr Ayshford Sandford and the Reverend H. H. Winwood in 1866', particularly as he describes 'creeping through a narrow entrance with an outlook to the sea on a precipitous side of a quarry'. The passage led to a chamber of 'considerable horizontal extent', which Dawkins called the 'Fairy Chamber', as it consisted of 'a rich red crystalline pavement, perfectly horizontal and studded here and there with round bosses, either red, or snow white'. From the roof hung stalactites 'offering the same beautiful contrast of colours, forming a delicate canopy of tassels ...' He then notes that:

> As I broke my way into some of the unexplored recesses, through the thickly planted straw-shafts, and scene after scene of fairy beauty, unsullied by man, opened upon my eyes, the ringing of fragments on the crystalline floor that accompanied almost every movement made me feel an intruder, and sorry for the destruction.

Another important cave, Coygan Cave, was likewise unfortunately close to an active quarry, and blasting in 1971 caused a roof collapse which sealed it permanently. Surprisingly it was then de-scheduled as an Ancient Monument and eventually completely destroyed by quarrying. Daylight Rock Fissure, which,

FIG 203. Nanna's Cave, a shallow rectangular opening only 5 m deep, is the most accessible cave on Caldey Island, although visitors are dissuaded from exploring because of safety concerns. (Jonathan Mullard)

as its name suggests, penetrates the spine of the cliff, includes a short tunnel which has yielded teeth from woolly rhinoceros, horse, and woolly mammoth. *The* collection of bones, again gnawed by hyena, included lion, cave bear, reindeer, ox or bison, and giant deer *Megaloceros giganteus*, sometimes known as the Irish elk.

Ogof y Benglog was also discovered by Van Nedervelde. An excavation in 1969 revealed a female human skull which had entered the small cave, possibly by rolling downslope from an extension lost to quarrying. Mel Davies, an enthusiastic cave explorer (1969a), noted that 'much probably remains to be discovered in the cave', which is only a few metres from Ogof yr Ychen.

The Eel Point Caves were first excavated around 1840 when 'abundant remains' of mammoth, rhinoceros, hyena, lion, bison, bear, deer and hippopotamus were found. Once again though the record of hippopotamus must be treated with suspicion, the rest of the animals representing a glacial fauna. When the cave was re-examined in 1950 by Grimes and Van Nedervelde quarrying had once again destroyed most of the cave, but in 1986 more bones were discovered including three individual rhinoceroses.

OGOF GOFAN

Ogof Govan is the most interesting of the caves situated in the sheer face of the Castlemartin cliffs. Located on the west side of Saddle Head, the main entrance is under an overhang about halfway down the cliff, which here is around 40 m high. Both this entrance and a second more accessible one are therefore only accessible by experienced climbers and cavers. There are at least six separate chambers, with the two entrances connected by a passage. The main chamber, the sixth, has a ceiling covered with 'straws' and other formations, while in the centre of the floor there are several massive stalagmites. One is over 5 m high. The formations surround a 'crystal pool', containing emerald-green water (Fig.

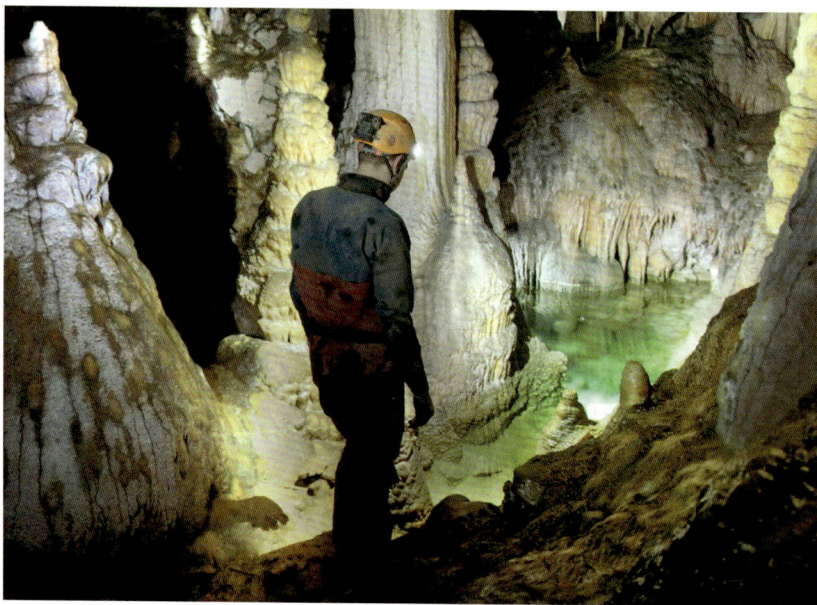

FIG 204. Ogof Gofan has been described as 'spectacularly well decorated', the 'crystal pool' of emerald-green water being surrounded by stalagmites, one of which is over 5 m high. (Stuart France/Cambrian Caving Council)

204). There are stalagmites in the connecting passage and, interestingly, several of these have been broken, or have had their tops chopped off. The pieces are still lying about in the cave but no longer fit on their stumps due to further growth. Davies (1969b) concluded from this that 'ancient man was either something of a vandal or deliberately tried to clear obstructions away to make the passage route more convenient'. As it would have been easy to throw the broken pieces out of the cave, it appears they may have been kept deliberately.

The large stalagmites and the other spectacular formations exist because of a major fault and a broken upper layer of rock, which allows plenty of water to percolate through the roof of the cave, calcium carbonate and other minerals being precipitated from the mineralised water. Excess water runs out of the cave through 'the Window', an opening about halfway up the cliff. There are not many situations like this on the Castlemartin Range and it is likely that both the specific geology and the cave are unique, although further explorations are continuing. The cave has been described by Dyfed Archaeological Trust as 'spectacularly well decorated, quite uncommon to the likes in the area and can be considered an archaeological site of national importance' (Houlston 2018). Access to the site by specialist caving groups is controlled by a permit system operated by the National Park Authority and the Ministry of Defence. The number of visiting cavers remains low but one visitor said that 'though the caving distance is relatively short, we all agreed that this is a very remarkable cave and the chambers within truly spectacular.'

Two years of funding by the Ministry of Defence, between 2016 and 2018, resulted in thirteen specialist visits to Ogof Gofan to collect biological and archaeological data. A bat research project did not find any hibernating bats in the main chambers, although there were a few individuals hidden in alcoves in the passage linking the two entrances. It seems that while many lesser horseshoe bats visit the cave during the night for feeding and temporary roosting it is not as important for hibernation as was once thought, with only three bats recorded during the visits.

OTHER CASTLEMARTIN CAVES

There are four other caves in the Castlemartin cliffs which have been explored, Ogof Pen Cyfrwy, Ogof Morfran, Ogof Garreg Hir and Ogof Bran Goesgoch. The first is situated further out on Saddle Head than Ogof Gofan and is therefore in a much more exposed position. As a result, most of its contents have long been washed out by the sea. In addition, a pothole inside the cave exits at sea level and there is evidence that water erupts though this during storms. In contrast, Ogof Morfran, a rock shelter about halfway down the cliff, has produced evidence that it was occupied for long periods of time. Excavations in the late 1960s and early 1970s recovered ox bones hacked and blackened by fire and a red deer antler which had its brow tine smoothed, as if it had been shaped and cut for use as a digging tool. Other finds included flint, charcoal, limpets, mussels, a sandstone saddle quern and fragments of Bronze Age pottery. In addition, there were Roman coins and part of a bronze pin dated to 268–274 CE. Even here though storm waves occasionally flood the cave and herring gulls build nests in it, causing a surprising amount of damage.

The difficulty in assessing Ogof Garreg Hir was noted at the beginning of this chapter, but excavations in the 1970s found two human bones, from the left side of a small adult, and the remains of red deer, roe deer, wolf, pig, hare and several species of birds. The bone assemblage is apparently unusual in that so many species were represented by so few bones. This suggest that there was once an outer passage, or chamber, which has been eroded by the sea. Finally, Ogof Bran Goesgoch was first discovered in 1969 but it was not until 1977 that three human bones dug up by burrowing rabbits were noticed on its floor.

Several other cave entrances are known, and some have been excavated without finding anything, while further caves have been detected from the sea following a traverse by boat. Although modern cavers have claimed these sites as new discoveries, the egg collectors of earlier centuries had somehow reached them all, Mel Davies noting that one elderly local man in 1969 had confirmed how widespread the practice had been (Davies 1989).

ISLAND CAVES

Owing to the difficulties of accessing caves on the islands to the west there have been few studies of their invertebrate populations, but in May 1929 William Bristowe explored Seal Hole on Skomer 'with the help of candles' (Bristowe 1929). While the sea washes right to the end of the cave, which is around 15 m deep, it does not reach the roof, and here Bristowe found a close relative of the orb-web cave spider, the orb-web spider *Meta merianae*. This species spins its web in shady, damp habitats, but is usually found concealed away from it during the day. The spider can be found in boathouses, cellars, abandoned rabbit and badger burrows, hollow trees and near the entrances of caves and culverts, particularly in the vicinity of water seeps (Harvey *et al.* 2002). Alongside the orb-web spider was the shore money-spider *Halorates reprobus*, one of which was observed eating a small sea slater *Ligia oceanica*. This money-spider has also been found in cormorant nests (Bristowe 1958).

There are no large caves on Skokholm but there are a few deep fissures in the cliffs into which the sea extends at high tide. In the darkest of these, in 1934, Bristowe found the orb-web cave spider, numbers of its white cocoons being suspended from the roof of the fissure. In the lighter, and shallower, fissures its place was taken by the orb-web spider and in the dampest ones the shore money-spider was present in abundance, 'its white cocoons occurring in hundreds on the walls' (Bristowe 1935).

CHAPTER 15

Trees and Woodland

Trees are the key to our survival. They came before us and we have used, abused and loved them for thousands of years. They have provided our sustenance – food, shelter, medicine and the air we breathe. They are our past and future.

Neil Sinden (1990)

FOLLOWING THE LAST GLACIATION, SOME 10,000 years ago, woodland expanded to cover most of Wales, but the story since has been one of continuous fragmentation and decline. Even before the trees returned human populations were well established, and in the Neolithic woodlands were already being actively managed and timber extracted. Indeed, one of the most marked effects of human activity upon the landscape that can be detected in pollen records is the clearance of woodland. In Pembrokeshire these records show evidence of extensive fellings in the Cleddau valley around 1450 BCE (Seymour 1985), while similar data at comparable dates at sites across mid and South Wales point to extensive deforestation during the middle Bronze Age (Walker *et al.* 2009). Some of the most important areas of surviving woodland cover the slopes of Cwm Gwaun (Fig. 205).

Timber was a valuable resource, and significant quantities of wood would have been required for building purposes such as the construction of timber roundhouses and other structures. In some instances, it is likely the demand for wood would have resulted in deliberate attempts at management, although coppicing may have occurred accidentally where felling had taken place and

FIG 205. Dense woodland on the steep slopes of Cwm Gwaun, with St Brynach's Church just visible through the trees. The valley contains some of the most extensive tracts of surviving semi-natural woodland in northern Pembrokeshire. (Jonathan Mullard)

browsing by animals was restricted, thereby allowing regrowth from the tree stumps (Caseldine 2018). Some idea of the amount of wood required to construct a roundhouse, and therefore the impact on local woodland, has been gained from experimental work at Castell Henllys, where the largest known roundhouse at the site has been reconstructed (Bennett 2001). All the wood used in the reconstruction was the result of coppicing. Around 90 hazel bushes were used in the construction of the wall and 34 oak trees used for the rafters, wall posts and ring beams, while 2,000 bundles of reed were used to thatch the roof (Fig. 206).

Gerald of Wales considered that there were three issues that resulted in the destruction of woodland in Wales:

First, the wood cut down was never copisid, and this hath beene a great cause of destruction of wood thorough Wales. Secondly, after cutting down of woodys, the gottys [goats?] hath so bytten the young spring that it never grew but lyke shrubbes. Thirddely, men for the monys destroied the great woddis that thei should not harborow theves.

FIG 206. The reconstructed roundhouse at Castell Henllys used 34 oak trees, 90 hazel bushes and 2,000 bundles of reed in its construction. The site is one of the most intensively studied hillforts in Britain and supports a continuing programme of research and excavation. (Jonathan Mullard)

During the Middle Ages, when the first reliable records begin, pleas and petitions concerning disputes over woodland were numerous in all parts of Wales, and in 1386 there are records of the felling of 3,000 green oaks, worth 200 marks (£133, equivalent to around £142,000 today), in the woods of Coydrath and Rodewood in Pembrokeshire (Owen 1918). There was similarly concern about the decay of the underwood there through lack of custody and enclosure.

Providing pasture for livestock used to be one of the most important functions of woodland, and indeed remains so in certain areas, such as the Forest of Dean and the New Forest, today. Where the lord of the manor was in full possession of the pasture it was customary to sell grazing rights for the year, usually for a cash sum. In 1326 the *Black Book of St David's: an extent of all the lands and rents of the Lord Bishop of St David's made by Master David Fraunceys* recorded that the Forest of Lloydarth, for example, covered 300 acres (121 ha) 'and they are able to keep there 20 mares in foal, 40 great beasts and 200 sheep, and the grazing of each great beast is worth 1d. and of every 10 sheep 1d. per annum' (Willis-Bund 1902). This heavy grazing was responsible for the degradation of the woodlands, the amount and quality of the herbage improving as the condition of the woods deteriorated. The annual harvest of fruit from the forest, both acorns and nuts, was worth two shillings in 1326, around £108 in today's currency. Sometime over the next 300 years, however, the remains of the ancient woodland became just a nuisance to be cleared.

Towards the end of the sixteenth century, the 'voyages of discovery' resulted in new trees being introduced to Wales from North America, including the 'firre tree'. Indeed, the first recorded introduction of exotic trees to Wales and the first record of planting and direct sowing of conifers relates to one 'Master Thomas Bowen of Trefloine in the County of Pembroke', who:

> had about fifteene or sixteene yeares past [c.1596/97] manie young and small plants of this kind brought him home by saylers from the Newfound land, with some of the earth wherein they did formerly grow, and planted them together with the said earth in convenient places about his house, where they have since so well prospered that many of them at this present [1612] are about foure foot in circuit, and also very high and tapering. And they will grow upon mountaines, gravellie soyles, or in good earth, either by planting the young tree, or sowing of the seed. (Church 1612)

Trefloyne Manor, near Tenby, is now a hotel and golf course, but no ancient trees appear to have survived.

A further drastic reduction in the area of woodland in the sixteenth and seventeenth centuries appears to have been due mainly to the well-documented and widespread 'assarting' of the remaining woodlands; that is, land clearance, sometimes with the aid of fire. As John Leland reported in the 1530s, 'the ground is sumwhat baren of wood as al Penbrookshire almost is, except where a few parkes be'. In this period there are increasing references to the fact that corn is now growing where good timber once stood, not least by George Owen, who noted in 1603 that 'This county groans with the general complaint of other counties of the decreasing of wood, for I find by matters of record that divers great cornfields were in times past great forests and woods.' In a list of 'Woods and Forests in times past and now destroyed and arable land' he mentions eight sites, including 'The Wood by Newe gall', 'Lloydarch Forest', 'Moelgrove' and 'Coedcadw'. While Owen's aim was to inform his readers about sources of timber 'for buildings and other necessities', in line with the practicalities of his era, it represents the most comprehensive audit of Pembrokeshire woodlands at that time.

During the reign of James I, a 'breviat' or survey was made of the woods in the royal forests, parks and chases in 1608. The king's forests in Pembrokeshire were estimated to contain 2,666 timber trees, 22,884 decaying trees and 21,032 saplings. The ratio of decaying trees to timber trees indicates that the condition of the forests was unsatisfactory. Even marking the timber trees with the king's initials did not make them safe from depredations, however, for in the Forest of Cilgerran even selected trees 'marked with the prince's name as special timber' were apparently illegally felled (Jones 1955). Similarly, another survey of a parcel of the Forest of Narberth, called Castle Lake Wood, reported that the 60 acres were 'well set with saplinge oaks .. among which .. are some dotard oaks of bigger proportion serving to noe other use but fireinge' (Owen 1914).

Before softwoods became generally available for pitwood at the end of the eighteenth century, all the timber used in the Welsh coalfields was hardwood, most of it acquired from the local area. For example, in 1713 in Pembrokeshire, 7 acres (2.8 ha) of 'fine wood ... which the colliers say is the best they ever saw for that use' were bought at £40 an acre for the use of Cardmaker's Pool coalpits. (Howells & Howells 1972). By 1763 though when a 74-gun ship was being built at Neyland the local woods were so denuded that they could not supply the timber for it. In the words of a shipwright working on the project, *'so effectually is this county stript of navy timber that we are not able to purchase so much as a futtock to put in her'* (Coed Coch Manuscript), a futtock being a slightly curved middle timber in a ship.

As a result of this destruction and a continuous demand for charcoal for the lead and iron industries, woodland in Pembrokeshire today only covers around 6 per cent of the land area. Extensive areas were felled during the First and Second World Wars, the regrowth creating even-aged stands with few veteran trees. Waldo Williams (1904–71), widely regarded as one of the finest Welsh-language poets Wales has ever produced, wrote his poem *Yr Hen Allt* (*The Ancient Wood*) about the recovery of the woodlands after the First World War. In this Williams notes that: 'The old wood, look, is growing again, On every side life is flooding back' (translation by Conran 1997). A census of woodlands in 1924 classed 43.8 per cent of the remaining Pembrokeshire woodlands as 'felled or devastated' (Linnard 1979). In the following decades the loss of the surviving native woodlands continued, mainly due to the development of conifer plantations – descendants perhaps of Master Thomas Bowen's trees. Examples include Ffynone and Cilgwyn Woodlands in the valley of the Afon Dulas, near Capel Newydd, which were planted in the early 1960s, following the destruction of the original oak woodland.

In the 1970s Dutch elm disease, caused by the fungus *Ophiostoma ulmi* and carried mainly by the elm bark beetle *Scolytus multistriatus*, affected an estimated 25 million elm *Ulmus* spp. trees in the United Kingdom. Wales, like the rest of southern Britain, was badly affected, and it is said that 90 per cent of the elms in Pembrokeshire died. Losing the majority of elms in the county dramatically changed the landscape forty years ago, and now there is another major threat from ash dieback. First confirmed in Britain in 2012, on ash saplings imported from nurseries in continental Europe, this is caused by another fungus, *Hymenoscyphus fraxineus*. There are currently no effective strategies for managing the disease, and the removal of trees in infected areas has little effect since the fungus lives and grows on leaf litter on the forest floor. Once a tree is infected the disease is usually fatal, although evidence suggests that older trees can survive infection for some time. Stands of ash on the base-rich soils in the south of the county are particularly at risk.

HOYLE CAVES WOODLAND

One of the sites potentially threatened by ash dieback is the woodland overlooking the Ritec valley, near Tenby (Fig. 207). Here the limestone slopes surrounding one of the main cave sites in Pembrokeshire, Hoyle's Mouth, support an extensive area

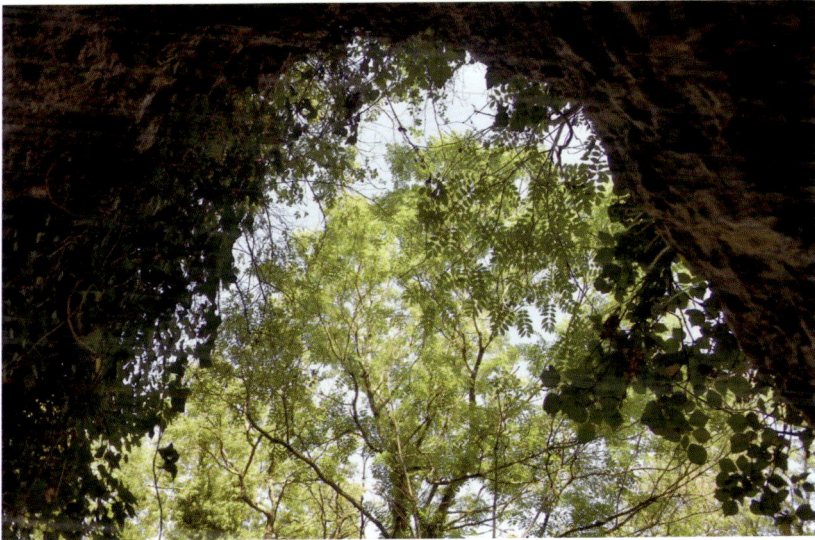

FIG 207. Ash woodlands, here viewed from inside Hoyle's Mouth Cave in the Ritec valley, are uncommon in Pembrokeshire. Confined to a few sheltered slopes, they are potentially threatened by ash dieback. (Jonathan Mullard)

of ash woodland with an understorey of hazel, elder, hawthorn *Crataegus monogyna*, holly and blackthorn. This type of woodland is reminiscent of the calcareous ash–hazel woods found in southern England and is uncommon in Pembrokeshire, being confined, as here, to a few sheltered slopes. Although its ancient origins are uncertain, there is written and cartographic evidence of open ash woodland in the vicinity in the mid and late nineteenth century. Edward Laws, for example, in *The History of Little England Beyond Wales* (1888), refers to a visit he made to the adjoining Longbury Bank Cave in 1878 with Professor George Rolleston (1829–81), Professor of Anatomy and Physiology at the University of Oxford.[*] They noted the trees and shrubs, which they regarded as an 'ancient flora ... unmixed with foreign importations'. Structurally the woodland here is diverse, with large ash trees mainly derived from overgrown coppice, and associated climbers of honeysuckle *Lonicera periclymenum* and traveller's-joy *Clematis vitalba*.

The ground flora includes bluebell, dog's mercury *Mercurialis perennis*, hairy brome *Bromus ramosus*, giant fescue *Festuca gigantea*, moschatel *Adoxa moschatellina*, lords-and-ladies *Arum maculatum*, sanicle *Sanicula europaea* and woodruff *Galium odoratum*. Orchids recorded from the site include common twayblade *Listera ovata*, early-purple orchid and common spotted-orchid *Dactylorhiza fuchsii*. Notable species are sweet violet *Viola odorata* and the three-cornered leek *Allium triquetrum*. The latter is an invasive species, originating from the Mediterranean, which derives its name from the triangular cross-section of the stem. It is now thoroughly naturalised and increasingly abundant and widespread in the milder areas of Britain.

OAK WOODLAND

The most common woodland habitat in the county is, however, oak woodland, the dominant tree being sessile oak. There are a variety of management approaches, with high forest, coppice with standards and old wood-pasture all represented. In general, this ancient semi-natural woodland only survived where the topography limited agricultural or forestry activities, examples being the small relict woodlands on the banks of the Cleddau and the hanging woodlands on the steeper slopes of Cwm Gwaun. Those that remain though are rich in species and designated as Sites of Special Scientific Interest, or National Nature Reserves. Nationally important epiphytic lichen communities occur here in the

---

[*] Rolleston's great-grandson is Frank Gardner, the BBC's Security Correspondent and current President of the British Trust for Ornithology.

FIG 208. The pollution-sensitive lungwort grows on trees in a number of woodlands, including here on the Stackpole National Nature Reserve. Bright green under moist conditions, lungwort becomes brown and papery when dry. (Jonathan Mullard)

clean oceanic air, with many species indicative of ancient woodland, including lungwort *Lobaria pulmonaria* (Fig. 208). This consists of an ascomycete fungus and a green alga in a symbiotic relationship with a cyanobacterium, an arrangement that involves members of three different kingdoms.

PENGELLI FOREST AND PANT-TEG WOOD

George Owen once owned Pengelli Forest, near Felindre Farchog, describing it as one of the nine forests and great woods remaining in the county (Fig. 209). Along with the adjoining Pant-teg Wood it still forms the largest block of ancient woodland in the county. The sinuous eastern boundary bank marks the original limit of the forest, which Owen said was enclosed with 'quicksett and pale rounde'. The canopy is a mixture of sessile oak and pedunculate oak

FIG 209. Pengelli Forest was once owned by George Owen. Due to felling during the Second World War the stands here are now even-aged, with few veteran trees surviving. The adjoining Pant-teg Wood is a sessile oak coppice. (Jonathan Mullard)

Quercus robur, downy birch, ash and alder, which dominates the streams and the waterlogged ground around their sources. Other species recorded from the site include aspen *Populus tremula*, wild cherry *Prunus avium*, wych elm *Ulmus glabra*, goat willow *Salix caprea* and native crab apple *Malus sylvestris*. There was a similar mix of species in Owen's time, and his description is the earliest known record of the habitat.

The Midland hawthorn *Crataegus oxycanthoides* grows in the forest; the three bushes found here are thought to be the only native specimens in West Wales. The species has, however, not been found in any other ancient woodland in the county and it is impossible to determine whether they are remnants of an original population, or more recent arrivals from seed carried by birds. It is very unlikely to have been deliberately introduced (Stephen Evans, personal communication). The *New Atlas of the British and Irish Flora* nevertheless shows it as not native west of Monmouthshire (Preston *et al.* 2002).

Notable plants in the ground flora include early dog-violet *Viola reichenbachiana*, woodruff, adder's-tongue and water avens *Geum rivale*. The site is rich in insects, with records of the oil beetle *Meloe proscarabaeus*, the dark bush-cricket *Pholidoptera griseoaptera* and the speckled bush-cricket *Leptophyes punctatissima*. Butterflies include purple hairstreak *Favonius quercus*, silver-washed fritillary *Argynnis paphia* and white-letter hairstreak *Satyrium w-album*, while pearl-bordered fritillaries *Boloria euphrosyne* occur in the adjacent fields.

The purple hairstreak is rarely seen, as it spends most of its time in the canopy, the main adult food source being honeydew. It is only driven down to seek fluid and nectar during prolonged droughts. One place in Pembrokeshire though where there is a good chance of seeing purple hairstreaks is the coast path above Goultrop Roads, south of Little Haven. Here oak trees grow on the cliff face, with the result that the canopy is almost level with the cliff top, making sightings relatively easy (Fig. 210). In 2017 twenty purple hairstreaks were recorded and, with care, it is possible to get good photographs of the insect. A similar approach at the Teifi Marshes Reserve again produces results. The trick at this site is to use the bird hide in the oak woodland to the south of the car park and climb to its upper storey. This provides a view down into the canopy, and reports into double figures have been received from this location in the past.

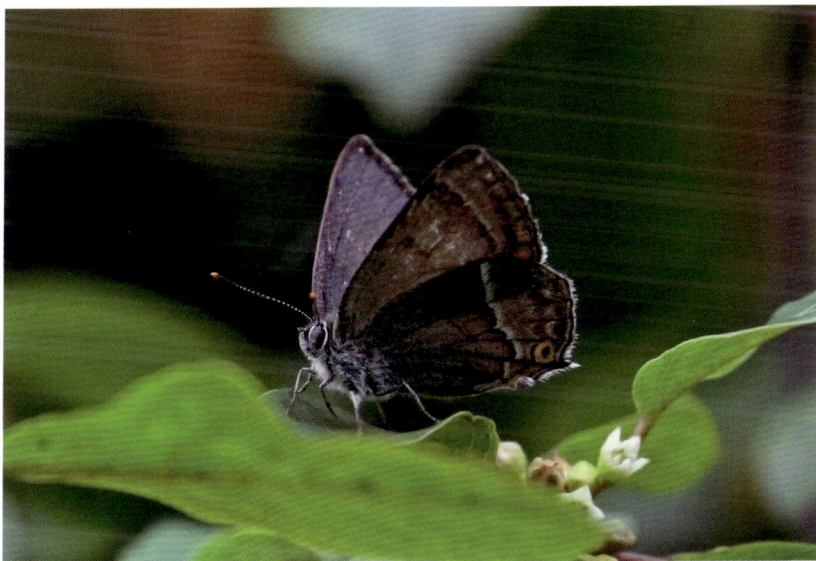

FIG 210. Purple hairstreaks are rarely seen at ground level, but there are two locations in Pembrokeshire where it is possible to ascend above the canopy and get close views of the insect. (C. J. Sharp)

Birds typically found here include redstart *Phoenicurus phoenicurus*, chiffchaff, wood warbler *Phylloscopus sibilatrix*, buzzard, tawny owl *Strix aluco* and sparrowhawk, with woodcock *Scolopax rusticola* visiting in winter. George Owen was particularly fond of woodcocks and relates how they were caught for food:

> *Their chief taking is in cockroades in woods, with nets erected up between two trees where in cock-shoot time (as it is termed), which is the twilight, a little after the breaking of the day and before the closing of the night, they are taken sometimes two, three or four at a fall. I have myself oftentimes taken six at one fall, and in one road at an evening taken eighteen, and it is no strange thing to take a hundred or six score in one wood in twenty-four hours if the haunt be good and much more has been taken, though not usually.* (Owen 1603)

Woodcocks are still common, migrants arriving in Wales at the end of October from breeding areas in England and the continent, though not in the numbers recorded by George Owen. The species is, however, under-recorded in Pembrokeshire and there are suggestions that there may be a small breeding population in the county (Green & Roberts 2004).

One species that does breed in Pengelli Forest is the rare barbastelle bat, the site having the distinction of being the only known nursery roost in Wales, although there are undoubtedly many others (Fig. 211). Most of the roosts are in cracks in trees, or behind flaking bark, often associated with large ivy *Hedera helix* stems. A dense understorey that includes holly is considered important for maintaining a favourable microclimate in the roosts. Once the young can fly it seems that the colony divides into smaller units and then reconvenes at a single roost in late July, sometimes in one of the roosts used before the young were born, but much remains to be learned about this enigmatic mammal. One threat to the barbastelle is that around half of the roosting sites and the majority of the foraging areas lie outside the boundaries of the National Nature Reserve, as the boundary was drawn up before the bats were discovered. There is therefore a risk of inappropriate land management damaging roosts and foraging areas.

A medium-sized bat, with a distinctive pug-shaped nose, the barbastelle is one of our rarest mammals. Its name is derived from the Latin, meaning 'star-beard'; it refers to the white hairs which stick out from the lower lip and body. Comparatively little is known about the barbastelle, the majority of United Kingdom winter records being of single bats in underground locations. As mentioned in Chapter 5, however, automated bat detectors on Skomer and Ramsey, where there is no woodland, have recorded this fast and agile animal, so we still have a lot to learn about its movements and habits. Barbastelles are in fact quite widespread in the county, and monitoring by placing bat

FIG 211. Pengelli Forest is currently the only known nursery roost in Wales for the rare barbastelle bat. The ears of the barbastelle are joined on the top of the head, a feature which distinguishes it from all other species of British bats. (Melvin Grey)

detectors in woodlands has shown that they are particularly numerous in south Pembrokeshire, particularly in the Canaston and Minwear area to the west of Narberth. In addition. barbastelles have been found along the north shore of Milford Haven and there is a recent bat-box record from the south side. It is thought they may feed on the coastal marshes. In north Pembrokeshire barbastelles have been observed foraging through gardens in Newport (Tom McOwat, personal communication).

Another rare tree-dwelling bat associated with old-growth woodland is Bechstein's bat *Myotis bechsteinii*. An individual of this species was found in the woods at Colby Woodland Garden, near Tenby, in 2009, but prior to that the nearest sighting of the species was in the Forest of Dean. It is found mainly in southern England, in Dorset, Wiltshire and Hampshire. Fossil records suggest it was once the most common bat species in Britain, but forest clearance and climatic changes have resulted in major population declines across its range. This species feeds in closed-canopy broadleaf woodland with a well-developed understorey, often close to water bodies, its diet consisting of craneflies, beetles and moths.

TY CANOL WOOD

This ancient woodland to the west of Pentre Ifan appears on maps dating from 1578 and was described by Owen as part of Cilrhydd, 'a fair wood and the best timber in the shire'. Evidence from large oak coppice stools, some more than 3 m in diameter, indicates that there may have been trees here in the Iron Age, or even earlier. It is now a National Nature Reserve, like Pengelli. In the northern, and lowest section, for the most part it is a typical Welsh upland oak wood, dominated by sessile oak with scattered specimens of ash, downy birch, hazel and rowan, though many of the largest trees were, again, felled in the First World War. In contrast, in the southern and central parts of the wood there are deep rocky gullies and the tree canopy consists largely of smaller, less well-grown sessile oaks and some coppiced trees. The wood is noted for the way it has grown over the numerous rocky outcrops and boulder fields (Fig. 212). On the very exposed southern edge, the canopy is no more than 2 m high in places.

FIG 212. Ty Canol Wood has grown over numerous rocky outcrops and is a superb example of wood-pasture. Sheep and cattle grazing are essential to maintain both the character of the woodland. Nearly 400 species of lichen and 125 mosses have been recorded from the area. (Jonathan Mullard)

The ground flora here is largely dominated by bilberry, with various ferns and bryophytes, including liverworts such as western earwort, straggling pouchwort, greater whipwort, prickly featherwort *Plagiochila spinulosa* and spotty featherwort *P. punctata*.

As wood-pasture, the site has a long history of grazing, with both sheep and cattle, which is essential to maintain both the character of the woodland and the species associated with this unusual site. The cattle create a higher browse line, which provides more opportunities for lichen growth on the tree trunks. The open, well-lit conditions are particularly favourable for 'old forest' lichen communities, and to date nearly 400 different species have been identified, including many rarities, making Ty Canol one of the best sites for lichens in Wales. The current total is much larger than the botanist Francis Rose recorded in 1975:

> In the small area of Ty Canol Wood of less than 100 acres, at least 170 lichen taxa are now known, of which at least 109 are found on the bark of trees as epiphytes, or on wood. At present only five woodlands in Wales are known with over one hundred epiphytic lichen taxa present within them; the others are the Gwaun valley woodlands, Coedmor woodland in Cardiganshire, Dynevor Park Woods in Carmarthenshire and Coed Crafnant in Merioneth. Experience of woodland lichen floras over Britain as a whole suggests that the primaeval oak forests probably had about 120–150 epiphytic lichen species per square kilometre at least, so Ty Canol Wood gives probably a better conception of the lichen flora of primaeval oak wood than the vast majority of woodland sites remaining in Wales.

Ty Canol is the 'type locality' for the crustose lichen *Schismatomma umbrinum*. Consisting of a thick, yellowish-brown thallus surrounded by a wide purplish-brown prothallus, it is abundant on deeply shaded vertical rock faces under the tree cover, thriving in these cool and humid old woodlands. Rare epiphytic species include string-of-sausage lichen *Phaeographis lyellii*, *Japewiella tavaresiana* and *Ochrolechia inversa*, a scarce species of moist woodland. Other notable lichens found here are *Cliostomum conranii*, *Gyalideopsis muscorum*, *Cetrelia olivetorum*, *Cladonia merochlorophaea*, *C. luteoalba*, *Lecidea fuliginosa* and *Parmelia britannica*. In addition, over 125 different mosses have been recorded from the reserve, including Killarney featherwort *Plagiochila killarniensis* and ghostwort *Cryptothallus mirabilis*. The two filmy-ferns, Tunbridge filmy-fern and Wilson's filmy-fern, are also present, along with hay-scented buckler-fern *Dryopteris aemula*.

Wilson's filmy-fern is a small plant which superficially resembles a tuft of moss rather than a fern. The individual leaves are very thin and transparent, and divided into parallel-sided, toothed segments. This species is frequent in western Scotland, the English Lake District, North Wales and southwest Ireland (Page 1988). The main factor controlling its distribution appears to be the number of rainy days per year (Richards & Evans 1972). Tunbridge filmy-fern is less widespread and abundant than Wilson's filmy-fern and has a fragmented distribution, which is presumably a relic of a more continuous past. The largest known population in the county is in Garn Wood in Cwm Gwaun. Populations though are dynamic and there seems to be a cyclical pattern of succession, where colonisation is followed by the gradual build-up of a mat of bryophytes, including the filmy-fern itself, which eventually fall off the rock surface under their own weight, leaving bare rock as a potential site for the cycle to start again. Most sites where the species occurs have a long history of continuous or dynamic tree cover, the filmy-fern re-colonising from local refuges after periods when the canopy has been open (Mullard 2014).

Important insects include two moths whose caterpillars feed on lichen: Brussels lace *Cleorodes lichenaria* and dotted carpet *Alcis jubata*. The latter mainly inhabits mature deciduous woodlands where lichen covers the trunks and branches of the trees, although it has colonised older conifer plantations that have a range of lichens. At rest, the moths are hard to see, protected by their superb cryptic camouflage. In contrast, the larvae are bright green with large black dots. The larva of a third scarce moth, the light knot grass *Apatele menyanthidis*, feeds on moorland plants like bog-myrtle and bilberry. Once again butterflies found here include purple hairstreak and silver-washed fritillary.

Surveys have shown that Ty Canol supports at least 28 species of land snails and nine species of slug. This richness is typical of old broadleaved woodlands in Wales, although the presence of hollowed glass snail *Zonitoides excavatus*, a known calcifuge, suggests that the base-poor soil restricts the diversity of snails here. A lack of available calcium is less significant for slugs, since the shell is reduced to a thin internal remnant. Their mobility can be remarkable. The tree slug *Lehmannia marginata*, for instance, climbs high into the canopy to graze algae from trunks and leaves. Other species recorded include the colourful, if less acrobatic, green-soled slug *Arion flagellus* and tawny soil slug *Arion owenii*, both of which were found in the reserve in 2012 (Ben Rowson, personal communication). Also present is the largest European land mollusc, the ash-black slug *Limax cinereoniger*, which can grow to more than 20 cm in length. Living on algae and fungi, it roams far from its habitual refuges at night, or in damp weather.

Large mollusc species are in a minority, however, and woodland faunas consist mainly of detritivores: snails found in leaf litter and decaying wood. Individuals can be numerous, but most are small and easily overlooked. They are often found by collecting leaf litter, sieving it, and inspecting the debris under magnification. In this way Ty Canol has been found to be home to the dwarf snail *Punctum pygmaeum*, two herald snails *Carychium* spp., three whorl snails, including the marsh whorl snail *Vertigo antivertigo* and a classic woodland microsnail, the prickly snail *Acanthinula aculeata*. This has a small conical shell, with four to five convex whorls, and conspicuous transverse ridges drawn out into long spines at their margin. This 'prickly' appearance giving it its common name. None of these species, however, exceeds 2.5 mm in length or width, so records of their presence are testament to dedicated searches by visiting naturalists (Ben Rowson, personal communication).

Remarkably, there appear to be no records from Ty Canol of two other small species, the plated snail *Spermodea lamellata* and the attractive chestnut-coloured English chrysalis snail *Leiostyla anglica*. The latter is usually found attached to twigs or larger branches in leaf litter on the ground. A 'near endemic' to Britain and Ireland, it is more common in Ireland than elsewhere in Europe (Kerney 1999). Subfossil shells reveal that both of these species were once widespread across southern Britain in the 'forest optimum' of the Boreal and Atlantic periods, 9,500 to 7,500 years ago. Subsequent climatic change and deforestation mean they are now known in Pembrokeshire from only a handful of ancient woodlands, including those in nearby Cwm Gwaun.

CWM GWAUN WOODLANDS

The three main areas of woodland, Garn Wood, Kilkiffeth Wood and Cwm Felin-ban, cover the steep valley sides in the middle reaches of the River Gwaun, where the river has carved a boulder-strewn gorge through the rocks. Overgrown sessile oak coppice dominates the tree canopy along with some hybrid oak *Quercus robur* × *petraea*. Downy birch and, in the more base-rich, or flushed, areas, ash and alder also occur here, along with several non-native species such as sycamore, beech *Fagus sylvatica* and hornbeam *Carpinus betulus*. The ground flora in all three woods contains species that are characteristic of ancient woodland sites in this part of Wales, such as woodruff, wood melick *Melica uniflora*, wood millet *Milium effusum*, common cow-wheat *Melampyrum pratense* and bearded couch *Agropyron caninum*.

Cwm Felin-ban, and an area in Kilkiffeth adjacent to some abandoned enclosures, are particularly notable for their rich lichen flora. At the latter site, bearded lichen *Usnea* spp. decorates hazels and other plants in quantities unusual even in West Wales (Fig. 213). Most of the Welsh population occurs in Pembrokeshire, but even here it is sparsely distributed and seldom encountered.

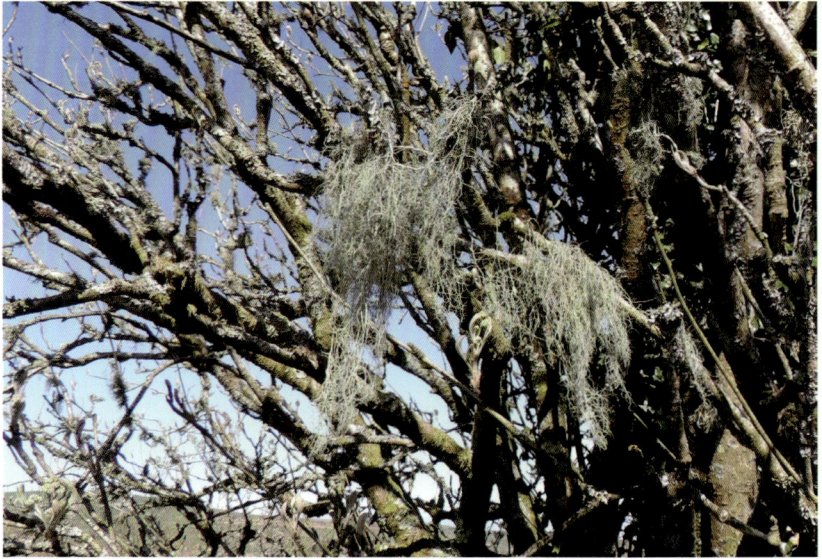

FIG 213. Bearded lichen decorates a rowan on the upper edge of Kilkiffeth Wood, Cwm Gwaun. Pembrokeshire contains most of the Welsh population of this species. (Stephen Evans)

Bearded lichen was included in Section 42 of the Natural Environment and Rural Communities Act, as a species of 'principal importance for conservation of biological diversity in Wales', because of its rarity and the significant loss of historic sites (Mullard 2014). On the larger oaks at Cwm Felin-ban other lichens, which thrive in the humid conditions, include *Catillaria pulverea, C. sphaeroides, Lecanactis premnea, Pannaria conoplea, Phyllopsora rosei, Leptogium lichenoides* and *Sticta fuliginosa*.

GALLT LLANERCH AND COED GELLI-DEG WOODS

These privately owned woodlands at the eastern end of Cwm Gwaun contain a number of veteran oak and ash trees, since part of the area escaped felling during the World Wars. It may be a relic of traditional 'coppice with standards', as described by George Owen. Mature sessile oak and hybrid oak dominate the canopy, and ash is abundant in places. Beech and sycamore are common in Coed Gelli-deg, however, and have the potential to change the character of the area, especially if the ash succumbs to disease. The shrub layer is dominated by

hazel and holly, with occasional specimens of rowan, goat willow and spindle
Euonymus europaeus. The ground flora includes many species typical of ancient
woodland, including moschatel, woodruff, sanicle and wood millet. More acidic
slopes have a mossy field layer with wavy hair-grass *Avenella flexuosa*, bilberry
and common cow-wheat.

The epiphytic lichen flora is particularly rich, with 21 species indicative
of ancient woodland. Lungwort is frequent on the ash trees, along with *Sticta
fuliginosa*, *S. limbata*, *S. sylvatica*, *Parmeliella* spp., *Dimerella lutea*, *Leptogium
teretiusculum*. *Gyalecta truncigena* and *Pachyphiale cornea* also occur here at their
only known location in the valley.

Dormice *Muscardinus avellanarius* are also present, the area to the north of
Mynydd Preseli, between Fishguard and Cardigan, containing the most westerly
population in Wales. The species has been recorded here since 1948, and dormice
are currently known from at least 24 separate sites, including Ty Canol Wood
(Jermyn *et al.* 2001). They are, however, still comparatively rare in Pembrokeshire
and threatened by the loss and fragmentation of habitat and poor woodland
management. This strictly nocturnal animal is largely confined to woods with a
rich understorey and areas of species-rich scrub. The physical structure of the
habitat is critically important. As dormice spend nearly all of their time in trees,
aerial pathways of interconnecting branches are essential to allow them to move
around their territories, but even in suitable habitats they occur at lower densities
than other rodents. Hazelnuts are an important source of food, as they enable
dormice to put on the necessary weight before hibernation.

HOOK WOOD

The ancient woodlands surviving on the Western Cleddau are rich in wildlife.
Hook Wood, in particular, contains many ancient woodland indicator plants
and is one of the most diverse oakwoods on the steep rocky shores of Milford
Haven (Fig. 214). George Owen described it as one of 'the best standinge woodes'
in Pembrokeshire. Yet again, however, the majority of the woodland was felled
during the First World War and, as a result it now consists mainly of even-aged
sessile oak coppice, which varies in height from about 15 m on the upper slope
to around 4 m on the low estuarine cliffs. A feature of the area is the occurrence
of discrete stands of aspen, alder, grey willow, goat willow and crab apple. Ash
occurs in a number of places on the upper slope, and in wetter areas lower down,
in combination with sessile oak, hazel and guelder-rose *Viburnum opulus*. More
open cliffs have ledges and crevices containing species such as wall pennywort

FIG 214. Hook Wood is one of the most diverse oakwoods on the steep rocky shores of the Western Cleddau. (Jonathan Mullard)

Umbilicus rupestris, tutsan *Hypericum androsaemum* and goldenrod, while ivy and field rose *Rosa arvensis* are common here.

Ferns are abundant and include broad buckler-fern *Dryopteris dilatata*, male-fern, scaly male-fern *D. pseudomas*, lady-fern, hard fern and soft shield-fern *Polystichum setiferum*.

CANASTON, PICKLE AND MINWEAR WOODS

Like the woodlands on the Western Cleddau, the first records of Canaston, Pickle and Minwear Woods come from George Owen. Situated at the head of the tidal section of the Eastern Cleddau, the area has been thickly wooded for centuries, much of it lying within the boundaries of the medieval Narberth Forest. In 1794, Charles Hassell recorded that most of the woodland was oak and managed for charcoal production and bark for tanning, but that good charcoal timber was running out. Indeed, estate maps of the time show areas of woodland that had been recently felled, thinned and coppiced. Hassall gives a detailed account of their management and the products that resulted:

The usual course of management in these woods, was, to thin the young shoots at the growth of three or four years, this was called waste weeding, and generally consisted of pruning away all the inferior shoots thrown out by the stumps or stools of the trees that had been felled, leaving the most vigorous shoots, to the number of four, five or six on a stool, to grow for a succeeding crop.
In about five years more, another weeding took place, and this was called a cordwood weeding, cuttings being come to a size fit for making charcoal.
The wood is left then to grow for a coppice of poles, and in about fifteen years more, the poles would be cut for the use of the neighbouring collieries, and would sell at about 7s 6d. a dozen, bark and cordwood included. The bark of these poles is either sold to the neighbouring tanners or exported to Scotland, where it finds a good market for the purpose of tanning neats and light hides.

Although up until the mid-nineteenth century the area of the woods had changed very little, during the twentieth century large tracts were unfortunately once again replaced by conifers. The sessile oaks that remain today therefore represent only a remnant of a much larger ancient woodland.

FIG 215. The main path through Minwear Wood National Nature Reserve. The site contains a number of notable tree species, including wild service tree. (Jonathan Mullard)

FIG 216. In Pembrokeshire, wild service tree is known as maple cherry, a name unique to the county. Once pollinated, the flowers develop into green-brown oval fruits. These are sometimes called chequers, since they are patterned with small pale spots when mature in mid to late autumn. (Tim Rich)

Minwear Wood has a well-developed understorey of holly, rowan and hazel, with bilberry, heather and brambles dominating the field layer (Fig. 215). Recently coppiced areas contain downy birch and grey willow, with alder near streams, while the woodland on the slopes above the Eastern Cleddau contains ash and beech, as well as oak. The ground flora is dominated by great wood-rush, along with broad buckler-fern, male-fern, scaly male-fern, hard fern and lady-fern.

The site contains a number of notable tree species, including crab apple, aspen and wild service tree *Sorbus torminalis*, the last species being typical of ancient and undisturbed lowland primary woodland (Fig. 216). Although it has a unique local name in Pembrokeshire, maple cherry, it was only formally recorded from the county in 1971 (Roper 1993). Since that date surveys have shown that it is comparatively widespread, the majority of trees growing, as here, in the oak woods on the low cliffs on the shoreline of Milford Haven and the Cleddau. There are a few small areas of wild service in the woods alongside the lower reaches of the Teifi and Nyfer, and a few inland records from north of Newgale.

In Pembrokeshire, the trees only fruit well after hot summers, since it is a warmth-loving species at the limit of its range in Britain. Since temperatures are generally too low for the seeds to ripen, it spreads principally by suckers.

BADGERS

Naturalists have known for centuries that the best place to look for a badger sett is within a wood but close to its edge, on a moderate convex slope where the soil is well drained and easily dug (Roper 2010). While badgers prefer deciduous woodland, or open grassland with some woodland, they can also be found in coniferous woodland and urban areas. Unless they are in dense woodland, where the lack of light prevents ground vegetation developing, the increased nutrient levels in the vicinity of setts means that they are often associated with elder, nettles, thistles, brambles and various grasses. Seeds of the grasses are spread around the entrance of the sett because of the badgers' practice of airing their bedding on the surface.

Badgers are instantly recognisable (Fig. 217) and yet rarely seen, being shy and nocturnal. Their distribution has historically been dependent on changing patterns of agriculture, the availability of grassland close to the sett, with good

FIG 217. Willow weaver Michelle Cain has created a giant badger sculpture at the Teifi Marshes Nature Reserve. (Jonathan Mullard)

populations of earthworms, being an important factor (Reason *et al.* 1993). Some badgers have been recorded as eating over two hundred earthworms in a single night. At the beginning of the twentieth century, badgers were probably rarer than they had been a hundred years earlier, almost certainly due to persecution by gamekeepers. Many gamekeepers lost their lives, however, during the First World War, and badger populations started to recover. At the last estimate, in 1995, there were 35,000 badgers in Wales (Harris *et al.* 1995). There have been no recent surveys but it seems likely that the population is at least stable.

Badgers now have legal status, for the first time in their long history, the Protection of Badgers Act 1992 giving them and their setts full protection, in England and Wales. Unfortunately, they continue to suffer from persecution by people who dig out setts in order to 'bait' the badgers against their dogs. This cruel and illegal activity leads to the death of badgers and sometimes the dogs as well. In addition, badgers face serious threats from increasing volumes of road traffic, urbanisation and ever more intensive agriculture. They also suffer localised culling because of bovine tuberculosis. Bovine tuberculosis is a major, and controversial, issue for farmers in Wales. A cull of badgers in Wales aimed at cutting tuberculosis in cattle was, however, replaced in March 2012 by a vaccination programme.

CONIFEROUS PLANTATIONS

British forests differ from those in the rest of Europe in that there is only one commonly planted native conifer, Scots pine *Pinus sylvestris*. (Our other native conifers are yew *Taxus baccata*, found mainly in churchyards and described in Chapter 17, and juniper, which in Pembrokeshire occurs only on Ramsey, as discussed in Chapter 6.) Much of the coniferous woodland in Wales consists instead of non-native species, mainly Norway spruce *Picea abies*, which provides two-thirds of the timber currently supplied to sawmills (Welsh Government 2018).

As well as the conifers in Canaston, Pickle and Minwear Woods, there are plantations on the southern slopes of Mynydd Preseli (Fig. 218), the largest of which is Pantmaenog Forest, although there are smaller areas around the nearby Rosebush reservoir. Pantmaenog is a privately owned forest, managed primarily for timber production, but it has now been opened to the public and there are several waymarked walking, mountain-biking and horse-riding trails.

There is, however, an increasing interest in returning these areas of conifer to native woodland. For example, Ffynone and Cilgwyn Woodlands in north Pembrokeshire, covering over 130 ha, are now owned by a workers' cooperative called Calon yn Tyfu Cyf/Growing Heart. Previously part of the Ffynone estate,

FIG 218. Mature conifer woodland at Llethrmawr on the southern slopes of Mynydd Preseli. (Jonathan Mullard)

prior to the 1950s much of this area was occupied by managed oak woodland over hazel coppice, with smaller areas of specimen plantings of various conifers around Ffynone Mansion. As mentioned in the introduction to this chapter, most of the oak woodland was felled following the break-up of the estate and replanted in the early 1960s with Douglas fir *Pseudotsuga menziesii*, Norway spruce, western hemlock *Tsuga heterophylla* and larch *Larix decidua*. The cooperative aims to progressively convert the woodland from the existing even-aged coniferous plantations, which yield an occasional large harvest of timber, to a mixed composition that will yield more frequent and varied outputs. The production of edible crops such as hazelnuts and mushrooms is a key priority, so we may have gone full circle, reflecting, perhaps, the fourteenth-century outputs from the Forest of Lloydarth.

The restoration by the National Park Authority of oak woodland on conifer woodland sites formerly owned and managed by the Forestry Commission, such as those in Cwm Gwaun, has benefited a range of species. Over the years the main aim of the woodland management has been on the restoration of native woodlands and public access. The National Trust has also been returning sites planted with conifers back to broadleaved woodland. The woodland at Little Milford, for example, is believed to date from at least the eleventh century, and oak was routinely coppiced here until the 1920s. After this date, however, large areas of the woodland were felled and replanted with conifers. The land was gifted to the National Trust in 1975, and in 2012 they harvested most of the conifers and replanted the cleared areas with a mix of broadleaved trees.

SQUIRRELS

The red squirrel is primarily an animal of northern conifer forests but in the past occupied deciduous woodland in the absence of any competing squirrel species (Morris 1993). Following the introduction of the grey squirrel in the nineteenth and twentieth centuries, red squirrels have now largely retreated back among the conifers, though even here greys can be found. The only record of red squirrels in Pembrokeshire to be found on Aderyn[*] dates from around 1950 in Ty Canol Wood. There have been no positive records of the species in Pembrokeshire since 1980, although there is sometimes confusion, as grey squirrels can have a red tinge to their coats. The red squirrels that have been introduced to Caldey are just managing to survive, thanks to supplementary feeding.

In 1945 there was still a large area of England and Wales where red squirrels were reported to be present, and in most of the Welsh counties free of grey squirrels they were widely distributed and increasing in numbers. Only in Pembrokeshire were they said to be less abundant than they had been and, in the absence of the grey squirrel, the decrease was attributed to the felling of forests during the Second World War, the decline of 'game preserves' and the wide use of steel traps intended for rabbits (Shorten 1954). The species was, however, said to be 'maintaining its status', with indications of a slight increase. They were seen fairly often in the lower part of Cwm Gwaun near Fishguard and in the vicinity of Newport and Nevern (Panting 1973). David Saunders reported one crossing the A40 south of Fishguard in March 1969 and there were two sightings in 1971, at Llawhaden and Saundersfoot.

Probably the earliest record of a grey squirrel in Pembrokeshire dates to January 1933, when someone who had formerly been a keeper at London Zoo, and was therefore familiar with the species, shot one in Llanycefn parish. This may possibly have been a captive specimen that had escaped. The previous year, in 1932, it had been made illegal to release a grey squirrel in Britain, but the damage was already done. The spread and establishment of the animal in Wales, however, dates almost entirely from an explosion of the population in 1945. Greys

---

[*] The Biodiversity Information and Reporting Database of Local Environmental Records Centres Wales, https://aderyn.lercwales.org.uk.

began to be seen in several places in Pembrokeshire around that date, having already travelled as far west as Treffgarne, Rudbaxton, Usmaston, Boulston and Cosheston. They may have been descendants of animals brought from London to Glamorgan in 1922 (Matheson 1955). Lockley reported that by 1961 the invaders had completely occupied West Wales, with breeding reported from near Aberystwyth and Pembroke (Lockley 1962). He was amazed by sightings of many young animals in quite unpromising habitats, such as high ground above the tree line, or rabbit burrows on sand dunes, as they dispersed in the autumn. The severe and prolonged winter of 1962/63 may have further assisted their spread, as there was another marked increase at that time and a corresponding decrease in red squirrels (Panting 1973). The extreme conditions had less effect on the grey, given the harsh winter weather in its native area, the northeast coast of North America. Grey squirrels are also blamed for the decline in red squirrels by competing for food and infecting them with the squirrel pox virus.

The Invasive Alien Species (Enforcement and Permitting) Order 2019 again makes the re-release of invasive species such as the grey squirrel illegal. It is too late, however, to make a major impact on the population by this method, with an estimated population of grey squirrels in Great Britain of some three million (Mathews *et al.* 2018). Stephen Harris, a well-known mammal expert, wrote some time ago that we need to accept that the grey squirrel is here to stay, and that the best place now for red squirrels is on small islands like Caldey.

DEER

Another species included in the 2019 Order is Reeves' Muntjac deer *Muntiacus reevesi*. Muntjac deer have been known in Pembrokeshire since the mid-1970s, but very few people have noticed them since they tend to be secretive, although they do have a loud barking call in the mating season. Two individuals arrived on Skomer in the early 1990s, possibly put there by someone who had injured animals and needed a safe place to release them. Neither of these survived more than a few months. A population in south Pembrokeshire is believed to have become established as a result of three animals which were said to have escaped from a former wildlife park across the border in Pendine (Chapman *et al.* 1994). They are solitary, with a home range of about 14 ha, and can breed throughout the year. Preferring dense habitats, such as conifer plantations, they are browsers, feeding on brambles, ivy, ferns, leaves and grasses.

Red deer and sika deer are often seen in the north of the county, especially, as mentioned in Chapter 11, on the Teifi Marshes. Both species are descended from animals that escaped, or were released, from another former wildlife park at Cilgerran. They are spreading slowly, but there are very few records of any deer away from that area. As previously described, the red deer on Ramsey are descended from animals farmed there in the 1980s, but there are still a few deer farms on the mainland. Sika are closely related to red deer, and hybrids can occur. Their preference for conifer plantations, especially the thick young stages, has been a major factor in their population growth and they can reach densities of up to 45 animals per km² in prime habitats. Sika can be a serious pest in commercial forestry plantations, scoring tree trunks and browsing young trees. The threat they pose to red deer, by hybridisation, is the major concern, however, and it is now illegal to transfer sika or red deer (which might in fact be hybrids) to islands that retain good populations of 'pure' red deer.

VETERAN AND NOTABLE TREES

There are a number of landscaped parks in Pembrokeshire, including Picton (Fig. 219) and Slebech, near Haverfordwest, and Hean Castle, near Saundersfoot, but they do not contain the really ancient trees found in parks in other parts of the United Kingdom. Perhaps, given the extent of woodland destruction in the past, this should not be a surprise, but there are a number of specimen trees noted by H. A. Hyde in his book on *Welsh Timber Trees* (1961) at Cefndyrys, Lawrenny, Pembroke Dock and Stackpole Court. Whether all the trees Hyde mentions are still standing is unknown, but at Stackpole he lists maritime pine *Pinus pinaster* and holm oak *Quercus ilex*, the latter represented by three trees, two of which were 24 m high. Colby Woodland Garden, now owned by the National Trust, was created by Samuel Kay, a wealthy pharmacist from Stockport who purchased the house here in 1873. The garden is home to the tallest Japanese redwood, actually a cedar *Cryptomeria japonica*, in Britain, which is 40 m high. There is also a 300-year-old black mulberry tree *Morus nigra* at St Brides Castle, with a girth of 4.55 m, but it is on private land and not accessible to the public. When King James I decided to produce his own silk, he ordered the aristocracy to plant mulberry trees to provide a supply of berries for the silkworms. He ordered black mulberry tree, but unfortunately the silkworms only eat the fruit of the white mulberry *Morus alba*!

There are other specimen trees in the towns of Pembrokeshire, but it is a declining resource. Wales is the first country in the world to undertake a complete canopy-cover study of all its urban areas, and in 2013 Pembrokeshire's urban cover was estimated to be 13.5 per cent, the fourth-lowest of all the counties (Natural Resources Wales 2016). Around 1,637 large trees were lost in the county between 2006 and 2013, many of them legacies of Victorian and Edwardian

FIG 219. Picton Castle sits at the centre of one of the three main historic parklands in Pembrokeshire, the others being Slebech and Hean Castle. The gardens and parkland here were once part of the larger Manor of Wiston. (Jonathan Mullard)

planting. As the largest settlement, Haverfordwest has the greatest extent of woodland and amenity trees and, perhaps not surprisingly, given its location in the windswept west, St Davids has the smallest amount.

In the wider countryside, the Woodland Trust's Ancient Tree Inventory lists over 600 trees in the county that are considered 'veteran' or 'notable'. The majority are clustered around the Eastern Cleddau, Saundersfoot and Tenby. They include a variety of species including ash and sessile oak. Veteran trees are defined as survivors that have developed some of the features found on ancient trees, but are only in their second, or mature, stage of life. Notable trees are mature trees which stand out in their locality because they are large in comparison with other trees nearby. None of these trees is as well-known, however, as the Crooked Oak.

The Crooked Oak

On the northern banks of the Eastern Cleddau, near the confluence with the Western Cleddau, stands the propped and gnarled 'crooked' oak, a tree with strong literary and artistic connections (Fig. 220). Waldo Williams often used to stay overnight in the nearby Croes Millin Chapel, sleeping on the hard benches

FIG 220. The Crooked Oak at Picton Point, a famous tree celebrated in poetry by Waldo Williams and painted by Graham Sutherland. In the past any oak such as this, with an L-shaped curve, was earmarked for making a boat's rudder. The unused 'rudder' of this great oak is now supported by a wooden brace. (Jonathan Mullard)

and rising early to walk down to the ancient oak on the riverbank. Here he could see and hear the birds on the estuary at dawn. The tree was celebrated in his poem *Y Dderwen Garn* (*The Crooked Oak*). This was the title which Williams gave to a 'half-rhyme' lyric registering his protest against a plan in 1959 to flood the area by building a dam on the upper reaches of the estuary. Thankfully the scheme never materialised. Williams obviously had a love of trees, since the only volume of poems he published during his lifetime was called *Dail Pren* (*The Leaves of the Tree*).

Other trees with natural crooks were used, in the past, to construct houses, furniture and farm equipment such as ox yokes – oxen being an important farm animal in Wales until the mid-nineteenth century. Suitably curved timber was also utilised in the construction of sleds and wheeled vehicles required for the all-important hay harvest, which was vital for keeping animals alive during the winter. Rich hay meadows were once therefore a critical part of the farm economy.

CHAPTER 16

Farmland Survivals

In some parts of the district the farming is good, but very frequently the farmer has favoured the botanist so far as to allow weeds to spread to a great extent over his arable land.

Charles Babington (1863)

ALMOST TWO-THIRDS OF THE SURFACE of Wales is covered by grassland, and Pembrokeshire is no exception. The long tradition of livestock production has resulted in a highly productive, but species-poor, grass sward across much of the county – a 'green desert' devoid of much wildlife interest. Slurry spraying adds to the problems since, as described in Chapter 10, it often results in nitrate pollution of both land and watercourses. Herb-rich hay meadows still exist, but it is notable that most of the meadows now designated as Sites of Special Scientific Interest form part of smallholdings owned and managed by former employees of conservation organisations. The remnants of the formerly extensive populations of the Tenby daffodil *Narcissus obvallaris*, once a characteristic flower of the fields in south Pembrokeshire, now only survive in marginal locations.

The county has been noted for its potato growing since the eighteenth century, and along the coastal strip and on the St Davids and Dale peninsulas there are extensive areas of arable cultivation (Fig. 221). The National Trust's Trehill Farm, adjacent to Marloes Mere, has potatoes as its main crop and is one of the few farms that manages to balance the requirements of arable farming with

FIG 221. Harvesting Pembrokeshire Early Potatoes on the Dale peninsula. Due to the beneficial effects of the Gulf Stream, there is a longer growing season here than in other parts of Britain. (Jonathan Mullard)

conservation. Much of the cultivated land in Pembrokeshire, like the grasslands, is intensively managed. There is still a distinctive arable flora in the south of the county, but it is certainly not as abundant as it would have been in the 1860s when Charles Babington visited the area (Babington 1863), and the survival of these rare plants is dependent on decisions made on an annual basis by farmers. Species may survive unseen in the seed bank for a number of years, but the continual use of herbicides will eventually result in their demise.

MEADOWS

Although, as described in Chapter 9, Castlemartin contains the largest areas of neutral grassland in Wales, some of the best surviving traditional meadows in Pembrokeshire surround the farmstead of Mountain Grove, to the west of Penffordd. Known as Mountain Meadows, the six fields here are extremely rich in species and were designated a 'Coronation Meadow' by Plantlife in 2012 (Fig. 222). In that year, one meadow in each county of the United Kingdom was selected to celebrate the sixtieth anniversary of the Queen's coronation. New meadows were created at 'recipient' sites in the same county, using seed from the relevant Coronation Meadow. In Pembrokeshire's case the recipient is Castle Meadow on Camrose Farm, which backs on to a motte and bailey. The funding which Plantlife used to set up the scheme ran out after three years, but a successor project, 'Magnificent Meadows', has been developed.

There are large populations of orchids at Mountain Meadows, including green-winged orchid, which was once frequent in the meadows of southwest Wales. It is now confined to this site, a few churchyards and the coastal grasslands. In addition, southern marsh-orchid, northern marsh-orchid and early marsh-orchid are abundant; along with heath spotted-orchid *Dactylorhiza maculata*, early-purple orchid, and common twayblade. In early summer the meadows are a poignant reminder of the richness of the countryside before agricultural intensification and how much we have lost.

In addition to orchids, the unimproved neutral grassland here contains sweet vernal-grass, bugle *Ajuga reptans*, glaucous sedge, rough hawkbit (which

FIG 222. Mountain Meadows, near Penffordd, are traditional meadows of exceptional diversity, with large populations of orchids. (Jonathan Mullard)

is almost dominant), oxeye daisy *Leucanthemum vulgare*, field wood-rush *Luzula campestris*, ribwort plantain *Plantago lanceolata*, bulbous buttercup *Ranunculus bulbosus*, meadow buttercup, red clover *Trifolium pratense*, yellow-rattle, crested dog's-tail *Cynosurus cristatus*, common knapweed, common sorrel *Rumex acetosa* and cock's-foot. In the less well-drained areas, there is devil's-bit scabious, whorled caraway *Carum verticillatum*, sneezewort *Achillea ptarmica*, tufted hair-grass *Deschampsia cespitosa*, meadowsweet, purple moor-grass, mat-grass and compact rush *Juncus conglomeratus*.

Adjoining the meadows there are a number of small overgrown pastures, and these, along with the overgrown hedgerows around the meadows, provide abundant shelter and food for invertebrates. A total of 24 species of butterfly occur here, including holly blue *Celastrina argiolus*, small pearl-bordered fritillary, silver-washed fritillary and the nationally scarce brown hairstreak *Thecla betulae*. A number of rare hoverflies have also been seen in the area, notably crimson-belted hoverfly *Brachypalpoides lentus*, grey-backed snout-hoverfly *Rhingia rostrata* and *Eupeodes bucculatus*, a poorly understood species. All these are abundant, highlighting the importance of this special area.

FIG 223. Orchid-rich meadows at Wyndrush Pastures, where meadow seed and honey are harvested, as well as the hay. (Matt Sutton)

Wyndrush Pastures, near Redberth, represent another significant group of meadows (Fig. 223). Here the neutral grassland is dominated by common bent, red fescue, sweet vernal-grass and Yorkshire-fog. Other plants include common knapweed, common bird's-foot trefoil and rough hawkbit. In the damper areas, greater bird's-foot trefoil, devil's-bit scabious, pale sedge *Carex pallescens* and zigzag clover *Trifolium medium* are locally abundant. In places, sharp-flowered rush and soft-rush dominate, and there are occasional patches of compact rush. Among the rushes there are herbs such as lesser spearwort *Ranunculus flammula*, common marsh-bedstraw *Galium palustre*, whorled caraway and purple-loosestrife. In places, the rushes grade into a tall-herb fen containing species such as meadowsweet, hemp-agrimony and wild angelica. Purple moor-grass also occurs, along with tawny sedge *Carex hostiana* and flea sedge *C. pulicaris*. There are numerous fungi, including the newly described purple-pink waxcap *Gliophorus reginae*.

As at Mountain Meadows, there is a rich variety of butterflies, including green hairstreak, and moths such as the nationally scarce Devon carpet moth *Lampropteryx otregiata*. There are significant numbers of bumblebees, including the rare shrill carder bee, also recorded from Castlemartin. Other notable insects include the small yellow-legged robberfly *Dioctria linearis* and the striped slender robberfly *Leptogaster cylindrica*. Despite the scrubby edges to the meadows and plenty of organic cowpats the hornet robberfly *Asilus crabroniformis* has not yet been recorded from Wyndrush Pastures but there is a known site near Newgale, which has produced counts of up to thirteen individuals.

The hornet robberfly may be the largest fly in the United Kingdom, being over 25 mm long when full grown with black and yellow markings that resemble a hornet (Fig. 224). Like other robberflies they are hunters, catching beetles, grasshoppers, wasps, and indeed other flies. The female lays her eggs on cow, horse or rabbit manure, the larvae feeding on dung-beetle grubs in the ground below. After three years they pupate and emerge as adults the following summer. Since the species relies on grazing animals and insect-rich habitats it is very vulnerable to changes in farming practices, which means that meadows such as these are probably the only place it will survive. A key issue is the use of general insecticides, such as avermectins for worming cattle, which also kill off the dung beetles the robberflies prey on.

Since 2006, the owners of Wyndrush Pastures, Matt Sutton and Vicky Swann, have followed a policy of using no lime or fertilisers and have not drained the fields. As a result, the previously grass-dominated sections have become rich in orchids and other meadow flowers. At the same time, however, there has been a decline in hay production and a consequent reduction in the number of calves

FIG 224. The hornet robberfly is a large predatory insect associated with grazing animals, particularly cattle, as their dung provides the habitat for the larvae. (Steven Falk)

or finished beef cattle they can sell. Similarly, the productivity of Mountain Meadows has apparently been depleted. Cattle reared exclusively on botanically rich pastures mature slowly, and although the finished animal provides high-quality meat, this is not rewarded by the mainstream agricultural system. As Matt and Vicky have noted, 'Keeping grazing animals is considered integral to good meadow management, but on a small scale this can be a financial liability as well as a large time commitment' (Sutton & Swann 2013).

Somerton Meadows, on the north side of the Castlemartin peninsula, cover most of a small stock farm that has been managed in a traditional way for many years. Here there is a herd of Dexter cows, an ancient Irish breed, and electric fencing is used extensively to control the grazing (Fig. 225). This sensitive management has produced a remarkable suite of herb-rich grasslands, which include locally uncommon plants such as downy oat-grass *Avenula pubescens,* quaking-grass, smooth-stalked sedge *Carex laevigata,* hay-scented buckler-fern, saw-wort and yellow oat-grass *Trisetum flavescens.* The site supports small breeding populations of marsh fritillary butterfly and, as at Wyndrush, the shrill carder bee.

The real glory of the meadows here though are the fungi. One of the best sites in Wales for grassland fungi, Somerton Meadows supports a diverse range of waxcaps Hygrophoraceae, coral fungi Clavariaceae, pinkgills Entolomataceae and earth tongues Geoglossaceae, as well as several *Dermoloma* species. Waxcaps, in particular, are regarded as nitrogen-sensitive organisms because their fruiting is inhibited by applications of fertilisers. They require low levels of soil disturbance and long periods of ecological continuity, so their presence here is a testament to the management regime (Ainsworth *et al.* 2013). Indeed, the owners, David and Holly Harries, are key members of the Pembrokeshire Fungus Recording Network.

In addition, the farm is the 'type locality' for two recently described grassland fungi, *Entoloma ochreoprunuloides* f. *hyacinthinum* and *Gliophorus europerplexus.* The latter is an indicator of mycologically rich grassland and was identified as a

FIG 225. Dexter cattle feeding on the rich grassland at Somerton Meadows. Electric fencing is used to manage the grazing, ensuring that both cattle and wildlife thrive. (Jonathan Mullard)

FIG 226. One of the recently described big blue pinkgills, *Entoloma ochreoprunuloides* f. *hyacinthinum*, at Somerton Meadows. This is now one of the best sites for grassland fungi in Wales. (David Harries)

result of a study of waxcap grasslands which was carried out by Kew (Ainsworth *et al.* 2013). The same research revealed that *Entoloma bloxamii*, the big blue pinkgill which is listed in Section 7 of the Environment (Wales) Act 2016 as 'an organism of principal importance for the purpose of maintaining and enhancing biodiversity in relation to Wales', was in fact a collection of several separate species. A later paper confirmed that there are four distinct species in the United Kingdom, three of which have been recorded in Pembrokeshire (Ainsworth *et al.* 2018). All of them produce robust fruiting bodies with a cap typically up to 80 mm across and coloured blue, or blue-grey (Fig. 226). To date there has been one record of *Entoloma ochreoprunuloides* f. *hyacinthinum* from Somerton farm, two records of *E. madidum* from Somerton Farm and Stackpole, and one record of *E. atromadidum* from Tyrhos chapel, Rhos-Hill.

Mountain Meadows, Wyndrush Pastures and Somerton Meadows are all designed as Sites of Special Scientific Interest, as recently as September 2017 in the case of Somerton, but there are also numerous other meadows in the county which are of interest. Probably the easiest to access is the former airfield outside St Davids owned by the Pembrokeshire Coast National Park Authority. Areas around the concrete runways have been developing as meadows for over twenty years now and the summer display of flowers is striking (Fig. 227).

Other smaller meadows include Cottesmore Park near Haverfordwest, with green-winged orchid and pale sedge, Forgotten Field, at the Park near Martletwy, and Blaencleddau near Mynachlog-ddu, which is rich in yellow-rattle and eyebright and has been under benign traditional management for over fifteen years (Matt Sutton, personal communication, 2018). Attempts are being made to restore a number of other meadows, and, indeed, create new ones using harvested seed. The owners of Wyndrush Pastures, for example, have, with grant aid from the Welsh Government's Sustainable Development Fund, bought their own brush-harvester to enable them to provide a mobile harvesting service (Sutton *et al.* 2013).

FIG 227. The meadows on the former St Davids airfield are easily accessible and a riot of colour during the summer months, with oxeye daisies, common spotted-orchids and southern marsh-orchids among other species. (Jonathan Mullard)

LOCUSTS?

There are older, and perhaps puzzling, records of the invertebrates that have been found in meadows. Probably the most notable, once again, is a description by George Owen of a strange event that happened at the beginning of June 1602:

> *There happened that suddenly, as if the same had fallen by a shower out of the air, a great piece of ground to the quantity of two hundred English acres was covered in a manner with a kind of caterpillars, or green worms, having many legs and bare without hair. They were found in such abundance that a man treading on the ground should tread upon twenty or thirty of them, and in this sort they continued for the space of three weeks or more, no man knowing how they came, nor any of the like sort ever seen in the county before or since, and being killed and opened, there was no gut or anything else within them but only grass which they had devoured. The place was on a hill in the parish of Maenclochog above Ffynnon Dewi. They were found, as it were with one accord, to go one way, namely, up the hill, and went over the hill a quarter of a mile and more, and as they went, did devour and consume the grass, that the ground appeared bare and red like fallow, and after they had continued there three weeks, there resorted thither an infinite number of sea mews and crows, as if all of many counties had been summoned thither, who in few days consumed them all, after they had consumed all the grass of the mountain. Also swine fed upon those worms eagerly, and waxed very fat.*

There is a possibility that these worms may have been locusts *Schistocerca gregaria*, Edward Lhuyd recording, some years later, in 1693 that 'There are Locusts lately come into Wales: whereof I have one sent me' (Gunther 1945). A locust was photographed in 2018 in a nature reserve managed by the Devon Wildlife Trust, so they do occasionally reach these shores even now, but never in the numbers described by Owen (Parker 2018). Owen would have been familiar with crickets and grasshoppers, however, so his description of the animals as caterpillars, or worms, is indeed mystifying.

MOLES

Originally inhabitants of deciduous woodland, moles *Talpa europaea* thrive in traditional pastures, but are found throughout mainland Britain wherever the soil is sufficiently deep for them to burrow. The word 'mole' is thought to derive from the Middle English word *mouldwarp*, which literally means 'earth-thrower'.

FIG 228. A mole in an improved pasture near St Davids. They are very vulnerable above ground, but can be seen more often than might be expected. (Jonathan Mullard)

There are estimated to be 3,250,000 moles in Wales, but this is an approximate figure since there are few estimates of density, and there are certainly no specific figures for Pembrokeshire (Harris *et al.* 1995). Of the 3,600 records, the majority appear to derive from sightings of molehills. Molehills though are not a reliable guide to population size, or location, but simply a sign of recent tunnelling activity. The impact of agricultural chemicals, particularly insecticides, on mole numbers is unknown. Sadly, moles still have no legal protection and are frequently regarded as pests, with many being killed each year.

Although moles spend almost all their time underground, digging out tunnels with their spade-like front paws and hunting earthworms, occasionally they can be seen on the surface (Fig. 228). A mole may move above ground when trying to detour around an underground obstacle, or when the ground is hard and invertebrate prey is scarce. Juvenile moles often surface as they abandon their mother's territory, leaving a fresh spoil heap with a noticeable hole at the centre.

POLECATS

Polecats range widely across habitats, from woodlands and farmland to coastal dunes. They are mainly nocturnal and solitary, with a diet that largely consists of rabbits and rats. They were once widespread, probably the third most common carnivore in Britain, with an estimated population of over 100,000 during the Mesolithic (Maroo & Yalden 2000). Interestingly, even though it must have been abundant, the polecat is not mentioned in the literature prior to the Norman conquest. The first reference in the Welsh language is found in *Llyfer Coch Hergest*, or the *Red Book of Hergest*, a large vellum manuscript dating from shortly after 1382 and one of the most important medieval manuscripts written in the Welsh language.

The polecat population underwent a severe decline during the eighteenth and nineteenth centuries, the animals being killed in large numbers by gamekeepers to protect poultry and game birds. The pioneering conservationist Charles Waterton (1782–1865) called it 'the stinking polecat, shunned by most people and persecuted by all' (Lovegrove 2007). By the early twentieth century the polecat was on the brink of extinction in Britain. From the 1930s onwards, however, the population began to slowly recover in Wales, expanding from a relict population around Aberdovey. Its return was almost certainly due to a reduction in gamekeepers during the First World War (Croose 2016). Between 1959 and 1962 there were 312 records from Wales and the English borders, only one of which was in Pembrokeshire, near Mynachlog-ddu. The species is now, however, considered to be well-established and widespread in the county. Despite this welcome change polecats are probably under-recorded, as they are elusive. Most sightings tend to be of animals killed on roads, or seen briefly as they cross roads at night.

TENBY DAFFODIL

The Tenby daffodil was first identified in 1796 by Richard Salisbury. Avoided by many of his contemporaries, partly because he refused to adopt the Linnaean system, Salisbury was nevertheless a dedicated botanist who contributed significantly to the development of the science. In his *Prodromus stirpium in horto ad Chapel Allerton vigentium* (1796), an initial publication intended as the basis for a later book, he wrote:

7.N. 1-florus: corollae lacciniis ½ longioribus, rectis, ovatis, interioribus multo angustioribus valde imbricates: coronâ infundibuliformi basi cylindraceá, 6-fidá, repando-dentatá, superne plicatá.

Pseudo Narcissus luteus. Besl. Hort. Eystt. Tert. Ord. mala.

Communicavit cl. Curtis.

Unfortunately, Salisbury gives no location for the find and, given the reference to Curtis, the specimen was presumably received from William Curtis, demonstrator of plants at the Chelsea Physic Garden from 1771 to 1777. Where Curtis obtained it is unknown. This distinctive plant, with short stiff stems and small yellow flowers in which the petals are at right angles to the trumpet, differed so much from the wild daffodil *Narcissus pseudonarcissus* ssp. *pseudonarcissus* Salisbury was familiar with in England that it was obviously a separate subspecies. The Tenby daffodil is only around 30 cm tall, and the yellow flowers are about half the size of those of the average daffodil. The flower has a neat trumpet and a ring of pointed petals, each with a small white tooth at the end. This tooth is perhaps the easiest way to distinguish it from other daffodils (Fig. 229).

FIG 229. The neat trumpet surrounded by a ring of pointed petals, each with a small white tooth at the end, identifies this as a Tenby daffodil. This is a cultivated specimen in the author's garden. (Jonathan Mullard)

The earliest specimen we have of the Tenby daffodil is stored in the herbarium of the Natural History Museum in London. It almost certainly originated from plants grown by the Reverend W. T. Bree in his garden at Allesley Rectory, near Coventry, from 'roots found apparently wild near Tenby, in Pembrokeshire, by the late Joseph Boultbee, Esq.' (Syme 1851). In a report published in *The Phytologist* in 1844 under the heading 'A new British Narcissus?' Bree noted that:

> *The late Mr Haworth, to whom I sent it in 1830, considered it new to Britain, and recorded it in the Philosophical Magazine under the name of Ajax lobularis. It is a highly ornamental species, a free flowerer, and increases readily. I think it may fairly be considered a native plant, (unless, indeed, it should be held to have been introduced by the Romans), for it is not likely that it should have been the outcast of a garden, being, as I believe, so little, if at all, known in the gardens, till of late years distributed by me among various private friends and public institutions, to all of whom it appears to have been previously unknown.*

Forty years later another observer recorded that Tenby daffodils could be seen for 20 miles from Tenby to Haverfordwest, and were also abundant around St Brides Bay (Baker 1884). A second sheet, from the herbarium of a Miss Moseley, gives the source of the plant as 'the Warren, Tenby received from Mr. Bree'. As far as is known this is the only reference to the site from which Mr Bree's friend collected the bulbs (Fred Rumsey, personal communication, 2016). As described in Chapter 9, the Warren, which stretches between Tenby and Giltar Point, is now mainly a golf course but Tenby daffodil still occurs on the northern slopes of the Point, perhaps survivors of the original population (Fig. 230). In March 1940 H. W. Pugsley noted daffodils 'in one hilly field near Tenby', and J. A. Whellan recorded them in 1941 'on a grassy limestone hill near Tenby', which might possibly be the same location.

Once widespread in the fields around Tenby, the daffodil unfortunately became a fashionable plant and, as a result, the population was greatly reduced by commercial bulb collectors in the nineteenth century. The curator of Tenby Museum, for example, reported in 1893 that:

> *From enquiries, I gather that up to 1885, a steady trade had been done by people here in the bulbs, men being sent into the country districts by one man here systematically to hunt for them. On some fields belonging to Holloway Farm, which you no doubt remember is just outside Tenby on the Marsh Road, the daffodil grew very abundantly: the owner, a man named Rees, learning of the value of the flowers at Covent Garden, sold the bulbs, the entire crop on fields, for £80. (Jones 1992)*

FIG 230. Tenby daffodils on the north side of Giltar Point, which is perhaps the 'grassy limestone hill' where they were recorded in the 1940s. They may be a relic of the original population. (Tim Rich)

The 'one man', the most ruthless plunderer of the plant, was Mr Shaw, an ambitious nurseryman. He was responsible both for popularising the name 'Tenby daffodil' and for bringing it to the brink of extinction in the wild. A paper read to the Cardiff Naturalists' Society by Charles Vachell in 1894 summarises his activities well:

> So delighted were the wholesale bulb dealers with the new flower that orders for the bulbs arrived in rapid succession. Mr Shaw was enabled to engage a staff of collectors who scoured the greater part of South Pembrokeshire for several seasons in a vigorous attempt to meet the phenomenal demand. Considerable quantities were found by the south side of the Haven, even as far as Castlemartin, but by far the largest quantities were obtained around Narberth, Clynderwen, Llanycefn and Maenclochog. As a rule, the farmers on whose land they grew regarded them as little better than weeds and readily parted with them for a trifle and sometimes for nothing, though Mr Shaw's men, as a result of three days' excursion, often brought him a heaped cart load, which he sold for £160 or more. So well did he keep his secret that he had a complete monopoly of the trade until the supply was practically exhausted.

The flower's curiosity value, and price, was increased by the development of various historical myths: the bulbs had been traded by Phoenician sailors for a cargo of anthracite, brought over by Flemish settlers in the early twelfth century, or planted by the monks on Caldey. Described in the *New Atlas of the British and Irish Flora* (Preston *et al.* 2002) as a 'plant of uncertain origin' and in other floras as 'introduced, long naturalised' (Stace 2019), it is now thought to be derived from the Spanish daffodil *Narcissus pseudonarcissus* ssp. *major* by cultivation in the medieval period. Spanish daffodils were certainly being cultivated in British gardens by 1629 and recorded growing in the wild in 1813. Plants closely resembling the Tenby daffodil occur in the wild in Spain, but hybridisation between other subspecies of daffodil complicates the situation. The only true native species of British daffodil appears to be the wild daffodil, although exactly where this originates from is not clear either.

Many of the colonies that survived the exploitation of the nineteenth century were sadly destroyed by intensive agriculture, but some plants remained. It is now naturalised once again in hedge-banks, churchyards and other refuges. The Tenby daffodil's spread was assisted by a 'bizarre sequence of events in the 1970s' (Mabey 1996). A boy from Essex, on holiday in the area, walked into the tourist information centre and asked where he could buy Tenby daffodil bulbs to take home to his aunt. None of the staff there had ever heard of the flower but a plumber who happened to be working in the office overheard the conversation and later brought in *The Readers Digest Book of British Flowers*, which provided a clear description of the plant. Realising the commercial potential, the tourist office persuaded a local nursery to source cultivated stocks and featured it in the next 'Tenby in Bloom' celebrations. In the spring the whole area is now covered with the bright blooms of Tenby daffodils.

Increasing numbers of daffodil cultivars are, however, becoming naturalised as a result of widespread municipal plantings, especially 'Primrose-peerless', 'Nonsuch' and varieties of 'Pheasant's eye' *Narcissus poeticus*. Despite the connection, daffodils have only been the national flower of Wales since the nineteenth century, and in the early twentieth century they were promoted by Lloyd George, who thought them more attractive than leeks.

ARABLE PLANTS

Arable plants, the 'weeds' in the crop, share the same ecological niche as the plants with which they coexist, thriving on regular disturbance and a short growing season. Many arable plants were included in Babington's list of 550 plant species found in south Pembrokeshire, published in 1863, but 150 years later the situation is very different. Populations are greatly reduced, and some species are now on the verge of extinction. During the twentieth century the amount of arable land in Wales reduced by around 65 per cent, and today it is estimated that only 11 per cent is under arable cultivation: a loss of 106,400 ha since 1970 (Shellswell 2015). This decline has been the result of a change from mixed farming to a mainly livestock-based economy relying on grass silage. The costs of arable farming have also increased, making it uneconomic for smaller farms. The increasing use of herbicides and fertilisers, alongside more efficient seed cleaning, have likewise played their part in reducing the populations of arable plants. Finally, there is now a greater proportion of autumn-sown cereals, compared with the spring-sown crops that many species historically relied upon for their survival.

Even the crops in the remaining fields where arable plants grow have changed significantly over the last hundred years, and this may have affected the surviving populations. Samples of ancient thatched roofs, which have preserved large quantities of medieval, and later, cereal crops, contain 'landrace' mixtures of bread wheat *Triticum aestivum*, rivet wheat *T. turgidum*, rye, barley and oats *Avena* spp. (Letts 2001). These grew to 1.8 m or more in height, much taller than modern varieties. Landraces evolve over the centuries when crops are grown in varied

FIG 231. Huge populations of corn marigold can be found in certain fields on the Dale and St Davids peninsulas and around Castlemartin. This field, owned by the National Trust, is adjacent to Marloes Mere. (Jonathan Mullard)

conditions from seed saved from the previous year's crop. The result is that every plant in a landrace is slightly different from its neighbour. Medieval crops were very uneven in height, ripening time and other traits. This diversity ensured that a proportion of the crop always set seed, regardless of drought, frost, or disease. Unfortunately, landraces across Britain were lost during the nineteenth century as farmers adopted the new cultivars created by plant breeders. Perhaps arable plants such as corn marigold *Glebionis segetum* grew much taller when they had to regularly compete with wheat nearly 2 m high. We cannot, however, easily recreate those early crops to find out.

Corn marigold is a feature of the slightly acidic sandy and loamy soils found in coastal fields, and specific farms on the Dale and St Davids peninsulas and around Castlemartin have long been noted for having fields with huge populations of this plant (Fig. 231). Cornflower *Centaurea cyanus* has been recorded regularly from a field near Lawrenny, and narrow-fruited cornsalad *Valerianella dentata* occurs nearby. One of the most recent finds was of hairy-fruited cornsalad *V. eriocarpa* in an arable field near St Ishmael's. Over 300 plants were recorded on the margins of a wheat crop, which probably means that the population is self-sustaining, germinating from the soil seed bank each year. There is only one other extant population known for hairy-fruited cornsalad, in North Wales, and it is not on arable land, so this find was extraordinary. The other Pembrokeshire record dates back to 1975, when it was recorded in cereal fields east of Fishguard. Unfortunately, hairy-fruited cornsalad has not been found at that location since, despite many repeat visits, but until recently the field margins there still supported species such as weasel's-snout, field woundwort *Stachys arvensis* and sharp-leaved fluellin. The county similarly supports the majority of the United Kingdom population of small-flowered catchfly.

The St Ishmael's area is a well-known location for arable plants, with some of the most important populations of small-flowered catchfly and annual knawel *Scleranthus annuus* in Wales. In addition, corn chamomile *Anthemis arvensis* has been found here. In contrast, corn spurrey *Spergula arvensis* and field woundwort are widespread across Pembrokeshire, and Wales is probably the best location for these plants, which are sparsely distributed elsewhere.

In contrast to the species described above, prickly poppy *Roemeria argemone* has always been a rare plant in Pembrokeshire, with only two records. It is associated with very lime-rich or sandy soils, and this may be the limiting factor in its distribution. The last known population near St Davids was on a roadside bank and is likely to be an ancient relict of the arable land managed organically behind Whitesands Bay. Rough poppy *Roemeria hispida* is also rare, with a small concentration of records on the St Davids peninsula, including some recent records on the seaward margin of an arable field alongside the coast path. This population is almost certainly of natural origin and has continued to survive through germination from the soil seed bank. There is only a single record, in 1851, for corn buttercup *Ranunculus arvensis*, which was found by Babington during a visit to the Rector of Gumfeston and his family.

Similarly, there are only four records from the county for shepherd's-needle *Scandix pecten-veneris*. It is primarily an autumn-germinating plant that tends to be found on the heavier clay soils, which are comparatively rare in Pembrokeshire. Three are prior to 1950 and the latest, from 2010, is probably a recent introduction, as the plant was found growing in a reseeded chicken run. In contrast, all of the records of henbane almost certainly represent plants of native origin, or at least long-established specimens. Many of the records, however, are on sand dunes, such as Freshwater West and Stackpole, or from sandy arable land behind the dunes.

The importance of arable habitats for bryophyte conservation in Britain is still not fully recognised, but the Survey of Bryophytes of Arable Land carried out by the British Bryological Society between 2003 and 2005 provided useful information, now supplemented by Sam Bosanquet's work on the mosses and liverworts of the county (Bosanquet 2010). The core of the arable 'assemblage' consists of the following mosses: rough-stalked feather-moss *Brachythecium rutabulum*, crimson-tuber thread-moss *Bryum rubens*, violet-tuber thread-moss *B. violaceum*, field forklet-moss *Dicranella staphylina*, serrated earth-moss, common feather-moss *Kindbergia praelonga*, Swartz's feather-moss *Oxyrrynchium hians*, cuspidate earth-moss *Phascum cuspidatum*, delicate earth-moss, common pottia, cylindric ditrichum and the stubble and beardless mosses *Weissia* spp.

Much more survey work needs to be undertaken, especially in some of the inland areas of Pembrokeshire. Even though the individual arable plants are annuals, the communities themselves are surprisingly stable, fields with a long history of cultivation having the richest flora (Wilson & King 2003). Seeds do not necessarily, however, germinate every year that the field is cultivated and can remain dormant, and undetected, in the soil until conditions are suitable. This can make surveying difficult, and repeat visits are often necessary to determine the presence, or absence, of species.

CLODDIAU

The word *cloddiau* can be variously translated as 'walls', 'banks', 'dykes', 'earthworks' or 'hedgerows', or any combination of these. The reason for building *cloddiau* in the first place was to enclose land; the Welsh words *cau* ('closed') and *cae* ('field') stem from the same root (Elias *et al.* 2016). The term is mainly associated though with raised earth banks, usually faced with stone. Found mainly in Pembrokeshire and the western coastal districts of Wales, they are a characteristic feature of the landscape (as illustrated in Fig. 8). Where other suitable materials were scarce, these structures enabled the creation of fields. Cloddiau are normally lower than a drystone wall, the majority being less than a metre in height, but they typically have a hedge on top and a ditch on one side, created by the removal of soil for the core. Unlike in a drystone wall, the stones are placed with their longest side vertical, the stone casing protecting the bank from erosion by livestock or weather. The compaction and integrity of the core are crucial to the structure's longevity.

It is said that over 80 per cent of flowering plant species in Wales have been recorded from cloddiau. While this is somewhat of an exaggeration, if the circumstances are right, they can indeed support a rich flora. Cloddiau on the coast, for instance, contain sea campion, spring squill, thrift and kidney vetch, while in upland areas plants such as harebells *Campanula rotundifolia*, heather, bilberry and western gorse predominate. In sheltered inland areas plants such as red campion, lesser stitchwort *Stellaria graminea* and bluebell occur. Today they are important for small mammals, birds, reptiles and insects, providing essential corridors through otherwise inhospitable agricultural land.

ADDERS

Gorse- and bracken-covered *cloddiau* are a significant habitat for adders. The 'Adders are Amazing!' project organised by Amphibian and Reptile Groups of the United Kingdom (ARG) is working with communities on the St Davids peninsula, one of Wales's key strongholds for the adder, to ensure a sustainable future for this snake (Fig. 232). The Pembrokeshire Adder Coordinator, Sam Langdon, is using volunteers to undertake detailed surveys of adder habitats and, especially,

FIG 232. A female adder basking on dead grass warmed by the sun. Only a metre or so from the coastal footpath on Pen y Cyfrwy, south of St Davids, this individual was unnoticed by the many walkers who passed close by, due to the way it blended in with its background. (Jonathan Mullard)

hibernation sites. Protecting these locations is critically important, because adders are at their most vulnerable when hibernating. They are often communal sites with large numbers of adders using the same place year after year. There is a major hibernation site on the edge of the road leading to Dowrog Common, but many of the locations used by adders are on non-designated sites, hence the need for community involvement and support. The findings of the project will be used to inform further initiatives and help reverse the negative attitude that people have towards adders and our other native snakes, grass snakes and smooth snakes *Coronella austriaca* (though the latter only occurs naturally on heathlands in Dorset, Hampshire and Surrey).

One of the biggest threats that adders in Pembrokeshire face is disturbance by walkers and their dogs. It is therefore important for dog owners to keep their dogs close to them when they are in suitable adder habitats. ARG have adopted the slogan 'Stop, Stand back, and Smile' (SSS), to remind people not to approach adders and to give them space. Stepping back as soon as you find an adder allows it to relax again, and it is then likely to stay put rather than head for the nearest cover as fast as it can. There are many comments on social media from people saying they will never visit Pembrokeshire again when they see posts about adders, but seeing an adder should be viewed as a special event, rather than something to be frightened of.

The specific sites discussed above represent fragments of a previous, and richer, landscape. Like the cliffs and sand dunes covered in previous chapters, they act as refuges for wildlife from intensive agriculture. But these are not the only areas to hold on to an older flora and fauna. Churchyards and burial grounds were often created by the enclosure of ancient pastures, and can be remarkably abundant in species as a result.

CHAPTER 17

Cathedral, Churches and Chapels

West Wales is a heartland of gorgeous religious buildings, as well as spiritual revolutions. Non-conformism exploded here in the 19th century, seeing preachers amassing huge crowds at our lovely Welsh chapels. But there are also older places ... islands inhabited by monks ... and ... our stunning cathedral in the United Kingdom's smallest city.

Visit Wales Website

H IDDEN IN THE VALLEY OF the River Alun, the location of St Davids Cathedral was chosen in the hope that the church would be overlooked by seaborne raiders, but it was ransacked at least seven times. Even as late as 1767 great parts of the cathedral were roofless, and a traveller, signing himself T. K. D., wrote to the *Gentleman's Magazine* in disgust:

Adjoining to the palace also appears venerable vestiges of an Abbey, now entirely unroofed in every part, exposing to the wind and rain the sculptured resting places of the worthies of past ages, so that I had the pleasure of beholding the Maidenhair, Hartstongues and Spleenwort, plentifully flourishing from between the Knights Templars legs and mitred noddles of long forgotten Bishops, decorating with living green, the monumental effigies of those whom time long since had crumbled into nothing.

FIG 233. St Davids Cathedral was built in 1181 on the alleged site of the saint's sixth century monastic settlement. The upper part of the tower supports a prolific growth of the lichens *Roccella phycopsis* and *R. furciformis*, which similarly occur on exposed sea cliffs. (Jonathan Mullard)

While ferns no longer grow inside the cathedral, the sheltered location has influenced the wildlife which can be found on, and around, this ancient building (Fig. 233).

In addition to the cathedral, and the various churches mentioned in earlier chapters, there are numerous other ecclesiastical sites in Pembrokeshire. Most of the churches and their surrounding churchyards date back hundreds, or even thousands, of years and, apart from burials, are rarely disturbed, the soil being mostly free from fertilisers and other agricultural chemicals. Churchyards and their mosaic of tall and short grassland, trees, boundary features, and the building itself, provide a haven for a variety of birds, bats and other wildlife. Chapels and their associated burial grounds can again be rich in species.

Probably the only religious site in Pembrokeshire not to have any specific wildlife associated with it is the tiny St Govan's Chapel on the Castlemartin Range, wedged, as it is, in a fissure on the cliffs (Fig. 234).

Most of the churches and chapels in Pembrokeshire have been visited by naturalists keen to record their plants and lichens, so there are comprehensive species lists available. The work started as a result of the Network Research Project on Churchyards and Other Burial Grounds, initiated by the Botanical Society of Britain and Ireland in the 1980s. Since that date Stephen Evans, the County Recorder, has been undertaking new, and repeat, surveys, particularly visiting chapel burial grounds, which were not well covered in the initial project. Members of the British Lichen Society have also investigated a number of sites in the county, as has the Pembrokeshire Fungus Recording Network. The National Park Authority now helps with the management of several churchyards and chapel burial grounds in order to maintain the diversity of wildlife in these areas. Three features are especially important: the presence of yew trees, lichens and species-rich grassland.

FIG 234. The little medieval chapel of St Govan's is located on a rocky shelf between high cliffs on the Castlemartin Range. (Jonathan Mullard)

YEW TREES

The yew is probably the most long-lived tree in northern Europe, and individual specimens are said to be over 1,000 years old. Richard Mabey, in *Flora Britannica* (1996), notes that:

> what sets yews most decisively apart from other trees in Britain is the remarkable and probably unique association they have with ancient churches ... Yews of great ages are rare outside churchyards and no other type of tree occurs so frequently inside the church grounds ... I do not know of any similarly exclusive relationship between places of worship and a single tree species existing anywhere else in the Western world.

Many of these yews seem to be much older than the church, and since the planting of wild tree species was very rare before the Middle Ages, it suggests that the original sacred sites were located close to existing yew trees. Yews are important for a wide variety of wildlife, and an ancient tree can support hundreds of different species, including mosses, lichens, birds and small mammals. The seeds develop within a bright red cup, the aril, that resembles a berry and are dispersed by birds, especially blackbirds *Turdus merula* and other thrushes.

Probably the most famous yew tree in Pembrokeshire is the 'bleeding yew' in Nevern churchyard (Fig. 235). An avenue of ancient yews lines the path from the churchyard gate to the porch of St Brynach's church. While most are unremarkable, the second tree on the right from the gate is the legendary bleeding yew, so called because blood-red sap continuously runs out through splits in the end of a branch that has been cut off close to the trunk. While at times sap flows from all yew trees, and especially so in spring when it rises, the flow from the bleeding yew has continued for as long as can be remembered. I first visited Nevern in the 1970s, and the yew was bleeding as profusely then as it does today. In fact, there is virtually no difference between photographs taken over the years. Unsurprisingly, there are a number of legends associated with the bleeding, which attracts Christian and Pagan pilgrims alike. The Christian version is that the tree bleeds every year in sympathy with Christ, while some say it bleeds for the wrongful hanging of a young man many years ago. Others say it will bleed until there is a Welsh prince installed in the now ruined Nevern Castle and yet another legend says it will bleed until the world is at peace.

Other notable yews, though not bleeding ones, can be found at Cilrhedyn, Freystrop, Nash, Little Milford and St Dogmaels. Cilrhedyn church is closed and

FIG 235. The 'bleeding yew' in Nevern churchyard is remarkable for the bloodlike fluid that continuously oozes from a wound on the tree. (Jonathan Mullard)

FIG 236. The yew in Nash churchyard divides into eight branches a short distance above the ground, almost certainly a result of damage when it was a young tree. (Jonathan Mullard)

in ruins but just inside the churchyard gate there is a single yew and beyond that a group of three yews, including one that is unusually tall for a churchyard yew. It is a sturdy tree, two thick upright parallel branches developing from the bole at a height of about 4 m. The most unusual feature of the yew in Freystrop churchyard is that after it became hollow it partially collapsed, leaving a standing section and a fallen section. A branch from the fallen side has subsequently fused with a standing branch to create a link between the two separated fragments. In contrast, at Nash a short bole divides into eight substantial branches, forming a circle around a central platform (Fig. 236). At Little Milford the yew is considered to be at least 500 years old and envelops a child's gravestone from 1819. The St Dogmaels yew is situated outside the church door. Unfortunately, here the crown of the tree has been drastically pruned and a high wall built around the tree itself, so its initial appearance does not reflect its probable age.

In addition to these, generally healthy, yew trees there are the sad remains of others, the most striking of which is on display in a covered part of the ruined church at Slebech. A notice here states that it is 'a relic of the famous yew which grew in Slebech churchyard. Its circumference was 13 feet 8 inches (4.2 m) and the Cambrian Archaeological Association on their visiting tour expressed their opinion that it was at least 800 years old.' As they become hollow and layer as they grow, it might never be possible to determine accurate ages for our oldest yews, but the Ancient Yew Group has devised a set of protocols to help determine an individual yew's status as ancient, veteran, or notable. Under this system an ancient yew is a tree which is a minimum of 800 years old, with no upper limit, a veteran yew has a minimum age of 500 years, but may be up to 1200 years old, and a notable yew has a minimum age of 300 years, but may be up to 700 years old (Mullard 2014).

To date, no record has been found showing precisely where the Slebech tree was located, its former appearance, or when and why it was felled. The history of the church, on the north bank of the Eastern Cleddau, is, however, 'bound up with the eccentric and high-handed Baron de Rutzen, a native of Courland (Latvia) who married the rich Slebech co-heiress in 1822' (Lloyd *et al.* 2010). De Rutzen considered that the ancient church brought the local people too close to the nearby Slebech Hall and built a new church on the busy new turnpike

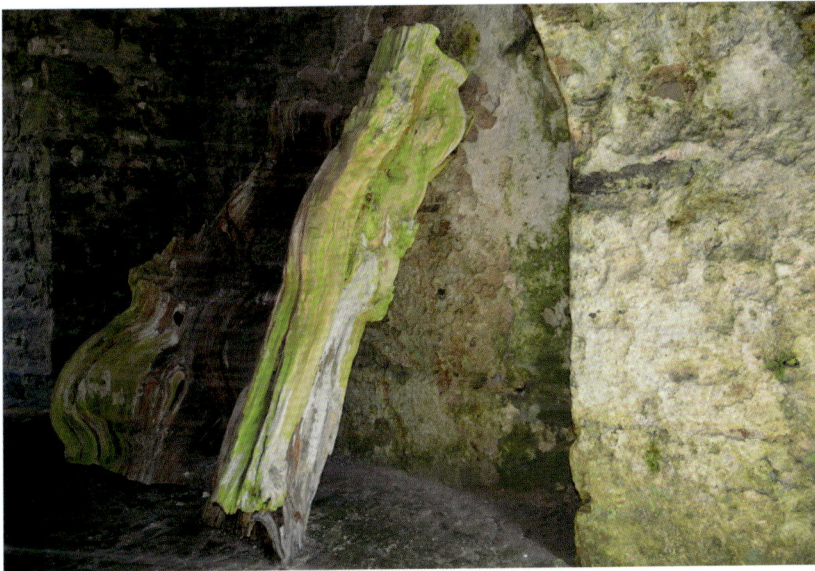

FIG 237. The remains of the famous yew which once grew in Slebech churchyard. It may have been felled on the orders of the eccentric Baron de Rutzen in 1844, when he stripped the roof from the church to render it unusable. (Jonathan Mullard)

road to serve them, together with the neighbouring parishes of Minwear and Newton North. When the replacement was completed in 1844, he stripped the roofs off the three medieval churches, without permission, to prevent them being used. The intervention of the Ecclesiastical Court resulted in the eventual reinstatement of Minwear, but Slebech and Newton North remained as empty shells. It is not too far-fetched therefore to connect the tree's demise with the activities of De Rutzen in 1844 (Fig. 237).

At Bridell, in North Pembrokeshire, all that remains of the yew are a few fragments of the stump on the south side of the church. Here there are no clues as to the reason for its destruction.

CUCKOO STONES

The churchyard at Nevern is famous for the Nevern Cross, one of Wales's most outstanding early Christian monuments; constructed in the late tenth, or early eleventh, century (Fig. 238). Each year on 7 April, St Brynach's feast day, it was said that the first cuckoo of the year would arrive from Africa, alight on the top of the great cross and sing to announce the arrival of spring. So regular was its appearance that during the medieval period the priest supposedly would not begin to say mass until it arrived. One year, during stormy weather, however, it did not arrive on time, so the priest and his congregation waited for several hours. Then, at last, the cuckoo fluttered down to perch on the cross, but it was clearly exhausted, and before it could utter a note it fell dead. Those who witnessed its death were convinced that it had flown through bad weather, determined to sing at Nevern as its ancestors had done before. George Owen considered that this 'vulgar tale, although it concerns in some sort church matters, you may either believe or not, without peril of damnation'. In Wales, it is said to be unlucky to hear the cuckoo before 5 April, but that it will bring prosperity for the whole year to someone who hears it on 28 April. Cuckoos usually start to arrive in Pembrokeshire around the second week of April, with some individuals remaining until the second week of September.

There are many locations in Britain said to be the traditional arrival point of the cuckoo in spring, but the Nevern Cross must be one of the most spectacular. Other locations tend to be prominent rocks, or boulders, such as Cuckoo-rock near Mount Sion Covert on the Castlemartin Training Area. In Ireland and Scotland, they were known as Gowk Stones, or Stanes, 'gowk' being an old name for the bird (Anon. 1889). Occasionally they are prehistoric standing stones, such as the Cuckoo Stone near Stonehenge. 'Celtic' mythology is full of references to

FIG 238. Nevern Cross, one of the most outstanding early Christian monuments in Wales and a noted 'cuckoo stone'. Among the lichens on the cross, when this picture was taken, is common goldspeck lichen. Unfortunately, this lichen flora has now disappeared, being replaced by a number of fast-growing species. (Jonathan Mullard)

cuckoos. Its call was believed to beckon souls, the bird travelling between the worlds of the living and the dead. The first appearance of a cuckoo was once commonly to bring about a 'gowk storm', a furious spring storm. So perhaps the Nevern cuckoo was the source of its own demise. Their departure, in late summer, is less dramatic than their arrival, commencing in late July and ending in mid-September, although one individual in 1954 was reported at Little Milford as late as 21 and 22 December.

The Reverend Murray Mathew recorded that the cuckoo was particularly common along the banks of the Daucleddau. Nearly a hundred years later the Pembrokeshire Breeding Birds Survey of 1984–88 found that cuckoos were still most frequent in the south of the county, with an estimated population of around 210 breeding females (Lovegrove *et al.* 1994). Today it is estimated that there are 90 birds in the county during the summer, but the species does not seem to be doing very well and the population has declined significantly over the last few years. In fact, if the situation continues the cuckoo may no longer be a herald of spring. A decline in the availability of insect food for the adults, declines in the populations of host species such as the meadow pipit, the fact that these hosts are breeding earlier because of climate change, and illegal trapping of migratory birds along the Mediterranean coast are all taking their toll. It is, on reflection, surprising that any of them survive at all.

Sometimes other species of cuckoo find their way to Pembrokeshire. As mentioned in Chapter 2, the first yellow-billed cuckoo recorded in Britain was shot at Stackpole in the nineteenth century, and a great spotted cuckoo *Clamator glandarius* was photographed in Penally on 19 March 2014. The latter species normally migrates from Africa to southern Spain but warm weather apparently caused it to bypass continental Europe. British Birds Rarities Committee records show that the great spotted cuckoo has been seen in Britain 46 times in all, but this was only the fourth time it had turned up in Wales.

CHURCHYARD LICHENS

The Nevern Cross, like many of the gravestones in the churchyard, used to be covered with a rich coating of lichens. When the photograph in Fig. 238 was taken species such as common goldspeck lichen *Candelariella vitellina* covered large areas of the cross head. Along with several other species it was responding to the nutrients produced by bird excrement and pollen from the yew tree which surrounds the head of the stone – an unusual, but well-documented, source of eutrophication. The result of rain-wash down the monument can be seen quite clearly. The lichen community on the head of the cross is now, however, mainly composed of fast-colonising, and fast-growing, species, probably owing to the continued increase in the supply of nutrients. Other ancient stones in the churchyard still support a more mature lichen community.

As described in previous chapters, there is an extensive range of suitable habitats for lichens in Pembrokeshire, but, even so, the religious buildings are of significant importance since some species in the county have only been recorded from these structures, such as *Limonaea sorediata* on the cathedral. This is a species found elsewhere on vertical walls of acid rocks near the sea, but it always avoids direct sun and rain, preferring to grow under overhangs and in cavities. Interestingly, the first bryophyte known to have been collected in the county, in 1838, slender stubble-moss *Gyroweisia tenuis*, was found by an unknown person on the cathedral. It is often found at the base of church walls on damp, shaded stone or brickwork. The specimen is now in the herbarium at the National Museum in Cardiff.

The considerable age of many of these sites is especially important as many lichens are slow-growing, and long-lived, and require the ecological continuity associated with these ancient structures. The variety of substrates and aspects also produces a rich diversity of species and, given the easy accessibility, they have been well studied. Provided there is sufficient light and moisture, lichens are able to colonise surfaces unsuitable for flowering plants. In churchyards, individual lichens may be almost as old as the gravestones upon which they grow. Aspect is a key element, and the species found on the north and south sides of a church or chapel are often completely different. The expanses of generally dry, vertical stonework too play a part in determining the species found here. The upper parts of church towers, in particular, provide a habitat that differs from the rest of the building in experiencing an abrupt wetting and drying cycle – and as a result there is sometimes a spectacular growth of shrubby lichens here (Gilbert 2000). As mentioned previously, on the tower of St Davids Cathedral there is a prolific growth of *Roccella phycopsis* and *R. furciformis*, both of which are normally confined to exposed outcrops overlooking the sea (see Fig. 233). The presence of trees favourable to lichen colonisation, such as ash, sycamore, elder, oak and hawthorn, helps increase the number of species to be found.

For its size, Wales supports the greatest known diversity of lichen species in the world, but the number of people who study them is diminishing rapidly. Plantlife Cymru hopes to inspire a new generation of lichenologists through the Cennad scheme. The title plays on *cen* (Welsh for lichen) and *cennad*, meaning messenger or envoy. Radio Wales 'Country Focus' broadcast in October 2018 highlighted the lichens on and around the cathedral, in particular *Roccella phycopsis*.

FIG 239. Stackpole Elidir churchyard has more recorded lichens than any other ecclesiastical site in Pembrokeshire, including St Davids Cathedral. (Jonathan Mullard)

Key sites for lichens include Stackpole Elidir with 160 recorded species (Fig. 239), St Davids Cathedral with 114 species and Angle with 112 species. Stackpole Cheriton churchyard contains a rare lichen, *Leptorhapis maggiana*, which was first recorded in 1995 growing on hazel. The number of species found, however, depends not only on the factors described above, but also on how many lichenologists have visited these sites, and how experienced they are. It seems to be the case with lichens that the more you look, the more you find. In Pembrokeshire a total of 305 lichens have been recorded from churchyards out of a grand total of 838 species for the county as a whole. Twenty-three species have been recorded only from churches, including *Verrucaria macrostoma* f. *furfuracea*, a churchyard specialist found on basic building stone or mortar, *Arthonia lapidicola*, a lichen specialising in sloping window sills subject to iron runoff from protective grills, *Lecanora antiqa*, which responds to base-rich stonework and mortar, and *Psilolechia leprosa*, which is copper-resistant and always found on basic stone or mortar associated with lightning conductors and under copper window grills (Ivan Pedley, personal communication).

SPECIES-RICH GRASSLAND

In the porch at Nevern Church there is a plant list and management plan for the churchyard, a result of the survey work in the 1980s mentioned at the start of this chapter. A total of 156 different species of plants have been recorded including primrose *Primula vulgaris*, early-purple orchid, rough hawkbit and ferns such as maidenhair spleenwort *Asplenium trichomanes*, lady-fern and scaly male-fern (Fig. 240). There has been too much frequent short mowing in places since then, but the area behind the church still retains its original meadow flora and few herbicides are used. Because of this, Nevern churchyard is frequently suggested as a place to visit if you are interested in the wildlife of churchyards. Illustrations of it feature in Francesca Greenoak's influential publication *God's Acre: the Flowers and Animals of the Parish Churchyard* (1985). These include a pencil drawing of the ancient cross, a sketch of herb-Robert *Geranium robertianum* and hart's-tongue fern *Asplenium scolopendrium* on a grave, a pygmy shrew hiding amongst other ferns, and honeysuckle. There is also a sketch of a butterfly and snail carved on a tomb in the churchyard.

There are many other botanically diverse churchyards, such as Manorbier, which contains pyramidal orchid, but unfortunately those in more urban settings are often over-manicured and suffer from herbicide use. In contrast, some churchyards in remote rural settings have become undermanaged with

FIG 240. A total of 156 species of plants have been recorded from 'God's Acre' at Nevern, including the yellow-flowered rough hawkbit. (Jonathan Mullard)

FIG 241. Llanion Cemetery on London Road, Pembroke Dock is one of a number of cemeteries that are now being managed for their wildlife. (Trevor Theobald)

scrub and tall vegetation dominant, and quite a few churches have closed, or been converted to dwellings. Taken as a whole they are, however, still a huge resource for our native flora, especially meadow flowers that have disappeared from the intensively farmed grasslands that are all too frequent across much of inland Pembrokeshire. Among the most interesting plants still to be found in churchyards and burial grounds are green-winged orchid, greater butterfly-orchid *Plantanthera chlorantha*, twayblade, dropwort *Filipendula vulgaris* and sweet violet. There are, of course, wonderful displays of primroses and bluebells in spring, and in places there are still reasonable stands of cowslip. In June 2018 Trevor Theobald, the ecologist for Pembrokeshire County Council, counted about 800 southern marsh-orchid flowering spikes and about 300 twayblade flowering spikes in Llanion Cemetery, London Road, Pembroke Dock (Fig. 241). They were in an area where the frequency of grass-cutting had been reduced about three years previously.

FIG 242. The distinctive violet coral has been recorded from Crymych Cemetery. It is one of the 'fairy clubs' typical of old unfertilised grasslands. (Sam Bosanquet)

One of the other features of these old landscape fragments is the fungi they support. Members of the Pembrokeshire Fungus Recording Group have been recording waxcaps in Pembrokeshire for a number of years and have found five churchyards that contain seventeen or more species of waxcaps – a remarkable number for a relatively small area. On the whole though churchyards tend to be relatively species-poor, given the high levels of nutrients in the soil. Chapel burial grounds are often better for fungi, as they are usually smaller and maintained more intensively, leading to areas of nutrient-poor grassland. In addition, many of them are in the north of the county on less fertile soils. An interesting find in the autumn of 2016 was a collection of the white form of pink waxcap *Hygrocybe calyptriformis* at Siloh Chapel's burial ground in Tufton, to the north of Llys-y-frân reservoir. This colour variant is mentioned in the standard guides to waxcaps, but is unusual and quite striking when first spotted.

The best sites for fungi tend to be the large cemeteries, six cemeteries in Pembrokeshire having cumulative records of eighteen or more waxcap species. The most outstanding of these is Crymych Cemetery in the northwest of the county, where 23 species of waxcap have so far been recorded, along with the uncommon, and beautiful, violet coral *Clavaria zollingeri* (Fig. 242). This produces conspicuous tubular, purple to pinkish-violet fruit bodies that can grow up to 10 cm tall and 7 cm wide.

INSECTS

In contrast to the data on plants and lichens, there is relatively little information on the insects to be found in churchyards and burial grounds in the county, but the elms in Pwllcrochan churchyard, on the Angle peninsula, are a good site for the white-letter hairstreak butterfly. This redundant church stands in a picturesque valley in an abandoned village adjacent to the Texaco oil refinery and the former Pembroke Power Station. It has recently been refurbished by the Prince's Trust Cymru and Texaco and converted into a field studies centre for schools (Fig. 243). This location can, however, be very noisy if the waste gas flares are operating on the adjacent refinery.

One of our more elusive butterflies, the white-letter hairstreak flies high in the tree tops, often appearing as a dark speck against the sky. It gets its name from the letter 'W' that is formed by a series of white lines found on the underside of the hindwings. Elm is its only foodplant and the butterfly therefore suffered as a result of Dutch elm disease in the 1970s and early 1980s. Locations such as Pwllcrochan are critical for its survival.

FIG 243. St Mary's church at Pwllcrochan, with a backdrop of flare pipes at the nearby oil refinery. The building has recently been refurbished by the Prince's Trust Cymru and Texaco as a field studies centre. The elms surrounding the churchyard support a colony of white-letter hairstreak, which is probably Pembrokeshire's most under-recorded butterfly. (Jonathan Mullard)

BATS

Churches are important for bats and some have provided safe roosting sites for many generations, at least 60 per cent of pre-sixteenth century churches being estimated to contain bat roosts. Despite some concerns about damage to furnishing and fittings, when they roost and fly inside buildings, bats help to control insect populations, catching the adults of wood-boring insects, such as deathwatch beetle *Xestobium rufovillosum* and common furniture beetle *Anobium punctatum*. At least fifteen species of bat have been recorded in Pembrokeshire, twelve of which appear to be resident or breeding, and many of these use the county's churches.

A good example is St Nicholas's Church, which is situated in the centre of the village of New Moat. Mainly rebuilt in the 1880s, although the medieval tower remains, it supports at least six species of bats, including Natterer's bat which roosts in the eaves, whiskered *Myotis mystacinus* or Brandt's bats *M. brandtii* (they are very similar, only being separated as distinct species in 1970), brown long-

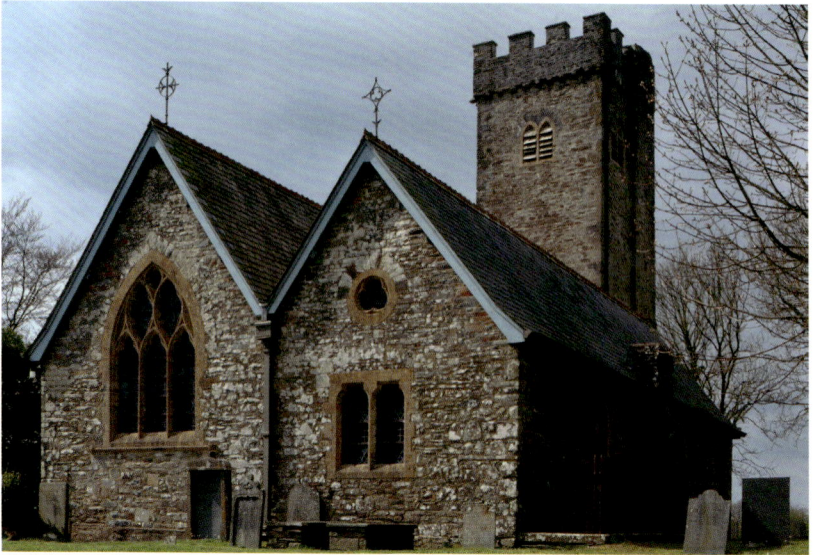

FIG 244. St Nicholas's Church in New Moat supports at least six species of bats, including greater horseshoe bats which use the crypt as a hibernation site. (Jonathan Mullard)

eared bats and both common pipistrelle and soprano pipistrelle (Fig. 244). There is also an unusually well-preserved crypt which is used as a hibernation site by greater horseshoe bats. The wooden door to the crypt has rotted away from the bottom, allowing the bats to access the area, and plans for its replacement include leaving an aperture to ensure they are undisturbed.

Greater horseshoe bats also roost in the vaulted cellars of the cloisters in St Davids Cathedral, which are not open to the public, and they have been recorded in the Bishop's Palace, too. Other ecclesiastical sites important for greater horseshoe bats include Nevern church, while lesser horseshoes can be found at Stackpole and Gumfreston. Important though churches such as New Moat are as hibernation and roost sites for bats, the main nursery sites are to be found elsewhere.

Bat nursery sites

While Pembrokeshire is one of the remaining strongholds for the greater horseshoe bat in Britain, there are only three known nursery roosts: in the loft of the clock tower at Stackpole, in the stable yard loft at Slebech, and in Felin Llwyngwair, a disused mill building east of Newport (Fig. 245). Though the greater horseshoe bat may once have been relatively common there was a substantial reduction in its range during the twentieth century. The fact that they feed entirely on insects, which we now know have declined significantly due to the agricultural chemicals sprayed on farmland, suggests that their food supply has been seriously affected (Yalden 1999). Pollution by agricultural organic matter and ammonia in water bodies has also been suggested as an important factor in the decline of bat populations. Bob Stebbings, a well-known authority on bats, has argued that the overall population of greater horseshoes in the United Kingdom dropped, between 1950 and 1980, from at least 58 nursery roosts, and perhaps 330,000 bats, to only twelve sizeable colonies and around 4,000 bats – a reduction of around 98 per cent (Stebbings 1988). The relative abundance of large insects in Pembrokeshire is almost certainly one of the factors in their continuing success here.

The colony of greater horseshoe bats in the clock-tower loft at Stackpole was first discovered in 1977, although judging from anecdotal information supplied by the Cawdor family and local people, it was a long-established breeding colony. The Welsh population is at the extreme northwestern extremity of the species' range, so the loft is important in concentrating a geographically dispersed

FIG 245. Pembrokeshire is one of the remaining strongholds for the greater horseshoe bat in Britain, but there are only three known nursery roosts in the county (Melvin Grey)

population at a single breeding site. Cellars in the nearby walled garden and old redundant heating ducts provide important transitory roosts, which the bats utilise in the spring and autumn. Surveys have established that the number of greater horseshoes at Stackpole has increased from around 150 bats in the 1980s to over 800 in 2017, so the population is extremely healthy. There are strong links between the bats at this site and Orielton, where greater horseshoes were first recorded in the 1950s by Lockley. In 1982 and 1983 movement of adults and young from Stackpole to Slebech was recorded, confirming the interdependence of the three sites.

The breeding colony in the stable yard loft at Slebech was discovered in 1982, but there had been reports of bats in the outbuildings there during the Second World War. The loft consists of two interconnected chambers, one being used for access by the bats and the other for breeding. The colony appears to be well established, with 25–29 young being successfully reared each year between 1983 and 1985, compared with 50–61 for the same period at Stackpole. The tunnels and cellars adjacent to the loft at Slebech provide an important winter hibernation site which is used by the bats from October to April.

Felin Llwyngwair is a mill built around 200 years ago. Once again there are good feeding habitats in the vicinity, including pastures with tall wooded hedges, the wooded valley of the Afon Nyfer and Carningli Common. Records of bat numbers indicate that this site was first used by greater horseshoe bats as an intermediate roost, breeding not being confirmed until July 2000, when 53 adults were found with around 30 young. The bats appear to use the loft space throughout the year, with numbers peaking during the summer. There are often wide fluctuations in numbers, however, which indicates that other, currently unknown roosts in the vicinity are used when weather conditions dictate.

Like the greater horseshoe bat, the population of lesser horseshoe bats in Britain has suffered such a serious decline that the species is now restricted to the west of Britain, with the main concentrations in Wales and southwest England. Once again, the importance of the buildings in the Stackpole estate for bats is highlighted by the fact that, at one time, the largest known nursery roost of lesser horseshoe bats in Pembrokeshire could be found here in the coach-house. The bats were first noticed here in June 1987, when 32 individuals were recorded. From that date the number of adults increased steadily, a total of 85 bats being counted in 1993, 99 in 1994, 103 in 1995, 102 in 1996, and 141 in 1997. It should be noted though that these numbers were recorded when work was being done to convert the main buildings at Stackpole into flats. During the following years, as they became more confident that the situation in their previous roost had stabilised, most, if not all, of the females using the coach-house moved back to Stackpole. As mentioned earlier, there is evidence of similar linkages between sites used by greater horseshoe bats.

Another large nursery roost for lesser horseshoe bats can be found in the stable block and cellars at Orielton. The first records here date to the late 1960s, a maximum of 45 bats being recorded between 1980 and 1985. Regular monitoring since 1993 showed that there was a general upward trend until 2000, when numbers reached 130. The roost has, however, declined in recent years, for reasons that are still unclear.

The largest breeding colony of soprano pipistrelle bats known in Pembrokeshire, comprising over 1,000 individuals, is located within another property on the Orielton estate. Smaller colonies, still measured in the hundreds, are, however, not unusual. All are associated with water and woodlands. These large colonies often split: for example, half the bats from Orielton eventually moved to another building nearby. In contrast, common pipistrelle colonies are generally much smaller.

Analysis of the weekly 'emergence counts' and the daily maximum and minimum air temperatures collected at Orielton for over twenty years has shown that year-to-year changes in the populations of pipistrelle bats could be predicted from the air temperature (Andrews *et al.* 2016). Adult non-biting midges *Chironomus* spp. constitute the main food item of pipistrelle bats, and their larval development and emergence times are dependent on temperature, thus providing clear evidence of the direct link between food availability and bat population size.

CHAPTER 18

Castles and Palaces

*For although all or most of the castles are ruined and uncovered, some for divers
hundred years past, yet are all the walls firm and strong and nothing impaired ...*
                                                              George Owen (1603)

RELIGIOUS BUILDINGS ARE NOT THE only historic structures that
support a range of plants and animals. Within Pembrokeshire there
are also a number of large and imposing castles, among the finest
to be seen anywhere in Britain. Most were built after the Norman incursion into
southwest Wales after 1093, but extensively remodelled, completely rebuilt, or
abandoned in the following centuries. For wildlife they act as artificial cliffs and
caves and have the potential to host a wide variety of species. The management of
both the structure itself and its surroundings can, however, be unsympathetic. To
date there have been few biological surveys of these sites, but Carew and Cilgerran
Castles support a range of species that could be found on similar sites elsewhere, if
the flora and fauna were conserved and encouraged. The remains of Tenby Castle
and the Castle Rocks are likewise of interest. Llawhaden Castle was a fortified
bishop's palace rather than a true castle, while the Bishops Palace, adjacent to St
Davids Cathedral, is a remarkable survival from the early fourteenth century.

CAREW CASTLE

Carew Castle, which replaced an earlier stone keep, was built on a limestone
bluff overlooking the Carew River around 1270. Constructed from locally sourced

Carboniferous Limestone, the medieval fortification was later transformed into an early Tudor palace, with elaborately carved Bath stone windows replacing most of the original ones (Lloyd *et al.* 2010). An Elizabethan façade was then grafted on top of this. From the late seventeenth century Carew was let to tenants, who robbed much of the dressed stone, and, over the following centuries, it gradually became derelict. In 1983, the National Park Authority leased the castle and its surrounds and began a programme of consolidation. This work was completed in 1998 and the castle and associated structures such as the boundary walls are now maintained as a 'partially restored ruin'. Most of the site is accessible to the public, and it is now a popular visitor attraction.

Shortly after taking over the site, the Authority became aware of the nature conservation interest of the site, and extensive biological surveys were undertaken in order to inform the consolidation work (Fig. 246). The limestone walls, together with the boundary walls, provide an extensive open habitat that has been colonised by species ranging from damp, crevice-dwelling ferns and mosses to species more typical of open rock faces and bare ground. Typical species recorded on the vertical wall surfaces include ferns such as rusty-back

FIG 246. The National Park Authority has undertaken extensive biological surveys of Carew Castle and, as a result, it is probably the best-studied historic site in Pembrokeshire. (Pembrokeshire Coast National Park Authority)

fern *Asplenium ceterach*, wall-rue *A. ruta-muraria*, maidenhair spleenwort, hart's-tongue and southern polypody *Polypodium cambricum*. Alongside these can be found wall pennywort, pellitory-of-the-wall *Parietaria judaica*, black medick *Medicago lupulina* and perennial wall rocket *Diplotaxis tenuifolia*. The last is scarce in Pembrokeshire, known from only two sites in the county, Carew Castle and Pennar Park, Pembroke. The wall walks and wall tops support a flora that includes grasses such as the nationally scarce compact brome *Bromus madritensis*, alongside mullein *Verbascum thapsus*, herb-Robert, common mallow, dandelion *Taraxacum* agg., thistles *Carduus* spp., Alexanders and ivy. On more open wall tops, biting stonecrop and mosses such as wall screw-moss *Tortula muralis* occur frequently.

Carew Castle is the Pembrokeshire stronghold for the southern, or Welsh, polypody, which grows in profusion on the castle walls and on the boundary wall alongside Mill Lane (Fig. 247). This winter-green fern has been described by Chris Page, the author of *Ferns*, New Naturalist 74, as the 'species par excellence of old castles especially in south and west Britain and especially of castles that have fallen into a reasonable state of dereliction exposing the lime mortar of their

FIG 247. The southern, or Welsh, polypody grows in profusion on the walls of Carew Castle and on the boundary wall alongside Mill Lane. The species is native to southern and western Europe, and there are many recorded cultivars. (Jonathan Mullard)

original construction and the lime-rich rubbly fill of their thick wall interiors' (Page 1988). Page cites Carew Castle as a 'good example' of such a site, but noted at the time of his visits that 'teams of workers seemed to be especially employed to go around pulling off this relatively harmless plant'. Following initial cleaning and repair work though the general approach to the subsequent maintenance of the castle and associated stone structures has been to allow shallow-rooting plants, such as ferns, to recolonise the site. Today there is ongoing interest in the different varieties of southern polypody that have been found at Carew, especially on the boundary wall.

As well as occurring on wall tops and ledges, compact brome has been recorded on, and at the base of, the south-facing castle walls among the limestone rubble. There are only a few records of the species in south Pembrokeshire, and elsewhere there are only isolated examples in southern England and South Wales. Compact brome is a species of Mediterranean origin but has been known at Carew Castle for a long time. At one stage it was considered to be native here, although the current thinking is that it is introduced. It has also been found on the wall on the north side of the mill pond and on the bank on the north side of Castle Lane. Another species associated with the grassland among the limestone rubble at the base of the south-facing castle walls is grey sedge *Carex divulsa*, which has a restricted, but increasing, distribution in south Pembrokeshire.

Several species of local or regional interest have been recorded in the grassland around the castle over the years, including fiddle dock *Rumex pulcher*. Carew Castle is probably the best site in Pembrokeshire for this species, which has a restricted distribution in the county. The population here comprises many hundreds of individual plants. It occurs in the semi-improved neutral grassland in front of the castle entrance and between the southeast and southwest towers. The grassland likewise supports one of the largest populations of spotted medick in the county. A total of 30 species of grassland fungi, including 21 waxcaps, three pinkgills, four spindles or clubs and two earth tongues have also been recorded here, making it one of the best grassland fungi sites in the county. Species of particular note include dingy waxcap *Hygrocybe ingrata*, citrine waxcap *H. citrinovirens* and short-spored earth tongue *Trichoglossum walteri*. The last of these is a rarely recorded species in the United Kingdom, and its taxonomic status is currently under review. It looks like a black finger sticking out of the ground, up to 9 cm high.

Information on invertebrates is much more limited than on other aspects of the natural history. Moth-trapping carried out in the late 1980s, however, confirmed that the castle supported good populations of species such as the muslin footman *Nudaria mundana*, a lichen-feeding moth that is closely associated with the lichen-encrusted walls. Cockchafer *Melolontha melolontha* and ground beetles together with large noctuid moths, such as large yellow underwing *Noctua pronuba*, occur in the grassland and are important prey species for the greater horseshoe bats that roost in the castle.

Ten species of bats are known to use the castle. Within the ruins there is an enormous range of undisturbed roosting and hibernation sites. These include several large capped and uncapped chimneys, a complex of tunnels, crevices in barrel-roofed and vaulted rooms and undercrofts, and numerous crevices in the walls of passages and rooms at different levels. Bats present include the two horseshoe bats and crevice-dwelling species such as whiskered, Brandt's, Daubenton's and Natterer's bats, soprano and common pipistrelle and brown long-eared bats. The castle is particularly important for greater horseshoe bats and is used during the spring and autumn by a significant proportion of the south Pembrokeshire breeding population. Individual bats may only stay for a night or two before moving on to other intermediate roosts or maternity roosts and hibernation sites in south Pembrokeshire. The link between the castle and the maternity roosts was first established through radio-tracking work carried out by Bob Stebbings in the late 1970s and early 1980s. Serotine bat *Eptesicus serotinus*, an infrequent visitor to the county (although it may be also be resident) has been recorded in the castle on a few occasions. An eleventh species, noctule bat, is regularly seen foraging above the castle and in the surrounding area, although they do not roost within the castle walls.

Several species of birds nest in holes and other features in the stonework, as well as in the scrub and woodland in the castle grounds. These include kestrel, tawny owl, barn owl (which breeds virtually every year), swallow, wren, blue tit *Cyanistes caeruleus*, robin, pied wagtail *Motacilla alba* and blackbird. Buzzards have nested occasionally in the castle, and peregrines regularly hunt over the area during the autumn and winter.

CILGERRAN CASTLE

As described in Chapter 11, Cilgerran Castle has arguably the most dramatic location of any of the Pembrokeshire castles, situated high above the Cilgerran Gorge. Prior to the current building, there was an earth and timber castle here, but it was destroyed by Welsh forces. The existing masonry dates from around the thirteenth century. The walls facing the cliff edge were the least heavily fortified, with the defences concentrated on those that faced inland. Today the castle is largely a ruin, though two substantial towers remain (Fig. 248).

FIG 248. Cilgerran Castle, overlooking the Afon Teifi on a rocky promontory, has large expanses of wall where the inner core is exposed, creating a habitat for a variety of plants. (Jonathan Mullard)

The site is owned by the National Trust but in the guardianship of Cadw, the historic environment service of the Welsh Government. Unfortunately, from the naturalist's point of view at least, the walls of these guardianship sites are often too well pointed and cleaned to support a large variety of plants, but at Cilgerran Castle there are large expanses of walls where the inner core is exposed, creating an artificial rockface. This has been colonised by a variety of plants, including several stands of autumn hawkweed *Hieracium sabaudum*. This is the commonest many-leaved hawkweed found in South Wales (Stace 2019). It is a feature of the lower reaches of the Teifi, occurring in both the Llechryd and Teifi gorges. Other locations in the county are few and far between, but it has been recorded from Treffgarne Gorge, Giltar Point and the old Mathry Road Station at Letterston (Stephen Evans, personal communication). Wood meadow-grass *Poa nemoralis* and common mallow are also found on the walls, together with rusty-back fern.

The grass inside the castle walls is maintained as a short sward characteristic of amenity grassland. Species diversity is low in these areas, with white clover *Trifolium repens*, false oat-grass and perennial rye-grass *Lolium perenne* dominating the area. Other species present include slender trefoil *Trifolium micranthum*, another species tolerant of grazing, mowing and heavy trampling, common whitlowgrass *Erophila verna*, Danish scurvygrass, creeping buttercup *Ranunculus repens*, black medick, daisy *Bellis perennis*, dandelion and smooth meadow-grass. Areas of the castle grounds that are paved, or bare ground, and low-level walls are very sparsely vegetated, with species that are typically found in dry, nutrient-poor habitats. These include smooth sow thistle *Sonchus oleraceus*, herb-Robert, ivy-leaved toadflax *Cymbalaria muralis*, eastern pellitory-of-the-wall *Parietaria officinalis*, red valerian *Centranthus ruber* and dove's-foot crane's-bill *Geranium molle* (Taylor 2015a). This is not a particularly exciting list from the botanist's viewpoint. Cadw is, however, considering changing its approach to vegetation management on historic sites to allow a more diverse range of plant species to become established, and this is to be encouraged.

Birds such as grey wagtail *Motacilla cinerea* and goldfinch *Carduelis carduelis* have apparently been observed nesting in the castle walls between the east and west towers, and there are house martin nests in the walls of the moat. A barn owl roosts in one of the many gaps in the wall but there appear to be no bat roosts, although common pipistrelle and soprano pipistrelle have been recorded foraging in the general area (Taylor 2015a).

TENBY CASTLE AND WALLS

The walls surrounding the historic core of Tenby are some of the most important surviving medieval city walls in Britain, but only isolated stretches of the castle walls remain, perched above the cliffs. Given the castle's position on the headland, a complete circuit of walling was probably considered unnecessary. Inside, at the highest point of the headland, is a watchtower, formed by two small towers joined together (Fig. 249). The exposed masonry here supports a range of plants including wall pennywort and maidenhair spleenwort. As described in Chapter 1, Beatrix Potter also mentioned wallflowers, as well as wild cabbage, in her letters, and both are still abundant on the walls here. Wallflowers have been cultivated since medieval times, and the species was first recorded as a wild plant by William Turner, the 'father of English botany' as long ago as 1548, when he wrote that it 'groweth upon the walles' (Pearman 2017).

FIG 249. Tenby Castle watchtower is situated at the highest point of the headland and supports a range of plants including wall pennywort, maidenhair spleenwort and wallflower. (Jonathan Mullard)

LLAWHADEN CASTLE

A fortified palace of the bishops of St Davids, Llawhaden Castle was totally rebuilt in the fourteenth century. It was abandoned in the sixteenth century and some of the stone removed for local building projects (Fig. 250). Once again, the walls here are over-maintained, at least as far as higher plants are concerned, but rue-leaved saxifrage *Saxifraga tridactylites* has been recorded from the site. This is a winter-annual that occurs in dry, open habitats and on artificial structures such as mortared walls. Southern polypody grows at the top of the well and is visible through the iron safety grid. It was once a feature of the Castle Pound but is no longer present there. There are still some fine stands of this fern, however, on walls elsewhere in the village (Taylor 2016).

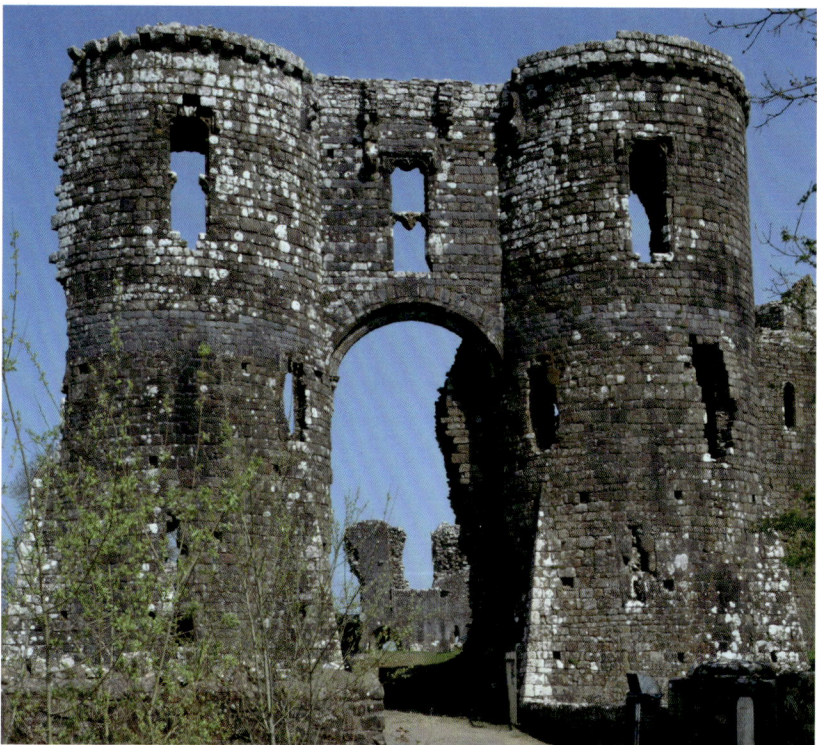

FIG 250. Llawhaden Castle, a fortified bishop's palace, was abandoned in the sixteenth century. The gatehouse is the largest surviving section. (Jonathan Mullard)

FIG 251. The rare false oxlip, a hybrid between primrose and cowslip, occurs in large clumps on the bank of the moat at Llawhaden Castle. (Jonathan Mullard)

The amenity grassland is again heavily managed, but there are some interesting remnants of unimproved neutral grassland on the sides of the dry moat and castle surrounds. This is a relic of what would once, before agricultural 'improvements', have been a widespread and common habitat. In total, 105 species have been recorded here, and in late spring these areas are a very attractive sight with large areas of bluebells and early-purple orchid, along with oxeye daisy, primrose and cowslip. Also present are large clumps of false oxlip *Primula × polyantha*, a hybrid between the last two species (Fig. 251). False oxlip is generally a much larger plant than the cowslip and should not be confused with the true oxlip *Primula elatior*. The latter is a very rare plant, only found in the woodlands in East Anglia and, to some extent, Buckinghamshire, where primroses are scarce.

ST DAVIDS BISHOP'S PALACE

In the valley of the River Alun in Pembrokeshire strange animals, monsters and mythical beasts wait silently for people to notice their presence. Carved in stone in the second quarter of the fourteenth century for the Bishop's Palace in St Davids, the now eroded corbels are the most interesting collection of sculpture from that period anywhere in Britain. Lions, dogs, a cat, an owl, monkey and a baboon jostle with mermaids and their companions for attention. The interior corbels are almost all human heads, in the courtyard there are animals and, on the east façade, grotesques and hybrid creatures look out across the valley. So visitors approaching from the cathedral were confronted by the forces of darkness. Inside the courtyard were normal creatures, though not local species, apart, perhaps, for the owl. The carvings in the interior of the building itself represented order and light (Coldstream 1999). These lavish decorations, and the conspicuous chequerboard stonework, are all testament to the wealth and status of the medieval bishops.

FIG 252. A symbol of the wealth and status of the medieval church, the Bishop's Palace in St Davids today supports a range of wildlife, including up to four species of bats. (Jonathan Mullard)

FIG 253. Jackdaws nest in the cavities on the walls of both the Bishop's Palace and St Davids Cathedral. (Jonathan Mullard)

Although the building that stands today is largely the result of building work in the late thirteenth and fourteenth centuries, the palace dates back to the sixth century. Ruinous since the seventeenth century, it hosts a wide range of wildlife that, for the most part, escapes the notice of the many visitors to this guardianship site (Fig. 252).

Edwin Lees in *The Botanical Looker-out* (1842) noted that 'the ruined palace of St Davids glows with the crimson flowers of the wall germander *Teucrium chamaedrys*'. This creeping evergreen perennial is no longer present but it was historically used as a medicinal herb for the treatment of gout, so it may have been deliberately planted. The dominant plant on the site today is another non-native, ivy-leaved toadflax. It is now considered naturalised, having been established in the United Kingdom for several hundred years. Ivy-leaved toadflax is thought to have been introduced into gardens prior to the seventeenth century, and records from the wild date from 1640. John Goodyer, the seventeenth-century botanist noted in 1621 that 'Cymbalaria ... runneth and spreadeth on the ground and clymeth and hangeth on walls even as Ivie or Chickweed doth, the branches are verie small round and smooth, limmer and pliant' (Gunther 1922).

In this sheltered spot ivy-leaved toadflax is accompanied by wall-rue, pellitory-of-the-wall and maidenhair spleenwort. These species are beginning to re-establish themselves and are being protected, rather than removed as they once would have been. Maidenhair spleenwort, in particular, is common on the sides of the 'thirty-nine steps', the main stone steps leading down to the nearby cathedral from the bell tower, so there are plentiful sources of seed nearby.

Today the choughs that once nested in the ruins of the Bishop's Palace have been replaced by numerous jackdaws, whose nests occupy many of the larger holes (Fig. 253). The buildings also host a number of bat species. Greater horseshoe bats roost in the latrine pit in the northwest section of the undercroft below the Great Chapel and the southwest chimney below the Bishop's Chapel, while Daubenton's bats roost in the apex of the arched ceiling in the eastern part of the undercrofts. Sometimes the bats can be heard during the day, and their presence is indicated by droppings and grease staining around the entrances to crevices. It is likely that common and soprano pipistrelles also use the site for roosting, as they are often seen around the buildings after sunset (Taylor 2015b).

CHAPTER 19

Conserving Pembrokeshire

One of the penalties of an ecological education is that one lives alone in a world of wounds. Much of the damage inflicted on land is quite invisible to laymen. An ecologist must either harden his shell and make believe that the consequences of science are none of his business, or he must be the doctor who sees the marks of death in a community that believes itself well and does not want to be told otherwise.

Aldo Leopold (1949)

L OCKLEY MOVED TO NEW ZEALAND in 1970 because of his frustration at being unable to protect the Pembrokeshire landscape from the development of oil refineries and power stations. These days, however, the pressures on the county's landscape have, in the main, global, rather than local, origins and, as such, are more difficult to address. There is, however, no excuse for inaction. We are probably the last generation that can stop devastating changes to the climate that will have impacts on both humans and the natural environment on which we depend for our inspiration and survival.

THE CLIMATE CRISIS

In a coastal area, one of the main factors driving changes will be sea-level rise, a result of the warming atmosphere and seas. The expansion of seawater as it gets warmer is responsible for the biggest change in sea levels. Oceans absorb more than 90 per cent of the heat that results from greenhouse gases, and the five years

FIG 254. The shingle ridge at Amroth currently protects the houses and shops on the opposite side of the beachfront road from flooding. If sea levels rise, settlements such as this are at risk. (Jonathan Mullard)

before 2019 were the five hottest on record (Cheng *et al.* 2019). The current rate of ocean warming has been calculated as the equivalent of five Hiroshima-size atomic bombs exploding every second (Abraham 2019).

According to the government's Climate Change Risk Assessment, sea levels in the United Kingdom are rising by approximately 3 mm a year, and it is expected that they will be at least a metre higher than modern levels by the end of the century (Committee on Climate Change 2016). Even if we manage to limit global warming to 2 °C, the target for the current climate negotiations, sea levels may still rise at least 6 m above their current height, radically reshaping coastlines and affecting millions of people in the process (Dutton *et al.* 2015). Antarctica alone has the potential to cause more than 1 m of sea-level rise by 2100 and more than 15 m by 2500, if emissions continue unabated (DeConto & Pollard 2016).

Even a small rise will have catastrophic effects on Pembrokeshire, resulting in the disappearance, or major relocation, of habitats such as beaches, sand dunes and shingle banks. Some coastal communities will become unviable as a result

(Fig. 254). Offshore, many of the small islets will disappear beneath the waves and a new geography will arise. The areas most immediately at risk include the low-lying land between Tenby and Penally, the shingle beach at Newgale and the Cleddau estuaries. As described in Chapter 9, Pembrokeshire County Council is already looking at the options for realigning the road at Newgale further inland, well away from the shingle beach.

A warming sea also means that the weather is becoming stormier. As described in Chapter 3, in the autumn of 2017 Storms Ophelia and Brian resulted in the Green Bridge of Wales losing a huge section of its seaward leg (Fig. 255). In April 2019 Storm Hannah reached Pembrokeshire, with the St Ann's weather station logging gusts of up to 130 km per hour. Huge seas battered Skokholm during the day, forcing the vast majority of cliff-nesting birds to return to the ocean and ride out the storm (Fig. 256). This left the guillemots and razorbills that were already incubating eggs (the earliest ever recorded on Skokholm) in a difficult position. Deprived of the protection of hundreds of other birds and exposed on all sides, they were unable to prevent the gulls coming in and taking their eggs. While the warmer spring weather is

FIG 255. The Green Bridge of Wales after the storms of autumn 2017 removed a huge section of its seaward leg. Compare this with the photograph in Fig. 38, which was taken in 2016. (Jonathan Mullard)

FIG 256. Huge waves batter the seabird colonies on Skokholm during Storm Hannah in April 2019. (Richard Brown and Giselle Eagle)

encouraging the birds to lay earlier, the breeding season is now overlapping with increasingly unpredictable weather. This combination does not bode well for the future. It is not only nesting birds that are being affected by the increasing number, and severity, of storms now reaching Pembrokeshire. Other species, such as the seal pups killed in 2017, are likewise very vulnerable. The increasing severity of our weather, in both spring and autumn, will continue to have substantial effects on animal populations.

Global warming is likewise impacting on the distribution of both animals and plants, with dramatic changes predicted over the course of the twenty-first century. In general, species are moving towards the poles and, on land, up to higher elevations where the temperature is lower. In Britain, the distributions of spiders, ground beetles, butterflies, grasshoppers and their allies have already shifted to higher elevations, moving upwards at a rate of around 11 m per decade, and to higher latitudes at 17 km per decade, but there is substantial variability across and within various groups (Chen *et al.* 2011). Since different species are reacting in different ways, the interactions between them are changing

(Tylianakis *et al.* 2008). While some species may be able to move in response to climate change, others may not, and they face extinction. In the seas, changes in the distribution and abundance of plankton and fish stocks are already affecting top predators such as seabirds.

Our protected areas will come under question, since, whether marine or terrestrial, they represent fixed lines on a map, which plants and animals do not respect. The Skomer Marine Conservation Zone, as described in Chapter 4, is currently at the northern limit for many warm-water species and the southern limit for some northern species. But with continually warming waters the species we are currently attempting to protect may in the future no longer be there. On land, National Nature Reserves may well remain, as areas managed for conservation, but sites designated due to the presence of particular species are very likely to lose their value, despite the best efforts of organisations such as the National Park Authority.

POLLUTION

Lockley was concerned about seabirds affected by ships illegally jettisoning oil around the islands but, as described in Chapter 8, plastic is now the main threat. The issue is not only that a number of the birds end up inextricably tangled in the plastic and die, but that they also accidently eat it. Plastic pollution is a global concern, with production increasing exponentially and concentrations in the oceans reaching 580,000 pieces per km^2 (Wilcox *et al.* 2015). So much plastic is present in the world's oceans that 90 per cent of seabirds consume it now, and it is predicted that virtually every bird will be consuming it by 2050. Sharp-edged plastic kills birds by perforating their internal organs, and the presence of plastic in the gut means that there may be little room left for food, which results in starvation.

Other forms of sea life are equally vulnerable, in particular from microplastics. These are fragments smaller than 5 mm, and in the marine environment they originate from a variety of sources, including fragmentation of larger plastic debris, pre-production pellets (nurdles) spilled during fabrication and transportation, outflow of wastewater containing microbeads from cosmetics and fibres from the washing of synthetic textiles, as well as road runoff containing fragments of vehicle tyres and marking paint. Their small size makes them highly likely to be ingested by a wide variety of marine animals, ranging from zooplankton such as copepods and other invertebrates such as shellfish, to vertebrates including both juvenile and adult fish, seabirds, seals and cetaceans. A

FIG 257. Volunteers clearing plastic and other debris from Bullslaughter Bay on the Castlemartin Range in April 2019. (Pembrokeshire Coast National Park Authority)

recent United Kingdom survey found microplastics in the digestive tract of every one of a sample of fifty marine mammals (Nelms *et al.* 2019).

The whole world is now affected by microplastics. They have been found in all habitats, including rivers, with around 80 per cent of marine microplastics originating from freshwater runoff. An initial survey of the Afon Cegin in North Wales, for instance, found an average of 76.9 pieces of microplastic in every litre of water, and the situation is likely to be similar for rivers in Pembrokeshire (Moore 2019). Recent research has found that half of the mayfly and caddisfly larvae surveyed in rivers in Wales contained microplastics (Windsor *et al.* 2019). These particles remained inside the animals as they transformed into adults, which then spread the pollution further as they flew, or were eaten by birds and bats.

There are larger distribution mechanisms at work as well. A study published in 2019 was the first to show that microplastic is raining down just as hard in remote environments as it is in urban areas, and that it can be blown long distances by the wind. It is therefore ubiquitous and impossible to remove (Allen *et al.* 2019). Humans are not exempt from the impacts. Microplastics were revealed in 2017 to be in tap water around the world, with the continuous fall of plastic onto reservoirs probably a major factor (Kosuth *et al.* 2018). Major changes are therefore needed in the way we use and dispose of plastic products. Public concern is on the rise, and an increasing number of people in Pembrokeshire, as elsewhere, are keen to help (Fig. 257), but far more radical solutions are needed.

The extensive slurry spreading within the catchment of the Milford Haven Waterway is another source of pollution, increasing nitrogen levels in the river systems, resulting in excessive algal growth and affecting freshwater life. Ammonia emissions from livestock manure are also one of the most potent air pollutants in the United Kingdom. A recent report to Defra stated that ammonia and nitrogen pollution, mostly from farms, is harming more than 60 per cent of the land area and affecting the most sensitive habitats for plants and wildlife (Hall *et al.* 2019). These areas are receiving concentrations of ammonia above the critical level for lichens, mosses, liverworts and similar plants. Just over half the land in Wales is affected, with Natural Resources Wales having identified a number of sites where lichens that are sensitive to nitrogen are dying off (Wasley *et al.* 2019). This may have wide-ranging effects, since lichens provide habitats for invertebrates, which in turn are eaten by other animals such as bats. As described in Chapter 17, pollution by agricultural organic matter and ammonia in water bodies has been suggested as an important factor in the decline of bat populations. Amazingly, lichens have been estimated to absorb half of the rainfall in woodlands, preventing flooding and storing water for release in dry periods. The fact that the water in Milford Haven contains too much nitrogen, and indeed phosphorus, means that it fails to meet statutory water quality standards. To address the situation the Pembrokeshire Coastal Forum has set up the Ecosystem Enterprise Partnership – Ecobank (EEP-Ecobank). This is developing a 'nutrient based trading scheme' to pay land managers for undertaking actions that benefit the natural environment (Pembrokeshire Coastal Forum 2019). It remains to be seen how successful this approach is.

Freshwater invertebrates are being subjected to an increasingly diverse cocktail of chemicals. Scientists from King's College London and the University of Suffolk recently even found cocaine, and other illicit drugs such as ketamine, in all the shrimp they tested in a rural area of eastern England (Miller *et al.* 2019). While concentrations of the drugs were low, contamination of the water courses, and our drinking water, by drug waste is an increasing problem. No studies of this kind have yet been undertaken in Wales, but given the widespread use of recreational drugs it is likely that at least some watercourses in Pembrokeshire are similarly affected. The effect of these chemicals on freshwater invertebrates, already under stress from other sources of pollution, is unknown.

CONSERVATION AND CARBON

We are now firmly in the Anthropocene, a new geological epoch that
commenced with the start of significant human impact on the Earth's geology
and ecosystems (Davison 2019). No formal date has yet been agreed by scientists,
but at present it is considered likely that it began around 1945–50, around the
time of the first large-scale nuclear tests and the development of consumer
plastics. Our war on biodiversity forms one of the primary characteristics of
this new epoch, and we are causing what has been described as Earth's sixth
major extinction. Humans are unique in the history of life on Earth, as we are
a globally distributed 'superpredator', relentlessly preying on the adults of
other apex predators, with widespread impacts on food webs across the world
(Darimont *et al.* 2015).

Most climate scientists agree that it is now too late to prevent 1.5 °C, or
more, of global warming by cutting our production of greenhouse gases. Even
if we reduced our emissions to zero tomorrow, it appears that we will probably
overshoot this crucial temperature limit. To prevent disaster, we need to
'decarbonise' our economy as soon as possible and capture the carbon dioxide
that has already been emitted. The industries surrounding Milford Haven
have their part to play here. It has been suggested that around one-third of the
greenhouse gas mitigation required between now and 2030 can be provided by
carbon capture through protecting and restoring ecosystems (Monbiot 2019).
By restoring and re-establishing forests, peatlands, saltmarshes, natural seabeds
and other ecosystems, large amounts of carbon could be removed from the air
and stored. At the same time, the protection and restoration of these ecosystems
would help minimise the sixth extinction.

People do not realise what we have already lost, which is why I have included
so many examples from the past in this book. In 1995 the renowned fisheries
expert Daniel Pauly came up with the term 'shifting baseline syndrome'
to explain our generational blindness to environmental destruction. Each
generation takes what it sees as it grows up as the norm, unaware of the way
things used to be. Pauly made the point that if you want to fight the loss
of memory and knowledge about the past, you have to rely on historical
information. He has stated that we have to rid ourselves of the notion that 'the
past is a provider of anecdotes and the present is a provider of knowledge'. The
writings of many of the people mentioned in this book may appear anecdotal,
but they should be treated with great respect. It is worth repeating here Robert
Drane's observations of the birds on Grassholm, written 127 years ago:

The first wonder was the number of birds. I know of no term or numeral that can adequately convey a conception of it – 'millions' will not do ... Let us say miriads, for that implies indefinitely great number, number without ascertainable limitation.

While Grassholm today is still a spectacular site, it is obviously not what it was, even allowing for the change in the species occupying the island. Records such as Drane's 'miriads' of puffins, and George Owen's description of 'the great abundance of herrings' around the Pembrokeshire coast, among others, tell us that we have to radically reset our mental baseline.

From the records left by George Owen we also know the locations of former woodlands, which are now farmland. While it is not possible to dramatically reduce plastic pollution without a global solution, the restoration of these destroyed Pembrokeshire woodlands would make an important local contribution to both carbon storage and wildlife conservation (Fig. 258). Indeed,

FIG 258. Re-establishing woodland in Pembrokeshire could help offset the county's contribution to climate change. Ancient woodlands such as these at Penlan Uchaf in Cwm Gwaun cannot be replicated, but new woodlands are possible and desirable. (Pembrokeshire Coast National Park Authority)

FIG 259. Tourists boarding a local bus at Melin Tregwynt woollen mill in 2013. The visitor economy is highly dependent on the landscape and wildlife of Pembrokeshire, as indicated by the graphics on this bus. (Jonathan Mullard)

increasing woodland cover is included among the top five priorities for the environment in Wales, which is one of the least wooded countries in Europe. Only about 15 per cent of the land area here is covered by woodland, compared to a European average of 37 per cent (Welsh Government 2018). The Environment (Wales) Act places a duty on Welsh Ministers to ensure that by 2050 net emissions are at least 80 per cent lower than the baseline set in legislation. Creating new woodland and managing existing woodlands are key parts of the approach, so as to protect, and increase, the woodland carbon sink.

Important as woodlands are, however, carbon accumulates most rapidly in vegetated coastal habitats, such as saltmarsh and seagrass beds, where it can be captured forty times faster per ha than in tropical forests (Duarte *et al.* 2005). Conserving the already extensive area of seagrass in Milford Haven and encouraging its expansion, both here and elsewhere, would make a reasonable impact for a relatively modest input of time and resources. Offshore there are concerns about the impact of trawling and dredging on carbon in seabed sediments. A recent paper suggests that while the mean annual burial rate for organic carbon is 2 kg per ha, trawlers might cause the resuspension of 18 kg per ha per year (Luisetti *et al.* 2019). Additionally, trawling first destroys, then prevents the re-establishment of the biotic crusts that once covered much of the shallow seabed around our coasts. Many of these crusts were composed of filter-feeders, including oysters and other bivalves that extracted carbon from the water, fixing it as calcium carbonate and in the process building substantial reefs. When these were damaged, not only was this carbon storage potential lost, but the underlying sediments became exposed to disturbance and, potentially, enhanced oxidation. Protecting areas of the seas around Pembrokeshire from trawling and dredging could therefore make a sizeable contribution to carbon capture.

The rise in sea level and more acidic oceans will threaten the coastal tourism infrastructure and natural attractions that are visited by around four and half million people each year. Climate change will also, as described, lead to changes in biodiversity, potentially affecting the eco-tourism on which so many businesses in Pembrokeshire depend (Fig. 259). The tourist industry in Pembrokeshire is aware of the need for an attractive and sustainable environment to attract visitors, but how will a rapidly changing climate affect business? With the coastline altered by rising sea levels and the seas themselves more acidic, will people still want to come?

While the current situation may appear daunting, the examples described above show that there are numerous opportunities for Pembrokeshire to lead the way in developing a sustainable future, protecting people while at the same time enhancing the wildlife of the county. But we do not have much time. There will be no sea parrots to amaze and delight further generations of visitors to Pembrokeshire if we do not act quickly to address the issues highlighted in this chapter.

The Plants and Animals of Little England

Included in Edward Laws' word list in *The History of Little England beyond Wales, and the Non-Kymric Colony settled in Pembrokeshire* (1888) are the names of a number of plants and animals. These are given below, together with the modern common and scientific names. As Laws points out, 'it must be understood that very many of them are not claimed as peculiar to Pembrokeshire', and indeed some are similar to those used in the English-speaking part of Gower (Mullard 2006). They are, however, words that were in common use in the county in the late nineteenth century.

Local name	Modern name	Scientific name	Note
Anny	Kittiwake	*Rissa tridactyla*	
Bluemorgan	Glaucous sedge	*Carex flacca*	bluish-grey leaves
Bog ginger	Water-pepper	*Persicaria hydropiper*	
Cracks	Wild plums	*Prunus domestica*	
Cutty moorcock	Water rail	*Rallus aquaticus*	'cutty' means small
Cutty wren	Wren	*Troglodytes troglodytes*	as above
Cutty evet	(Common?) newt	*Lissotriton vulgaris*	as above
Distel	Thistle	*various*	from the Dutch 'distel'
Eligug	Guillemot	*Uria aalge*	
Elver	Eel	*Anguilla anguilla*	
French cockle	Prickly cockle	*Acanthocardia echinata*	
Greybird	(Mistle?) Thrush	*Turdus viscivorus*	
Grey dullun	Dunnock	*Prunella modularis*	
Harfish	Razor shell	*Ensis arcuatus*	
Hen	Ocean quahog	*Arctica islandica*	

Local name	Modern name	Scientific name	Note
Keaks	Hogweed	*Heracleum sphondylium*	
Lady dishwash	Pied wagtail	*Motacilla alba*	See below*
Lapster	(Common?) Lobster	*Homarus gammarus*	pronunciation only
Leather mouse	Bat	*various*	
Limpin	(Common?) Limpet	*Patella vulgata*	
Liverocks	Reedmace	*Typha latifolia*	
Mealymoth	Lesser whitethroat	*Sylvia curruca*	'mealy mouth' – from the dark cheek feathers
Moory hen	Moorhen	*Gallinula chloropus*	
Moory pinnock	Meadow pipit	*Anthus pratensis*	
Mumruffin	Long-tailed tit	*Aegithalos caudatus*	
Nisbil	Hedgehog	*Erinaceus europaeus*	
Pardo	Great black-backed gull	*Larus marinus*	Welsh parddu = soot
Penny sow	Woodlouse	*Oniscus asellus?*	
Pyatt	Magpie	*Pica pica*	
St George's duck	Shelduck	*Tadorna tadorna*	
Sea parrot	Puffin	*Fratercula arctica*	
Sea pyatt	Oystercatcher	*Haematopus ostralegus*	
Seegar	Spiny lobster	*Palinurus elephas*	
Siggywiggy	Blue tit	*Cyanistes caeruleus*	
Silly willies	Common sandpiper	*Actitis hypoleucos*	
Sneak	Snail	*various*	
Spur	Common tern	*Sterna hirundo*	
Tosty	Cowslip	*Primula veris*	
Veer	Weasel	*Mustela nivalis*	
Woodcush	Woodpigeon	*Columba palumbus*	

* The origin of the washer names for pied wagtails is a mystery, but it may be because women once washed clothes and household utensils by a stream or village pump – the sort of places that the birds frequent. On Gower the bird is known as Lady Washdish.

Organisations and Contacts

THERE ARE NUMEROUS ORGANISATIONS CONCERNED with the natural history of Wales and the countryside in general, far too many to mention here. The following list therefore only covers those especially relevant to Pembrokeshire, or to topics covered in this book.

Amphibian and Reptile Conservation Trust
The Trust is committed to conserving amphibians and reptiles and saving the disappearing habitats on which they depend.
www.arc-trust.org

Ancient Yew Group
The aims of the group include raising public awareness of the national and worldwide importance of ancient yews. It has carried out extensive surveys, researching and collating all modern and historical references to yew trees across Britain.
www.ancient-yew.org

Botanical Society of Britain and Ireland
The BSBI County Recorder works with the National Trust, the Wildlife Trust of South and West Wales, the Pembrokeshire Coast National Park, Natural Resources Wales and the Ministry of Defence to survey the rarer plants of Pembrokeshire.
bsbi.org/pembrokeshire-v-c-45

British Bryological Society
The Society exists to promote the study of mosses and liverworts and has published the only modern field guide to mosses and liverworts of Britain and Ireland, enabling accurate identification in the field.
britishbryologicalsociety.org.uk

British Lichen Society
The Society's aims are to promote and advance the teaching and study of lichens, to encourage and actively support the conservation of lichens, and to raise public awareness of the beauty of lichens and of their importance as indicators of the health of our environment.
www.britishlichensociety.org.uk

Butterfly Conservation
The charity is devoted to saving butterflies, moths and their habitats throughout the UK. The South Wales branch runs guided walks, workshops, lectures and moth-trapping across the area.
butterfly-conservation.org/in-your-area/south-wales-branch

Defence Infrastructure Organisation
The work of the Defence Infrastructure
Organisation includes looking after the
Ministry of Defence's wider obligations as
steward of designated conservation areas,
protected species and heritage assets.
The DIO publishes *Sanctuary* magazine,
which includes information on wildlife
and conservation activities across the MoD
training estate.
www.gov.uk/government/publications/
dte-pembrokeshire-public-information-leaflet

Field Studies Council
The Council want to create a world
where everyone feels connected to the
environment so they can enjoy the benefits
it gives and make choices that help protect
it. They have two centres in Pembrokeshire,
at Dale Fort and Orielton.
https://www.field-studies-council.org/

National Museum Wales
The museum is an independent chartered
body and a registered charity which receives
its principal funding through grant-in-aid
from the Welsh Government.
museum.wales

National Trust
Founded in 1895 with the aim of saving
the nation's heritage and open spaces, the
National Trust protects and conserves a
wide range of properties and countryside in
Pembrokeshire.
www.nationaltrust.org.uk/days-out/regionwales/
pembrokeshire

Natural Resources Wales
A Welsh Government sponsored body,
responsible for the management of the
natural resources of Wales. Formed from
a merger of the Countryside Council for
Wales, Environment Agency Wales, and

Forestry Commission Wales.
naturalresources.wales

Pembrokeshire Birds
Pembrokeshire Birds is the website
for birders in Pembrokeshire to share
sightings.
pembsbirds.blogspot.com

**Pembrokeshire Coast National Park
Authority**
The National Park is managed by the Park
Authority, which has around 150 staff and a
committee of 18 Members.
www.pembrokeshirecoast.wales

Pembrokeshire Coastal Forum
The Coastal Forum works to protect the
coast and marine environments for current
and future generations to enjoy.
www.pembrokeshirecoastalforum.org.uk

Pembrokeshire Fungus Recording Network
The Pembrokeshire Fungus Recording
Network operates through email contacts
and updates to the website. There is also
a Facebook group, 'Finding Fungi in
Pembrokeshire', which posts interesting
finds and helps with identification.
www.pembsfungi.org.uk

Plantlife Wales
Plantlife works to raise the profile of plants
and fungi, to celebrate their beauty, and to
protect their future.
www.plantlife.org.uk/wales

RSPB Cymru
The RSPB (Royal Society for the Protection of
Birds) has been protecting wildlife in Wales
since 1911 and owns Ramsey and Grassholm,
together with a number of smaller islands.
www.rspb.org.uk/about-the-rspb/get-in-touch/
rspb-offices/wales

Shark Trust

The Shark Trust's vision is for a future where sharks thrive within a globally healthy marine ecosystem. It helps to protect shark species, or populations, threatened with extinction.

www.sharktrust.org

Vincent Wildlife Trust

The Trust's work is focused on British and Irish mammals, especially bats and mustelids, including badgers, pine martens and otters.

www.vwt.org.uk

Wildlife Trust of South and West Wales

The Wildlife Trust of South and West Wales is one of 47 Wildlife Trusts across the UK. It manages thirteen nature reserves in Pembrokeshire, including Skokholm and Skomer.

www.welshwildlife.org

References

Abraham, J. (2019). Our oceans broke heat records in 2018 and the consequences are catastrophic. *The Guardian*, 16 January.

Adams, R. (1972). *Watership Down*. Rex Collings, London.

Ainsworth, A. M., Cannon, P. F. and Dentinger, B. T. M. (2013). DNA barcoding and morphological studies reveal two new species of waxcap mushrooms (Hygrophoraceae) in Britain. *MycoKeys* 7, 45–63.

Ainsworth, A. M., Suz, L. M. and Dentinger, B. T. M. (2016). *Hohenbuehelia bonii* sp. nov. and *H. culmicola*: two pearls within the marram oyster. *Field Mycology* 17, 78–86.

Ainsworth, A. M., Douglas, B. and Suz, L. M. (2018). Big blue pinkgills formerly known as *Entoloma bloxamii* in Britain: *E. bloxamii* s. str., *E. madidum*, *E. ochreoprunuloides* forma *hyacinthinum* and *E. atromadidum* sp. nov. *Field Mycology* 19, 5–15.

Alexander, R. (2017). *Waterfalls of Stars: My Ten Years on the Island of Skomer*. Seren Books, Bridgend.

Allen, S., Allen, D., Phoenix, V. R. *et al.* (2019). Atmospheric transport and deposition of microplastics in a remote mountain catchment. *Nature Geoscience* 12, 339–344.

Altringham, J. (2003). *British Bats*. New Naturalist 93. Collins, London.

Andrews, P. T., Crump, R. G., Harries, D. J. and Andrews, M. M. (2016). Influence of weather on a population of soprano pipistrelle bats in West Wales, UK: a 20-year study estimates population viability. *Endangered Species Research* 30, 19–28.

Anon. (1841). [Article on Sir Roderick Murchison's 'Silurian system'.] *Edinburgh Review* 73, 1–141.

Anon. (1889). Cuckoo or Gowk? *Bruce Herald* Volume XX, Issue 2054, 12 April.

Anon. (1934). Notes: The VIII. International Ornithological Congress. *British Birds* 28, 73–74.

Archer-Thomson, J. and Cremona, J. (2019). *Rocky Shores*. British Wildlife Collection 7. Bloomsbury, London.

Arndt, E. A. (1989). Ecological, physiological and historical aspects of brackish water fauna distribution. In J. S. Ryland and P. A. Tyler (eds.), *Reproduction, Genetics and Distribution of Marine Organisms.* Proceedings of the 23rd European Marine Biology Symposium, Swansea, 5–9 September 1988. Olsen & Olsen, pp. 327–338.

Babington, C. C. (1863). On the botany of South Pembrokeshire. *Journal of Botany* 1, 258–270.

Babington, C. C. (1886). Pembrokeshire plants and the Rev. Mr. Holcombe. *Journal of Botany* 24, 22–23.

Bailey, J. H., Nelson-Smith, A. and Knight-Jones, E. W. (1967). Some methods for transects across steep rocks and channels. *Underwater Association Report* 1996-67, 107–111.

Baines, M. E. and Evans, P. G. H. (2012). *Atlas of the Marine Mammals of Wales,* 2nd edition. CCW Monitoring Report 68. Countryside Council for Wales, Bangor.

Baines, M. E., Earl, S. J., Pierpoint, C. J. L. and Poole, J. (1995). *The West Wales Grey Seal Census.* CCW Contract Science Report 131. Countryside Council for Wales, Bangor.

Baker, J. G. (1884). On the British daffodils. *Journal of Botany* 22, 193–195.

Baker, J. M. (1987). From chocolate mousse to acid drops: an introduction to FSC research. *Biological Journal of the Linnean Society* 32, 105–109.

Ballantine, W. J. (1961). A biologically-defined exposure scale for the comparative description of rocky shores. *Field Studies* 1 (3), 1–19.

Barbera, C., Bordehore, C., Borg, J. A. *et al.* (2003). Conservation and management of northeast Atlantic and Mediterranean maerl beds. *Aquatic Conservation: Marine and Freshwater Ecosystems* 13, S65–S76.

Barnes, R. S. K., Coughlan, J. and Holmes, N. J. (1973). A preliminary survey of the macroscopic bottom fauna of the Solent, with particular reference to *Crepidula fornicata* and *Ostrea edulis. Proceedings of the Malacological Society* 40, 253–275.

Barnett, R. J. (1986). Further records of *Elachista collitella* (Duponchel) (Lepidoptera: Elachistidae) *Entomologist's Gazette* 37, 240.

Barradell, M. G. (2009). Fine scale use of Ramsey Sound, Pembrokeshire, West Wales, by Harbour Porpoise (*Phocoena phocoena*). BSc Honours dissertation, Pembrokeshire College.

Barrett, C. G. (1878). Notes on Pembrokeshire Tineina. *The Entomologist's Monthly Magazine* 14, 268–272.

Barrett, J. and Yonge, C. M. (1958). *Collins Pocket Guide to the Sea Shore.* Collins, London.

Barrett-Hamilton, G. E. H. (1914). *A History of British Mammals.* Gurney & Jackson, London.

Bassindale, R. and Clark, R. B. (1960). The Gann Flat, Dale: studies on the ecology of a muddy beach. *Field Studies* 1 (2), 1–22.

Bates, D. E. B., Bromley, A. V. and Jones, A. S. G. (1969). The geology of the Bishops and Clerks Islands, Pembrokeshire. In A. Wood (ed.), *The Pre-Cambrian and Lower Palaeozoic Rocks of Wales.* University of Wales Press, Cardiff.

Bean, E. J. and Appleby, T. P. S. (2014). Guidelines for sustainable intertidal bait and seaweed collection in Wales: legislative review. A report to the Pembrokeshire Marine SAC Relevant Authorities Group. University of the West of England, Bristol.

Bendelow, V. C. and Hartnup, R. (1980). *Climatic Classification of England and Wales.* Soil Survey Technical Monograph 15. Soil Survey, Harpenden.

Bennett, P. (2001). Roundhouses in the landscape: recent construction work at Castell Henllys, Pembrokeshire. *CBA Wales Newsletter* 21, 13–15.

Benton, T. (2012). *Grasshoppers and Crickets.* New Naturalist 120. William Collins, London.

Beolens, B., Watkins, M. and Grayson, M. (2014). *The Eponym Dictionary of Birds.* Bloomsbury, London.

Berry, R. J. (1977). *Inheritance and Natural History.* New Naturalist 61. Collins, London.

Berry, R. J. (2009). *Islands.* New Naturalist 109. HarperCollins, London.

Berry, R. J. and Jakobson, M. (1975). Ecological genetics of an island population of the house mouse (*Mus musculus*). *Journal of Zoology* 175, 523–540.

Birkett, D. A., Maggs, C. A. and Dring, M. J. (1998). Maerl (Volume V). An overview of dynamic and sensitivity characteristics for conservation management of marine SACs. UK Marine SACs Project. Scottish Association for Marine Science.

Birkhead, T. (2016). Long lives, short memories: guillemots on Skomer since the 1930s. *Natur Cymru* 61, 10–14.

Bishop, G. M. and Earll, R. (1984). Studies on the populations of *Echinus esculentus* at the St Abbs and Skomer Voluntary Marine Nature Reserves. *Progress in Underwater Science,* 9, 53–66.

Blake, C. (2005). Use of fossil and modern coralline algae as a biogenic archive. PhD thesis, Queen's University Belfast.

Blanchard, M. (1997). Spread of the slipper limpet *Crepidula fornicata* in Europe: current state and consequences. *Scientia Marina* 61 (Supplement 9), 109–118.

Borrow, G. H. (1862). *Wild Wales: Its People, Language and Scenery.* John Murray, London.

Bosanquet, S. (2010). *The Mosses and Liverworts of Pembrokeshire.* Privately published.

Bowen, I. (1914). *The Great Enclosures of Common Lands in Wales.* Chiswick Press, London.

Boyce, D. C. (2004). A review of the invertebrate assemblage of acid mires. English Nature Research Reports 592. English Nature, Peterborough.

Bree, W. T. (1844). New British Narcissus? Art. XXIII – Variations Original and Select. *The Phytologist* 1, 61–62.

Bristowe, W. S. (1929). The spiders of Skomer Island (S. Wales). *Proceedings of the Zoological Society of London* 99, 617–622.

Bristowe, W. S. (1935). The spiders of Skokholm (S. Wales), with notes on a phalangid new to Britain. *Proceedings of the Zoological Society of London* 105, 233–239.

Bristowe, W. S. (1958). *The World of Spiders.* New Naturalist 38. Collins, London.

Brooke, M. (2010). *The Manx Shearwater.* T. & A. D. Poyser, London.

Brown, P. (2015). Weatherwatch. *The Guardian,* 21 September.

Brownlow, H. G. (1952). The design, construction and operation of Heligoland traps. *British Birds* 45, 387–399.

Bullimore, B. (2013). Milford Haven Waterway Environmental Surveillance Group: twenty years of partnership surveillance. *Porcupine Marine Natural History Newsletter* 24.

Bullimore, B. (2014). Monitoring Wales' only Marine Nature Reserve. *Natur Cymru* 50.

Bullimore, B. (2017). All at sea – marine conservation in Wales: gloom or hope? *Natur Cymru* 62.

Bunker, F. St P. D. (2011). Monitoring of a maerl bed in the Milford Haven Waterway, Pembrokeshire, 2010. A report to the Countryside Council for Wales by MarineSeen. CCW Contract Science Report 979.

Bunker, F. St P. D. and Camplin M. D. (2007). A study of the Milford Haven maerl bed in 2005 using drop down video and diving: a report to the Countryside Council for Wales by MarineSeen. CCW Contract Science Report 769.

Bunker, F. St P. D. and Holt R. H. F. (2003). Survey of sea caves in Welsh Special Areas of Conservation 2000 to 2002. A report to the Countryside Council for Wales by MarineSeen, Pembrokeshire. CCW Marine Monitoring Report 6. Countryside Council for Wales, Bangor.

Burton, M., Lock, K., Newman, P. and Jones, J. (2016). Skomer Marine Conservation Zone. Distribution and abundance of *Echinus esculentus* and selected starfish species 2015. NRW Evidence Report 158. Natural Resources Wales, Cardiff.

Burton, M., Lock, K., Newman, P. and Jones, J. (2017). Skomer MCZ Scallop Report. NRW Evidence Report 196. Natural Resources Wales, Cardiff.

Buxton, J. (1950). *The Redstart*. New Naturalist Monograph M02. Collins, London.

Buxton, J. and Lockley, R. M. (1950). *Island of Skomer: a Preliminary Survey of the Natural History of Skomer Island, Pembrokeshire*. Staples Press, London.

Cabot, D. (1999). *Ireland*. New Naturalist 84. HarperCollins, London.

Carey, D. A., Hayn, M., Germano, J. D., Little, D. I. and Bullimore, B. (2015). Marine habitat mapping of the Milford Haven Waterway, Wales, UK: comparison of facies mapping and EUNIS classification for monitoring sediment habitats in an industrialized estuary. *Journal of Sea Research* 100, 99–119.

Caseldine, A. E. (2018). Humans and landscape. *Internet Archaeology* 48. doi: 10.11141/ia.48.4.

Cefas/EA/NRW (2017). Salmon stocks and fisheries in England and Wales 2016. Centre for Environment, Fisheries & Aquaculture Science, Environment Agency and Natural Resources Wales.

Champneys, A. (2012). Factors affecting distribution and habitat selection of water shrews *Neomys fodiens*. PhD thesis, Nottingham Trent University.

Chandler, R. H. (1909). On some unrecorded erratic boulders in south Pembrokeshire. *Geological Magazine* 6 (5), 220–222.

Chapman, J. (1991). The later parliamentary enclosures of South Wales. *Agricultural History Review* 39, 116–125.

Chapman, N., Harris, S. and Stanford, A. (1994). Reeve's Muntjac *Muntiacus reevesi* in Britain: their history, spread, habitat selection and the role of human intervention in accelerating their dispersal. *Mammal Review* 24, 113–160.

Charles, B. G. (1992). *The Place-names of Pembrokeshire*. National Library of Wales, Aberystwyth.

Chen, I., Hill, J. K., Ohlemüller, R. *et al.* (2011). Rapid range shifts of species associated with high levels of climate warming. *Science* 333, 1024–1026.

Cheng, L., Zhu, J., Abraham, J. *et al.* (2019). 2018 continues record global ocean warming. *Advances in Atmospheric Sciences* 36, 249–252.

Church, R. (1612). *An Olde Thrift Newly Revived.* London.

Coghlan, A. (2016). Bacteria found to eat PET plastics could help do the recycling. *New Scientist*, 19 March.

Coldstream, N. (1999). From another world: the grotesque sculptures at St Davids. *Heritage in Wales*, Summer, 10–12.

Committee on Climate Change (2016). *UK Climate Change Risk Assessment 2017 Evidence Report.* www.theccc. org.uk/UK-climate-change-risk-assessment-2017.

Conder, P. (1990). *The Wheatear.* Christopher Helm, London.

Conolly, A. (1994). Castles and abbeys in Wales: refugia for 'mediaeval' medicinal plants. *Botanical Journal of Scotland* 46, 628–636.

Conran, T. (1997). *The Peacemakers.* Gomer Press, Llandysul.

Corbet, G. B. (1964). Regional variation in the bank vole *Clethrionomys glareolus* in the British Isles. *Proceedings of the Zoological Society of London* 143, 191–219.

Countryside Commission (1969). *Special Study Report Volume 2: Nature Conservation at the Coast.* HMSO, London.

Countryside Council for Wales (2009). Pembrokeshire Marine European Marine Site. Advice provided by the Countryside Council for Wales in fulfilment of Regulation 33 of the Conservation (Natural Habitats) Regulations 1994. CCW, Bangor.

Coutts, R. R. and Rowlands, I. W. (1969). The reproductive cycle of the Skomer vole (*Clethrionomys glareolus skomerensis*). *Journal of Zoology* 158, 1–25.

Crane, P. and Murphy, K. (2010). An early medieval settlement, iron smelting site and crop processing complex at South Hook, Herbrandston, Pembrokeshire. *Archaeologia Cambrensis* 159, 117–196.

Croose, E. (2016). *The Distribution and Status of the Polecat* (Mustela putorius) *in Britain 2014–2015.* Vincent Wildlife Trust, Ledbury.

Crothers, J. H. (ed.) (1966). *Dale Fort Marine Fauna (second edition).* Supplement to *Field Studies* 2. Field Studies Council.

Crump, R. G. and Emson, R. H. (1983). The natural history, life history and ecology of the two British species of *Asterina.* *Field Studies* 5, 867–882.

D., T. K. (1767). Account of a journey through North Wales 1767. *Gentleman's Magazine* 37, 589–590.

Darimont, C. T., Fox, C. H., Bryan, M. H. and Reimchen, T. E. (2015). The unique ecology of human predators. *Science* 349, 858–860.

Davies, M. (1969a). Cave sites (Ogof Gofan, Ogof Pen Cyfrwy, Ogof Morfran). *Archaeology in Wales* 9, 13–14.

Davies, M. (1969b). Ogof Gofan. *Cambrian Caving Club Journal* 4, 9.

Davies, M. (1981). Victorian naturalists in Tenby. *The Pembrokeshire Historian* 7, 16–24.

Davies, M. (1989). Recent advances in cave archaeology in southwest Wales. In T. D. Ford (ed.), *Limestones and Caves of Wales.* Cambridge University Press, Cambridge, pp. 79–91.

Davis, T. A. W. (1970). *Plants of Pembrokeshire.* West Wales Naturalists' Trust, Haverfordwest.

Davison, N. (2019). The Anthropocene epoch: have we entered a new phase of planetary history? *The Guardian*, 30 May.

Dawkins, W. B. (1874). *Cave Hunting: Researches on the Evidence of Caves Respecting the Early Inhabitants of Europe.* Macmillan, London.

Dawson, I. (2000). 66 years on. *Spider Recording Scheme Newsletter* 36, 6.

DeConto, R. M. and Pollard, D. (2016). Contribution of Antarctica to past and future sea-level rise. *Nature* 531, 591–597.

Defoe, D. (1742). *A Tour through the Whole Island of Great Britain*. London.

De la Beche, H. T. (1829). On the geology of southern Pembrokeshire. *Transactions of the Geological Society of London* (series 2) 2 (1), 1–20.

Dicks, B. (1987). The FSC Oil Pollution Research Unit – the 1980s and beyond. *Biological Journal of the Linnean Society* 32, 111–126.

Dix, T. (1866). A list of birds observed in Pembrokeshire. *The Zoologist* (series 2) 1, 132–140.

Dix, T. (1869). Ornithological notes from Pembrokeshire. *The Zoologist* (series 2) 4, 1670–1681.

Donovan, E. (1805). *Descriptive Excursions Through South Wales and Monmouthshire in the Year 1804, and the Four Preceding Summers*. London.

Down, C., Phillips, J., Ranger, A. and Farrell, L. (2002). A hemlock water dropwort curry: a case of multiple poisoning. *Emergency Medicine Journal* 19, 472–473.

Drane, R. (1893). Natural history notes from Grassholm. *Report and Transactions of the Cardiff Naturalists' Society* XXVI, 1–13.

Drane, R. (1898). A pilgrimage to Golgotha. *Report and Transactions of the Cardiff Naturalists' Society* XXXI, 38–51.

Drayton, M. (1612). *Poly-Olbion*. London.

Duarte, C. M., Middelburg, J. J. and Caraco, N. (2005). Major role of marine vegetation on the oceanic carbon cycle. *Biogeosciences* 2, 18.

Dutton, A., Carlson, A. E., Long, A. J. *et al.* (2015). Sea-level rise due to polar ice-sheet mass loss during past warm periods. *Science* 349 (6244), aaa4019.

Eastham, A. (2016). Goosey Goosey Gander with Jemima Shelduck in attendance: two Stone Age occupation caves in south Pembrokeshire. Pembrokeshire Historical Society. www.pembrokeshirehistoricalsociety.co.uk/goosey-goosey-gander-jemima-shelduck-attendance-two-stone-age-occupation-caves-south-pembrokeshire.

Edington, J. M., Morgan, P. J. and Morgan, R. A. (1973). Feeding patterns of wading birds on the Gann Flat and river estuary at Dale. *Field Studies* 3, 738–800.

Edwards, A., Garwood, P. and Kendall, M. (1992). The Gann Flat, Dale: thirty years on. *Field Studies* 8, 59–75.

Elias, T., Davies, J. H. and Roberts, D. (2016). Er clod i'n cloddiau cerrig. *Natur Cymru* 58.

Ellis, J. R., Milligan, S. P., Readdy, L., Taylor, N. and Brown, M. J. (2012). Spawning and nursery grounds of selected fish species in UK waters. Science Series Technical Report 147. Cefas, Lowestoft.

Emson, R. H. and Crump, R. G. (1979). Description of a new species of *Asterina* (Asteroidea) with an account of its ecology. *Journal of the Marine Biological Association of the United Kingdom* 59, 77–94.

Evans, F. (1989). A review of the management of lowland wet heath in Dyfed, West Wales. Contract Surveys 42. Nature Conservancy Council.

Evans, R. J., O'Toole, L. and Whitfield, D. (2012). The history of eagles in Britain and Ireland: an ecological review of placename and documentary evidence from the last 1500 years. *Bird Study* 59, 335–349.

Evans, S. (2013). The lesser butterfly-orchid/ tegeirian llydanwyrdd, *Platanthera bifolia*, in Pembrokeshire (v.c.45). *BSBI Welsh Bulletin* 92, 16–19.

Evans, S. (2018). Inland locations for vernal squill/seren y gwanwyn (*Scilla verna*) in Pembrokeshire. *BSBI Welsh Bulletin* 101, 24–28.

Evans, S., Warr, J. and Gomersall, F. (2012). Dale Fort, Pembrokeshire (v.c.45): BSBI Wales AGM 12th–14th August 2011. *BSBI Welsh Bulletin* 90, 18–20.

Evans, S. B. (1999). *Centaurium scilloides*. In M. J. Wigginton (ed.), *British Red Data Books: 1. Vascular Plants*. JNCC, Peterborough.

Evans, S. E. and Roberts, P. J. (2015). Welsh dune fungi: data collation, evaluation and conservation priorities. NRW Evidence Report 134. Natural Resources Wales, Bangor.

Fairlie, S. (2009). A short history of enclosure in Britain. *The Land* 7, 16–31.

Falconer, R. W. (1848). *Contributions Towards a Catalogue of Plants Indigenous to the Neighbourhood of Tenby*. Longman, Brown, Green and Longmans, London; R. Mason, Tenby.

Falk, S. J. and Crossley, R. (2005). *A Review of the Scarce and Threatened Flies of Great Britain, Part 3: Empidoidea*. JNCC Species Status Project.

Fenton, R. (1811). *A Historical Tour through Pembrokeshire*. London.

Fisher, J. and Vevers, H. G. (1943). The breeding distribution, history and population of the North Atlantic Gannet (*Sula bassana*). *Journal of Animal Ecology* 12, 173–213.

Fortey, R. (2000). *Trilobite!: Eyewitness to Evolution*. HarperCollins, London.

Freeman, M. (2019). Early tourists in Wales: 18th and 19th century tourists' comments about Wales. https:// sublimewales.wordpress.com.

Fretter, V. and Graham, A. (1981). The prosobranch molluscs of Britain and Denmark Part 6. *Journal of Molluscan Studies*, Supplement 9, 285–363.

Fullagar, P. J., Jewell, P. A., Lockley, R. M. and Rowlands, I. W. (1963). The Skomer vole (*Clethrionomys glareolus skomerensis*) and long-tailed field mouse (*Apodemus sylvaticus*) on Skomer Island, Pembrokeshire in 1960. *Proceedings of the Zoological Society of London* 140, 295–314.

Galperin, Y. and Little, D. I. (2014). Forensic evaluation of Milford Haven sediment hydrocarbon contamination. EGC Consulting Supplemental Report to Milford Haven Waterway Environmental Surveillance Group.

George, G. T. (2015). *The Geology of South Wales: a Field Guide*, 2nd edition. Geoserv Publishing, Maidstone.

George, T. N. (1970). *British Regional Geology: South Wales*. HMSO, London.

Gerald of Wales (1191). *Itinerarium Cambriae or Journey through Wales*. Translated by Lewis Thorpe. Penguin, Harmondsworth, 1978.

Gerarde, J. (1597). *The Herbal or Generall Historie of Plants*. London.

Gibson, E. (1695). *Camden's Britannia newly translated into English, with large additions and improvements*. Printed by F. Collins, for A. Swalle and A. and J. Churchil.

Gilbert, O. (2000). *Lichens*. New Naturalist 86. HarperCollins, London.

Gillham, M. E. (1953). An ecological account of the vegetation of Grassholm Island. *Journal of Ecology* 41, 84–99.

Gillham, M. E. (1955). Ecology of the Pembrokeshire islands: III. The effect of grazing on the vegetation. *Journal of Ecology* 43, 172–206.

Gillham, M. E. (1956a). Ecology of the Pembrokeshire islands: IV. Effects of treading and burrowing by birds and mammals. *Journal of Ecology* 44, 51–82.

Gillham, M. E. (1956b). Ecology of the Pembrokeshire islands: V. Manuring by the colonial seabirds and mammals, with a note on seed distribution by gulls. *Journal of Ecology* 44, 429–454.

Gilliland, P. M. and Sanderson, W. G. (2000). Re-evaluation of marine benthic species of nature conservation importance: a new perspective on certain 'lagoonal specialists' with particular emphasis on *Alkmaria romijni* Horst (Polychaeta: Ampharitidae). *Aquatic Conservation: Marine and Freshwater Ecosystems* 10, 1–12.

Godfrey, A. (2002). A reconnaissance of coastal soft cliff sites in south and south-west Wales to determine their importance for invertebrates. CCW Contract Science Report 506. Countryside Council for Wales, Bangor.

Goodman, G. T. and Gillham, M. E. (1954). Ecology of the Pembrokeshire islands: II. Skokholm, environment and vegetation. *Journal of Ecology* 42, 296–327.

Gosse, P. H. (1856). *Tenby: a Sea-Side Holiday.* Van Voorst, London.

Goudie, A. S. and Gardner, R. (1992). *Discovering Landscape in England and Wales.* Chapman & Hall, London.

Gough, H. J. (1971). A preliminary survey of the Collembola of Skokholm. *Field Studies* 3, 497–504.

Grall, J. and Hall-Spencer, J. M. (2003). Problems facing maerl conservation in Brittany. *Aquatic Conservation: Marine and Freshwater Ecosystems* 13, S55–S64.

Granmo, A., Rämä, T. and Mathiassen, G. (2012). The secrets of *Cryptomyces maximus* (Rhytismataceae): ecology and distribution in the Nordic countries (Norden), and a morphological and ontogenetic update. *Karstenia* 52, 59–72.

Green, J. (2002). *Birds in Wales 1992–2000.* Welsh Ornithological Society, Blaenporth.

Green, J. and Roberts, O. (2004). *Birding in Pembrokeshire.* Welsh Ornithological Society, Blaenporth.

Green, S. and Walker, E. (1991). *Ice Age Hunters: Neanderthals and Early Modern Hunters in Wales.* National Museum of Wales, Cardiff.

Greenoak, F. (1979). *All the Birds of the Air: the Names, Lore and Literature of British Birds.* Andre Deutsch, London.

Greenoak, F. (1985). *God's Acre: the Flowers and Animals of the Parish Churchyard.* E. P. Dutton, London.

Gregory, K. (1976). *First Cuckoo: Letters to The Times 1900–75.* Allen & Unwin, London.

Grimes, W. F. (1944). *Priory Farm Cave, Monkton, Pembrokeshire – with a report on the Human and Animal Remains by Lionel F. Cowley M.Sc. Archaeologia Cambrensis.*

Guardian (2012). 'Cutpurse wasp' wins species naming competition. *The Guardian,* Tuesday 24 July.

Gunther, R. W. T. (1922). *Early British Botanists and Their Gardens, Based on Unpublished Writings of Goodyer, Tradescant, and Others.* Oxford University Press, Oxford.

Gunther, R. W. T. (ed.) (1945). *Life and Letters of Edward Lhwyd 1660–1709.* Printed for the Subscribers, Oxford.

Gurney, S. (1857). The Cornish chough near Tenby. *The Zoologist* 15, 5790.

Gwynne, F. P. (1868). *Allen's Guide to Tenby.* W. Kent and Co., London; C. S. Allen, Tenby.

Hall, J., Rowe, E., Smith, R. *et al.* (2019). Trends report 2018: trends in critical load and critical level exceedances in the United Kingdom. Report to Defra under Contract AQ0843. CEH Project: NEC05708.

Hare, E. J. (2009). Island syndrome in rodents: a comparative study on island forms of the bank vole, *Myodes glareolus*. PhD thesis, Queen Mary University of London.

Harries, D. (2011). Willow Blister (*Cryptomyces maximus*) in Pembrokeshire. *Pembrokeshire Fungus Recorder* 2011 (3).

Harries, D., Hodges, J. and Theobald, T. (2013) The role of a local fungus recording network in conserving an internationally threatened species, *Cryptomyces maximus* (willow blister). *Fungal Conservation* 3, 7–11.

Harries, D., Hodges, J. and Theobald, T. (2015). An introduction to the sand dune fungi of West Wales. Pembrokeshire Fungus Recording Network.

Harris, M. P. (1965). Puffinosis among Manx shearwaters on Skokholm. *British Birds* 58, 426–434.

Harris, M. P. (1984). *The Puffin*. T. & A. D. Poyser, Calton.

Harris, S., Morris, P. Wray, S. and Yalden, D. (1995). A review of British mammals: population estimates and conservation status of British mammals other than cetaceans. JNCC, Peterborough.

Harrison, T. D. (1994). Notable Coleoptera from the Angle Peninsula, Pembrokeshire including *Harpalus melancholicus* Dejean (Carabidae). *Entomologist's Monthly Magazine* 130, 148.

Harvey, P. R., Nellist, D. R. and Telfer, M. G. (eds.) (2002). *Provisional Atlas of British Spiders (Arachnida, Araneae)*, Volumes 1 & 2. Biological Records Centre, Huntingdon.

Hassall, C. (1794). *A General View of the Agriculture of the County of Pembroke with Observations on the Means of its Improvements*. Board of Agriculture of Great Britain. London.

Hawkeswood, N. (2006). Restoring Cors Penally. *Sanctuary* 35, 54–55.

Haycock, A. (2016). A review of the status of wetland birds in the Milford Haven Waterway and Daugleddau Estuary, 2016. A report to the Milford Haven Waterway Environmental Surveillance Group.

Haycock, B. (2005). Marsh fritillary surveillance in Castlemartin Range. CCW internal document. Countryside Council for Wales, Bangor.

Haycock, B. (2017). Annual surveillance of chough populations in the Pembrokeshire Coast National Park in 2016: Castlemartin Coast SPA. A report to Natural Resources Wales and to the Pembrokeshire Coast National Park Authority.

Haycock, B. (2018). Annual surveillance of chough populations in the Pembrokeshire Coast National Park in 2018: Castlemartin Coast SPA. A report to Natural Resources Wales and to the Pembrokeshire Coast National Park Authority.

Haycock, B. and Hurford, C. (2006). Monitoring choughs *Pyrrhocorax pyrrhocorax* on the Castlemartin peninsula: a case study. In C. Hurford and M. Schneider (eds.), *Monitoring Nature Conservation in Cultural Habitats: a Practical Guide and Case Studies*. Springer, Dordrecht, pp. 247–258.

Hayward, P. J. (2004). *A Natural History of the Seashore*. New Naturalist 94. HarperCollins, London.

Hiscock, K. (ed.) (1998). Benthic marine ecosystems of Great Britain and the north-east Atlantic. JNCC, Peterborough.

Hiscock, K. (2018). *Exploring Britain's Hidden World: a Natural History of Seabed Habitats.* Wild Nature Press, Plymouth.

Hoare, R. C. (1793). *Journal of a Tour of South Wales.* London.

Holyoak, D. T. (2006). *Petalophyllum ralfsii.* United Kingdom Biodiversity Action Plan. Plantlife, Salisbury.

Houlston, L. (2016). Pembrokeshire – Castlemartin Range. *Sanctuary* 45, 97.

Houlston, L. (2018). Around the regions: Pembrokeshire – Castlemartin. *Sanctuary* 47, 91.

Howe, M. (2016). The end of the line for the strandline beetle? *Natur Cymru* 59, 10–13.

Howells, B. E. and Howells, K. A. (eds.) (1972). *Pembrokeshire Life: 1572–1843. A Selection of Letters.* Pembrokeshire Record Society, Haverfordwest.

Howells, M. F. (2007). *British Regional Geology: Wales.* British Geological Survey, Nottingham.

Howells, S. (1997). Geology and scenery of the Welsh islands. In P. M. Rhind, T. H. Blackstock and S. J. Parr (eds.), *Welsh Islands: Ecology, Conservation and Land Use.* Proceedings of the Welsh Islands Conference, Cardiff, November 1995. Countryside Council for Wales, Bangor, pp. 1–13.

Howells, S. (2008). Castlemartin ranges: rocks and landforms. *Sanctuary* 37, 12–15.

Humphery-Smith, C. R. (1971). Heraldry and the martyrdom of Archbishop Thomas Becket. *The Heraldry Society Coat of Arms* 85.

Hutterer, R., Ivanova, T., Meyer-Cords, C. and Rodrigues, L. (2005). *Bat Migrations in Europe: a Review of Banding Data and Literature.* Naturschutz und BiologischeViefalt 28. Federal Agency for Nature Conservation, Bonn.

Huxley, J. (director) (1934). *The Private Life of the Gannets.* Available on British Film Institute website: https://player.bfi.org. uk/free/film/watch-private-life-of-the-gannets-1934-online.

Hyde, H. A. (1961). *Welsh Timber Trees: Native and Introduced,* 3rd edition. National Museum of Wales, Cardiff.

Ingrouille, M. J. and Stace, C. A. (1986). The *Limonium binervosum* aggregate (Plumbaginaceae) in the British Isles. *Botanical Journal of the Linnean Society* 92, 177–217.

IUCN (2011). *Species on the Edge of Survival.* IUCN and HarperCollins, London.

Ixer, R. A. and Bevins, R. E. (2011). Craig Rhos-Y-Felin, Pont Saeson is the dominant source of the Stonehenge rhyolitic 'debitage'. *Archaeology in Wales* 50, 21–31.

James, J. W. C. (2008). Sand wave morphology and development in the Outer Bristol Channel (OBel) Sands. Marine and River Dune Dynamics Conference V, Leeds, April 2008.

Jangoux, M. (1987). Diseases of Echinodermata. I. Agents microorganisms and protistans. *Diseases of Aquatic Organisms* 2, 147–162.

Jehu, T. J. (1904). The glacial deposits of northern Pembrokeshire. *Transactions of the Royal Society of Edinburgh* 47, 53–87.

Jermyn, D. L., Messenger, J. E. and Birks, J. D. S. (2001). *The Distribution of the Hazel Dormouse* Muscardinus avellanarius *in Wales.* Vincent Wildlife Trust, Ledbury.

John, B. (2018). *The Stonehenge Bluestones.* Greencroft Books, Newport.

Joint Nature Conservation Committee. (2007). Second Report by the UK under Article 17 on the implementation of the Habitats Directive from January 2001 to December 2006. Conservation status assessment for: S1029: *Margaritifera margaritifera* – Freshwater pearl mussel. JNCC, Peterborough.

Jones, A. C. L., Halley, D. J., Gow, D., Branscombe, J. and Aykroyd, T. (2012). Welsh Beaver Assessment Initiative Report: an investigation into the feasibility of reintroducing European Beaver (*Castor fiber*) to Wales. Wildlife Trusts Wales.

Jones, B. L. and Unsworth, R. K. F. (2016). The perilous state of seagrass in the British Isles. *Royal Society Open Science* 3 (1), 150596. doi: 10.1098/rsos.150596.

Jones, D. (1992). *The Tenby Daffodil*. Tenby Museum.

Jones, F. (1954). *The Holy Wells of Wales*. University of Wales Press, Cardiff.

Jones, J. and Lock, K. (2014). Distribution and abundance of *Palinurus elephas* in Pembrokeshire. Marine Conservation Society/Seasearch.

Jones, J. and Whitmore, J. (2015). Distribution of non-native species on vertical artificial structures in Milford Haven. SEACAMS Project RD-067.

Jones, J., Burton, M., Lock, K. and Newman, P. (2016). Skomer Marine Conservation Zone sponge diversity survey 2015. NRW Evidence Report 159. Natural Resources Wales, Cardiff.

Jones, T. (1973). Perlau Taf. *The Carmarthenshire Historian, Carmarthenshire Community Council* 10.

Jones, T. I. J. (1955). *Exchequer Proceedings Concerning Wales in Tempore James I. Abstracts of Bills and Answers and Inventory of Further Proceedings*. University of Wales Press, Cardiff.

Judd, S. (2006). Status and distribution of the shieldbug *Odontoscelis fuliginosa* (L.)

and seedbug *Pionosomus varius* (Wolff) (Hemiptera: Heteroptera) associated with bare and partially vegetated dunes on the Castlemartin peninsula. *British Journal of Entomology and Natural History* 19, 97–103.

Kaiser, M. J., Bergmann, M., Hinz, H. *et al.* (2004). Demersal fish and epifauna associated with sandbank habitats. *Estuarine, Coastal and Shelf Science* 60, 445–456.

Kay, Q. O. N. (1998). A review of the existing state of knowledge of the ecology and distribution of seagrass beds around the coast of Wales. CCW Contract Survey FC 73-01-168. Countryside Council for Wales, Bangor.

Kay, Q. O. N. and John, R. (1995). The conservation of scarce and declining plant species in lowland Wales: population genetics, demographic ecology and recommendations for future conservation in 32 species of lowland grassland and related habitats. Science Report 110. Countryside Council for Wales, Bangor.

Kerbiriou, C., Gourmelon, F., Jiguet, F. *et al.* (2006). Linking territory quality and reproductive success in the Red-billed Chough *Pyrrhocorax pyrrhocorax*: implications for conservation management of an endangered population. *Ibis* 148, 352–364.

Kerney, M. P. (1999). *Atlas of Land and Freshwater Molluscs of Britain and Ireland*. Conchological Society of Great Britain and Ireland. Harley, Colchester.

Kindler, C., Chèvre, M., Ursenbacher, S. *et al.* (2017). Hybridization patterns in two contact zones of grass snakes reveal a new Central European snake species. *Scientific Reports* 7 (1), 7378. doi: 10.1038/s41598-017-07847-9.

Knight, G. T. and Howe, M. A. (2006). *The Conservation Value for Invertebrates of Selected Coastal Soft Cliff Sites in Wales.* CCW Contract Science Report 761. Countryside Council for Wales, Bangor/ National Museums Liverpool.

Kosuth, M., Mason, S. A. and Wattenburg, E. V. (2018). Anthropogenic contamination of tap water, beer, and sea salt. *PLoS One* 13 (4), e0194970.

Laimann, J. (2018). Research briefing: common land. National Assembly for Wales, Senedd Research 18-044.

Laws, E. (1888). *The History of Little England beyond Wales, and the Non-Kymric Colony settled in Pembrokeshire.* George Bell and Sons, London.

Laws, E. (1908). Bronze implements from the shores of Milford Haven. *Archaeologia Cambrensis* 8, 114–115.

Leach, A. L. (1918). Flint-working sites on the submerged land (submerged forest) bordering the Pembrokeshire coast. *Proceedings of the Geologists' Association* 29 (2), 46–64.

Leach, A. L. (1933). The geology and scenery of Tenby and the South Pembrokeshire Coast. *Proceedings of the Geologists' Association* 44 (2), 187–216.

Lees, E. (1842). *The Botanical Looker-out Among the Wild Flowers of the Fields, Woods, and Mountains of England and Wales.* Tilt and Bogue, London.

Lees, E. (1853). Notes on the localities of some Pembrokeshire plants, observed in May and June, 1853. *The Phytologist,* 4, 1013–1018.

Lemoine, P. (1910). Répartition et mode de vie du maerl (*Lithothamnium calcareum*) aux environs de Concarneau (Finistère). *Annales de l'Institut Océanographique, Paris* 1, 1–29.

Leopold, A. (1949). *A Sand County Almanac.* Oxford University Press, New York.

Letts, J. (2001). Living under a mediaeval field. *British Archaeology* 58, 11–13.

Lewis, M. (1988). Cambrian stratigraphy and trilobite faunas in southwest Wales. PhD thesis, University of Wales.

Lewis, S. (1833). *A Topographical Dictionary of Wales,* Volumes 1 & 2. London.

Linnard, W. (1979). The history of forests and forestry in Wales up to the formation of the Forestry Commission. PhD thesis, University of Wales.

Little, A. and Hiscock, K. (1987). Surveys of harbours, rias and estuaries in southern Britain: Milford Haven and the estuaries of the Cleddau. Nature Conservancy Council, CSD Report No. 735.

Lloyd, B. Diaries held at National Museum Wales.

Lloyd, B. (1944). The Boncath Kite. *Folklore* 55 (4), 166–167.

Lloyd, B. (1948). Notes on the flora of Pembrokeshire. Extracted from the Journals of Bertram Lloyd, F.L.S., M.B.O.U, F.R.E.S. by Sylvia Lloyd. *North Western Naturalist,* 23, 88–95.

Lloyd, T., Orbach, J. and Scourfield, R. (2010). *The Buildings of Wales: Pembrokeshire.* Yale University Press, New Haven, CT.

Lock, K. (2007). Milford Haven Waterway Seasearch 2007 Biodiversity Action Plan: species and habitat surveys. Seasearch and Marine Conservation Society.

Lock, K. (2011). Milford Haven Seasearch and native oyster surveys 2010 & 2011. Seasearch and Marine Conservation Society.

Lock, K. (2017). Native oyster *Ostrea edulis* Milford Haven waterway survey report. Seasearch and Marine Conservation Society.

Lock, K. and Bullimore, B. (2018). Seasearch surveys in Milford Haven: a twelve-year summary 2004–2015. Seasearch South and West Wales. www.seasearch.org.uk/downloads/Milford-Haven-2004-2015.pdf.

Lock, K., Stamp, T. and Wood, C. (2014). Seasearch Wales 2014 summary report. Seasearch and Marine Conservation Society.

Lock, K., Newman, P., Burton, M. and Jones, J. (2015). Skomer Marine Conservation Zone nudibranch diversity survey 2014. NRW Evidence Report 67. Natural Resources Wales, Cardiff.

Lockley, A. (2013). Island Child: My Life with R. M. Lockley. Gwasg Carreg Gwalch, Llanrwst.

Lockley, R. M. (1930). Dream Island. Witherby, London.

Lockley, R. M. (1934). Island Days. Witherby, London.

Lockley, R. M. (1936). Skokholm Bird Observatory. British Birds 29, 222–235.

Lockley, R. M. (1938). I Know an Island. George G. Harrap, London.

Lockley, R. M. (1942). A new way of running a farm. The Countryman 26, 37–45.

Lockley, R. M. (1943). Inland Farm. Witherby, London.

Lockley, R. M. (1946). The Island Farmers. Witherby, London.

Lockley, R. M. (1947). Letters from Skokholm. J. M. Dent, London.

Lockley, R. M. (1948). The Golden Year. Witherby, London.

Lockley, R. M. (1956). The outermost rocks of Wales. Nature in Wales 2, 257–266.

Lockley, R. M. (1957). Pembrokeshire. Robert Hale, London.

Lockley, R. M. (1962). The grey squirrel invasion in 1961. Nature in Wales 8, 36.

Lockley, R. M. (1964). The Private Life of the Rabbit. Andre Deutsch, London.

Lockley, R. M., Ingram, G. and Salmon, H. M. (1948). The Birds of Pembrokeshire. West Wales Field Society, Haverfordwest.

Lomolino, M. V. (1985). Body size of mammals on islands: the island rule reexamined. The American Naturalist 125, 310–316.

Long, D. and Williams, J. (2007). Juniper in the British Uplands: the Plantlife Juniper Survey Results. Plantlife, Stirling.

Lovegrove, R. (2007). Silent Fields. Oxford University Press, Oxford.

Lovegrove, R., Williams, G. and Williams, I. (1994). Birds in Wales. T. & A. D. Poyser, London.

Loxton, R. G. (1997). Invertebrates of the Welsh islands. In P. M. Rhind, T. H. Blackstock and S. J. Parr (eds.), Welsh Islands: Ecology, Conservation and Land Use. Proceedings of the Welsh Islands Conference, Cardiff, November 1995. Countryside Council for Wales, Bangor, pp. 70–78.

Luisetti, T., Turner, R. K., Andrews, J. E. et al. (2019). Quantifying and valuing carbon flows and stores in coastal and shelf ecosystems in the UK. Ecosystem Services 35, 67–76.

Lydekker, R. (1896). A Hand-book to the British Mammalia. Edward Lloyd, London.

Mabey, R. (1996). Flora Britannica. Sinclair Stevenson, London.

Mackie, A. S. Y., James, J. W. C., Rees, E. I. S. et al. (2006). The Outer Bristol Channel marine habitat study: summary document. Amgueddfa Cymru/National Museum Wales, Cardiff.

Malloy, A. (2019). Updated Statement 13:00: Oil Pollution Incident 03/01/2019. Port of Milford Haven.

Maroo, S. and Yalden, D. W. (2000). The Mesolithic mammal fauna of Great Britain. *Mammal Review* 30, 243–248.

Mason, R. (1868). *A Guide to the Town of Tenby and its Neighbourhood*, 5th edition. R. Mason, Tenby.

Mason, R. (1875). *A Guide to the Town of Tenby and its Neighbourhood*, 7th edition. R. Mason, Tenby.

Matheson, C. (1932). *Changes in the Fauna of Wales Within Historic Times*. National Museum of Wales, Cardiff.

Matheson, C. (1955). The rodents of Wales. *Nature in Wales* 1, 111–115.

Mathew, M. A. (1894). *The Birds of Pembrokeshire and its Islands*. R. H. Porter, London.

Mathews, F., Coomber, F. W., Wright, J. and Kendall, T. (2018). *Britain's Mammals 2018: the Mammal Society's Guide to their Population and Conservation Status*. Mammal Society, London.

McInnes, R. and Benstead, S. (2013). *Art as a Tool in Support of the Understanding of Coastal Change in Wales*. Crown Estate, London.

McKay, K. (1989). *A Vision of Greatness: the History of Milford 1790–1990*. Brace Harvatt Associates, Milford Haven.

McKie, R. (2018). Eerie silence falls on Shetland cliffs that once echoed to seabirds' cries. *The Observer*, 3 June.

Miles, D. (1987). *The Pembrokeshire Coast National Park*. David & Charles, Newton Abbot.

Miles, D. (1994). [Introduction and notes.] In *The Description of Pembrokeshire by George Owen*. Gomer Press, Llandysul.

Miller, T. H., Tiong Ng, K., Bury, S. T., et al. (2019). Biomonitoring of pesticides, pharmaceuticals and illicit drugs in a freshwater invertebrate to estimate toxic or effect pressure. *Environment International* 129, 595–606.

Minter, D. (2017). The Global Fungal Red List Initiative. IUCN and Species Survival Commission.

Mitchell, P. I., Newton, S. E., Ratcliffe, N. and Dunn, T. E. (2004). *Seabird Populations of Britain and Ireland: Results of the Seabird 2000 Census (1998–2002)*. T. & A. D. Poyser, London.

Monbiot, G. (2019). Averting climate breakdown by restoring ecosystems: a call to action. Natural Climate Solutions. www.naturalclimate.solutions/the-science.

Monks, N. (2017). The graptolites of Abereiddy Bay. *Deposits Magazine*. https://depositsmag.com/2017/05/04/the-graptolites-of-abereiddy-bay.

Moore, D. (2019). Microplastic found in Britain's 'most iconic' rivers and lakes – FoE. *Circular* (CIWM), 7 March. www.circularonline.co.uk/news/microplastic-found-in-britains-most-iconic-rivers-lakes-foe.

Morgan, G. (2013). Gannet cam: the Grassholm research story continues. RSPB. https://community.rspb.org.uk/placestovisit/ramseyisland/b/ramseyisland-blog/posts/gannet-cam-the-grassholm-research-story-continues.

Morris, M. C. F. (1897). *Francis Orpen Morris, a Memoir*. J. C. Nimmo, London.

Morris, P. A. (1993). *A Red Data Book for British Mammals*. Mammal Society, London.

Mullard, J. (2006). *Gower*. New Naturalist 99. HarperCollins, London.

Mullard, J. (2014). *Brecon Beacons*. New Naturalist 126. William Collins, London.

Murchison, R. I. (1839). *The Silurian System*. J. Murray, London.

Murchison, R. I. (1854). *Siluria: the History of the Oldest Known Rocks Containing Organic Remains, with a Brief Sketch of the Distribution of Gold over the Earth.* London.

Natural Resources Wales (2014). The Cleddau and Pembrokeshire coastal rivers abstraction licensing strategy: a licensing strategy to manage water resources sustainably. NRW, Cardiff.

Natural Resources Wales (2016). Tree cover in Wales' towns and cities: understanding canopy cover to better plan and manage our urban trees. NRW, Cardiff.

Natural Resources Wales (2018). Rare fan shell found in Welsh waters. Natural Resources Wales website, 11 October. https://naturalresources.wales/about-us/news-and-events/news/rare-fan-shell-found-in-welsh-waters.

Neale, J. J. (1896). Natural history notes. *Report and Transactions of the Cardiff Naturalists' Society* XXVIII.

Nelms, S. E., Barnett, J. Brownlow, A. et al. (2019). Microplastics in marine mammals stranded around the British coast: ubiquitous but transitory? *Scientific Reports* 9, 1075.

Newman, P., Lock, K., Burton, M. and Jones, J. (2015). Skomer Marine Conservation Zone annual report 2015. NRW Evidence Report 149. Natural Resources Wales, Cardiff.

Newman, P., Lock, K., Burton, M. and Jones, J. (2018). Skomer Marine Conservation Zone annual report 2017. NRW Evidence Report 250. Natural Resources Wales, Bangor.

Nicholson, E. M. (1929). Report on the 'British Birds' census of heronries, 1928. *British Birds* 22, 270–323, 334–372.

Nicholson, G. (1840). *The Cambrian Traveller's Guide and Pocket Companion.* George Nicholson, Stourport.

Niemann, D. (2013). *Birds in a Cage: Germany, 1941. Four P.O.W. Birdwatchers: the Unlikely Beginning of British Wildlife Conservation.* Short Books, London.

North, F. J. (1957). *Sunken Cities: Some Legends of the Coast and Lakes of Wales.* University of Wales Press, Cardiff.

Oates, M. (1999). Sea cliff slopes and combes – their management for nature conservation. *British Wildlife* 10, 394–402.

Olds, L., Chmurova, L., Dinham, C. and Falk, S. (2018). *Wales Threatened Bee Report.* Buglife Cymru, Cardiff. www.buglife.org.uk/wales-threatened-bee-report.

Owen, G. (1595). *A Pamphelett conteyninge the description of Milford havon.* London.

Owen, G. (1599). Treatise on clay marl. MS; in the Vairdre Book, National Library of Wales.

Owen, G. (1603). *The Description of Pembrokeshire.* Edited with an introduction and notes by Dillwyn Miles. Gomer Press, Llandysul, 1994.

Owen, H. (ed.) (1914). *A Calendar of Public Records Relating to Pembrokeshire.* Cymmrodorion Record Series No. 7, Volume II.

Owen, H. (ed.) (1918). *A Calendar of Public Records Relating to Pembrokeshire.* Cymmrodorion Record Series No. 79, Vol. III, p. 105.

Owen, H. W. and Morgan, R. (2007). *Dictionary of the Place-names of Wales.* Gomer Press, Ceredigion.

Owen, T. M. (1978). *Welsh Folk Customs.* National Museum of Wales, Cardiff.

Packe, C. (1743). *A New Philosophico-chorographical Chart of East Kent.* Canterbury.

Page, C. N. (1988). *Ferns.* New Naturalist 74. Collins, London.

Panting, P. J. (1973). Mammal field notes. *Nature in Wales* 13, 207–208.

Parker, C. (2018). Locust blown to Devon gives nature lover a shock. *The Times*, 24 October.

Parker Pearson, M., Pollard, J., Schlee, D. and Welham, K. (2016). In search of the Stonehenge quarries. *British Archaeology* 146, 16–23.

Parker-Rhodes, A. F. (1950). *Fungi, Friends and Foes.* Paul Elek, London.

Pauly, D. (1995). Anecdotes and the shifting baseline syndrome of fisheries. *Trends in Ecology and Evolution* 10, 430.

Pearman, D. (2017). *The Discovery of the Native Flora of Britain and Ireland.* Botanical Society of Britain and Ireland, Bristol.

Pembrokeshire Coastal Forum (2019). *Annual Report 2017–2018.* PCF, Pembroke Dock.

Pennick, N. (1987). *Lost Lands and Sunken Cities.* Fortean Tomes, London.

Penrose, R. S. (2019). *Marine Mammal and Marine Turtle Strandings (Welsh Coast): Annual Report 2018.* Marine Environmental Monitoring, Cardigan.

Penrose, R. S. and Gander, L. R. (2018). *British Isles and Republic of Ireland Marine Turtle Strandings and Sightings: Annual Report 2017.* Marine Environmental Monitoring, Cardigan.

Perkins, I. (2016). Castlemartin choughs are in a safe pair of hands. *Sanctuary* 45, 27.

Perry, S. L. (2015). *Living Seas: Future Fisheries.* The Welsh Fishing Industry. Report by the Wildlife Trust of South and West Wales. WTSWW Living Seas Report 1.

Picton, B. E. and Morrow, C. C. (2007). Encyclopaedia of Marine Life of Britain and Ireland. www.habitas.org.uk/marinelife.

Pierpoint C. (2008). Harbour porpoise (*Phocoena phocoena*) foraging strategy at a high energy, near-shore site in south-west Wales. *Journal of the Marine Biological Association of the United Kingdom* 88, 1167–1173.

Ponder, W. F. (1988). *Potamopyrgus antipodarum*: a molluscan coloniser of Europe and Australia. *Journal of Molluscan Studies* 54, 271–285.

Port of Milford Haven (2016). *Port of Milford Haven Annual Report 2016.* PoMH.

Potter, B. (1902) *The Tale of Peter Rabbit* Frederick Warne and Co. London

Potter, J. (2013). Field trip report: geology and churches in Pembrokeshire, part 1. *Magazine of the Geologists' Association* 12 (3), 15–16.

Powell, M. C. (1990). Leach's petrels in Wales, December 1989. *Welsh Bird Report* 4, 36–37.

Praeger, R. L. (1911). Clare Island Survey, Part X: Phanerogamia and Pteridophyta. *Proceedings of the Royal Irish Academy* 31, 1–112.

Preston, C. D., Pearman, D. A. and Dines, T. D. (2002). *New Atlas of the British and Irish Flora.* Oxford University Press, Oxford.

Prime, C. T. (1960). *Lords and Ladies.* New Naturalist Monograph M17. Collins, London.

Prosser, M. V. and Wallace, H. L. (2003). Milford Haven saltmarsh survey. Report to the Milford Haven Waterway Environmental Surveillance Group.

Pugsley, H. W. (1924). A new *Statice* in Britain. *Journal of Botany* 62, 129–134.

Pugsley, H. W. (1931). A further new *Limonium* in Britain. *Journal of Botany* 69, 44–47.

Pugsley, H. W. (1941). *Narcissus obvallaris. Journal of Botany* 79, 27.

Pulvertaft, D. (2011). *Figureheads of the Royal Navy.* Seaforth Publishing, Barnsley.

Raven, C. E. (1942). *John Ray, Naturalist: His Life and Works*. Cambridge University Press, Cambridge.

Raye, L. (2016). The forgotten beasts in medieval Britain: a study of extinct fauna in medieval sources. PhD thesis, Cardiff University.

Reason, P., Harris, S. and Cresswell, P. (1993). Estimating the impact of past persecution and habitat changes on the numbers of badgers *Meles meles* in Britain. *Mammal Review* 23, 1–15.

Rees, F. L. (1950). *List of Pembrokeshire Plants*. Tenby Museum and West Wales Field Society, Tenby.

Rees, G. (2005). The Strumble Head Story – so far. Pembrokeshire Birdwatchers Conference 2005/Pembrokeshire Bird Group website.

Rees, G., Haycock, A., Haycock, B. *et al.* (2009). *Atlas of Breeding Birds in Pembrokeshire 2003–2007*. Pembrokeshire Bird Group.

Reid, C. (1913). *Submerged Forests*. Cambridge University Press, Cambridge.

Rich, T. C. G. (2005). Could *Centaurium scilloides* (L. f.) Samp. (Gentianaceae), Perennial Centaury, have colonised Britain by sea? *Watsonia* 25, 397–401.

Rich, T. C. G., Evans, S. B., Evans, A. E. *et al.* (2005). Distribution of the western European endemic *Centaurium scilloides* (L. f.) Samp. (Gentianaceae), Perennial Centaury. *Watsonia* 25, 275–281.

Rich, T. C. G. and McVeigh, A. (2019) *Gentians of Britain and Ireland BSBI Handbook No. 19*. Botanical Society of Britain and Ireland

Richards, P. W. and Evans, G. B. (1972). Biological flora of the British Isles. No. 126. *Hymenophyllum tunbrigense* (L.) Sm.; *Hymenophyllum wilsonii* Hooker. *Journal of Ecology* 60, 245–268.

Riddelsdell, H. J. (1905). Lightfoot's visit to Wales in 1773. *Journal of Botany* 43, 290–307.

Robertson, J. (2007). Dinas Island Farm, Pembrokeshire – a golden future? *British Wildlife* 19, 20–27.

Roper, P. (1993). The distribution of the wild service tree, *Sorbus torminalis* (L.) Crantz, in the British Isles. *Watsonia* 19, 209–229.

Roper, T. J. (2010). *Badger*. New Naturalist 114. HarperCollins, London.

Rose, F. (1975). The vegetation and flora of Tycanol Wood. *Nature in Wales* 14, 178–185.

Ross, N. (2006). A re-evaluation of the origins of late Quaternary ramparted depressions in Wales. PhD thesis, Cardiff University.

Ross, N., Harris, C. and Brabham, P. J. (2007). Internal structure and origins of late Devensian 'ramparted depressions', Llanio Fawr, Ceredigion. *Quaternary Newsletter* 112, 6–21.

Rotheroe, M. (1993). The larger fungi of Welsh sand dunes. Cambrian Institute of Mycology, Lampeter.

Salisbury, R. A. (1796). *Prodromus stirpium in horto ad Chapel Allerton vigentium*. Linnean Society, London.

Salmon, H. M. and Lockley, R. M. (1933). The Grassholm gannets: a survey and census. *British Birds* 27, 142–152.

Salmon, M. A. (2000). *The Aurelian Legacy: British Butterflies and their Collectors*. Harley, Colchester.

Saunders, D. (1986). *The Nature of West Wales*. West Wales Trust for Nature Conservation; Barracuda Books, Buckingham.

Saunders, D. (2008). 'For five shillings this duck can be yours': a history of the Orielton Decoy. *British Wildlife* 20, 37–43.

Saunders, D. (2019). Orielton Duck Decoy, Pembrokeshire: a rich history. *Birds in Wales* 16 (1), 3–16.

Saunders, D. and Saunders, S. (1992). Blackburnian Warbler: new to the Western Palearctic. *British Birds* 85, 337–343.

Saxton, C. (1579). *An Atlas of England and Wales*. London.

Schwägerl, C. (2013). Reviving Europe's biodiversity by importing exotic animals. *Yale Environment 360*. https://e360.yale.edu/features/reviving_europes_biodiversity_by_importing_exotic_animals.

Scott, V. (1980). *Inferno 1940*. Western Telegraph, Haverfordwest.

Sea Mammal Research Unit (2017). *Scientific Advice on Matters Related to the Management of Seal Populations: 2017*. NERC Special Committee on Seals. SMRU, Aberdeen.

Sell, P. and Murrell, G. (2018). *Flora of Great Britain and Ireland: Volume 1. Lycopodiaceae – Salicaceae*. Cambridge University Press, Cambridge.

Seymour, W. P. (1985). The environmental history of the Preseli region of south-west Wales over the past 12,000 years. PhD thesis, University of Wales, Aberystwyth.

Shellswell, C. H. (2015). Wales' important arable plants. Plantlife Cymru, Cardiff.

Shorten, M. (1954). *Squirrels*. New Naturalist Monograph M12. Collins, London.

Sinden, N. (1990). *In a Nutshell: a Manifesto for Trees and a Guide to Growing and Protecting Them*. Common Ground, London.

Sneddon, P. and Randall, R. E. (1993). Vegetated shingle structures of Great Britain: Appendix 1. Shingle sites in Wales. JNCC, Peterborough.

South, R. (1980). *The Moths of the British Isles – First Series*. Edited and revised by H. M. Edelsten and D. S. Fletcher. Warne, London.

Sowerby, J. (1803). *Coloured Figures of English Fungi or Mushrooms*, Volume 3. London.

Spooner, B. and Roberts, P. (2005). *Fungi*. New Naturalist 96. HarperCollins, London.

Stace, C. (2019). *New Flora of the British Isles*, 4th edition. C&M Floristics, Penrith.

Stace, C. A., Preston, C. D. and Pearman, D. A. (2015). *Hybrid Flora of the British Isles*. BSBI, Bristol.

Stebbings, R. E. (1988). *The Conservation of European Bats*. Christopher Helm, London.

Steers, J. A. (1953). *The Sea Coast*. New Naturalist 25. Collins, London.

Stephens, N. (2014). Reconnecting our South Wales water voles. Report to Natural Resources Wales. Wildlife Trust of South and West Wales, Bridgend.

Stewart, M. (1933). *Ronay: a Description of the Islands of North Rona and Sula Sgeir, Together with Their Geography, Topography, History, and Natural History, Etc., to Which Is Appended a Short Account of the Seven Hunters, or Flannan Islands*. Oxford University Press, London.

Stewart, N. F. (2004). Important stonewort areas: an assessment of the best areas for stoneworts in the United Kingdom. Plantlife International, Salisbury.

Stiller, M., Baryshnikov, G., Bocherens, H. *et al*. (2010). Withering away: 25,000 years of genetic decline preceded cave bear extinction. *Molecular Biology and Evolution* 27, 975–978.

Stott, T. and Mitchell, C. (1991). Orielton Duck Decoy: the story of its decline. *Field Studies* 7, 759–769.

Sutton, M. and Swann, V. (2013). Harvest

time again: seed-collection and meadow restoration in Pembrokeshire. *Natur Cymru* 47, 24–27.

Syme, J. T. (1851). Notice of *Narcissus (Ajax) lobularis* Haw. *The Phytologist* 4, 156–157.

Taylor, R. (2014). Pembroke islands bat report 2014. BSG Ecology, Monmouth.

Taylor, R. (2015a). Cilgerran Castle phase 1 habitat and bat survey report. BSG Ecology, Monmouth.

Taylor, R. (2015b). St Davids Bishops Palace bat survey report. BSG Ecology, Monmouth.

Taylor, R. (2016). Llawhaden Castle phase 1 habitat and bat survey report. BSG Ecology, Monmouth.

Telfer, M. G. (2018). The status and distribution of the ground beetle *Harpalus melancholicus* at Stackpole Warren in 2017. NRW Evidence Report 247. Natural Resources Wales, Bangor.

Ternstrom, M. (1946). A brief history of the gannet colony on Lundy. *Report of the Lundy Field Society* 46, 39–42.

Thomas, H. H. (1923). The source of the stones of Stonehenge. *Antiquaries Journal* 3, 239–260.

Thomas, T. H. (1890–91). Visit to the Gannet settlement upon the island of Grassholm. *Report and Transactions of the Cardiff Naturalists' Society* 1890–91, 57–64.

Thompson, D. J., Rouquette, J. R. and Purse, B. V. (2003). Ecology of the Southern Damselfly. Conserving Natura 2000 Rivers. Ecology Series No. 8. English Nature, Peterborough.

Toulmin Smith, L. (ed.) (1906). *The Itinerary in Wales of John Leland in or about the years 1536–1539. Extracted from his MSS.* George Bell and Sons, London.

Tracy, J. (1850, 1851). Catalogue of Birds taken in Pembrokeshire, with Observations on their Habits, Manners, etc. *The Zoologist* 8, 2639–2642; 9, 3045–3049.

Turner, D. and Dillwyn, L. W. (1805). *The Botanist's Guide through England and Wales.* Phillips and Fardon, London.

Tylianakis, J. M., Didham, R. K., Bascompte, J. and Wardle, D. A. (2008). Global change and species interactions in terrestrial ecosystems. *Ecology Letters* 11, 1351–1363.

Unsworth, R. (2015). Seagrass meadows in Wales: vulnerable features and fish nurseries. *Natur Cymru* 55.

Unsworth, R., Collier, C., Waycott, M., Mckenzie, L. and Cullen-Unsworth, L. (2015). A framework for the resilience of seagrass ecosystems. *Marine Pollution Bulletin* 100, 34–46.

Vachell, C. T. (1894). A contribution towards an account of the narcissi of South Wales. *Report and Transactions of the Cardiff Naturalists' Society* XXVI, 81–94.

Vahed, K. (2019). The life cycle of the Scaly Cricket (or Atlantic Beach Cricket) *Pseudomogoplistes vicentae* (Gorochov, 1996). *Journal of Insect Conservation.* https://doi.org/10.1007/s10841-019-00187-1

Votier, S. C., Bicknell, A., Cox, S. L., Scales, K. L. and Patrick, S. C. (2013). A bird's eye view of discard reforms: bird-borne cameras reveal seabird/fishery interactions. *PLoS One* 8 (3), e57376.

Walker, M., Jones, S., Hall, J. and Sambrook, P. (2009). A preliminary palaeoecological record from Llyn Llech Owain, near Gorslas, Carmarthenshire. *Archaeology in Wales* 49, 53–58.

Wallace, I. (2003). The beginner's guide to Caddis (order Trichoptera). *Bulletin of the Amateur Entomologists' Society* 62, 15–26.

Walmsley, A. (2008). The Norfolk 'pingo' mapping project 2007–2008. Norfolk Wildlife Trust, Norwich.

Warren, M. and Wigglesworth, T. (undated). Silver-Studded Blue priority species factsheet. Butterfly Conservation.

Wasley, A., Heal, A. and Harvey, F. (2019). Ammonia pollution damaging more than 60% of United Kingdom land. *The Guardian*, Tuesday 18 June.

Webber, J. (2002). Conserving the coastal slopes 1999–2002. Pembrokeshire Coast National Park Authority.

Welsh Government (2018). *Woodlands for Wales: the Welsh Government's Strategy for Woodlands and Trees.* Welsh Government, Cardiff.

Welsh Ornithological Society (2017). Birds in Wales: rarity tracker 2017. https://birdsin.wales/rare-birds/rarity-tracker-2017.

Westerman, N. (2010). Tenby guidebooks. *Tenby Times*, January/February.

Whellan, J. A. (1941). Notes on the flora of south-west Wales. *North Western Naturalist* 16, 93.

Whitehouse, S. (2011). Moths in special habitats: sea-cliffs and headlands. Bird Guides. www.birdguides.com/articles/invertebrates/moths-in-special-habitats-sea-cliffs-and-headlands.

Widgery, J. (2000). Wildlife reports: grasshoppers and relatives. *British Wildlife* 12, 54–55.

Wight, M. (1948). Ramsey: a Pembrokeshire island. *Nature in Wales* 29, 512–514.

Wilcox, C., van Sebille, E. and Hardesty, B. D. (2015). Threat of plastic pollution to seabirds is global, pervasive, and increasing. *Proceedings of the National Academy of Sciences of the United States of America* 112, 11899–11904.

Williams, E. (2015). Newgale shingle bank adaptation plan. Pembrokeshire County Council.

Williams, M., Davies, J., Waters, R., Wilson, D., Wilby, P. and Scofield, D. (2004). Cardigan rock sequence revealed: graptolites are the key to reconstructing an area of complex Ordovician geology. *Earthwise* (British Geological Survey) 20, 12–13.

Willis, B. (1717). *A survey of the cathedral-church of St. David's, and the edifices belonging to it, as they stood in the year 1715.* Printed for R. Gosling, London.

Willis-Bund, J. W. (ed.) (1902). *Black Book of St. David's Church in Wales, Diocese of St. Davids.* Honourable Society of Cymmrodorion, London.

Willughby, F. (1678). *The ornithology of Francis Willughby of Middleton in the county of Warwick Esq, fellow of the Royal Society in three books*, edited by John Ray.

Willmot, A. and Moyes, N. (2015). *The Flora of Derbyshire.* Pisces, Newbury.

Wilmott, A. J. (1918). *Erythraea scilloides* in Pembrokeshire. *Journal of Botany* 56, 321–323.

Wilson, P. and King, M. (2003). *Arable Plants: a Field Guide.* WildGuides, Old Basing.

Windsor, F. M., Tilley, R. M., Tyler, C. R. and Omerod, S. J. (2019). Microplastic ingestion by riverine macroinvertebrates. *Science of the Total Environment* 646, 68–74.

Winwood, H. H. (1865). Exploration of the 'Hoyle's Mouth' cave near Tenby. *Geological Magazine* 2 (16), 471–473.

Wyn, H. (2000). *Battle of the Preselau: the Campaign to Safeguard the 'Sacred' Pembrokeshire Hills 1946–1948.* Clychau Clochog, Maenclochog.

Wyndham, H. P. (1775). *A Gentleman's Tour through Monmouthshire and Wales in the months of June and July, 1774*. Printed for T. Evans, London.

Yalden, D. (1999). *The History of British Mammals*. T. & A. D. Poyser, London.

Yarrell, W. (1839). *A History of British Birds*. Van Voorst, London.

Young, M. R., Cosgrove, P. J. and Hastie L. C. (2000). The extent of, and causes for, the decline of a highly threatened naiad: *Margaritifera margaritifera*. In G. Bauer and K. Wächtler (eds.), *Ecology and Evolutionary Biology of the Freshwater Mussels Unionoidea*. Springer-Verlag, Berlin, pp. 337–357.

Young-Powell, M. (2017). Breeding Dartford Warblers in Pembrokeshire in 2017. *Pembrokeshire Bird Group Report*.

Wilson, P. (2017) *Field Gentian* Gentianella campestris (L.) Boerner *in Pembrokeshire. Results and analysis of 2016 fieldwork*. The Species Recovery Trust

Ziuganov, V., Zotin, A., Nezlin, L. and Tretiakov, V. (1994). *The Freshwater Pearl Mussels and Their Relationships with Salmonid Fish*. VNIRO, Moscow.

Index

SPECIES INDEX

GENERAL INDEX

The New Naturalist Library